Lecture Notes in Geoinformation and Cartography

Series Editors: William Cartwright, Georg Gartner,Liqiu Meng, Michael Peterson

Emmanuel Stefanakis • Michael P. Peterson •
Costas Armenakis • Vasilis Delis (Eds.)

Geographic Hypermedia

Concepts and Systems

With 158 Figures

Editors:

Dr. Emmanuel Stefanakis
Department of Geography,
Harokopio University
70, El. Venizelou Ave., 17671 Athens,
Greece
Email: estef@hua.gr

Dr. Costas Armenakis
Centre for Topographic Information
Natural Resources Canada
615 Booth Street,
Ottawa, Ontario
Canada K1A 0E9
Email: armenaki@NRCan.gc.ca

Dr. Michael P. Peterson
Department of Geography/Geology
University of Nebraska at Omaha
Omaha, NE 68182 USA
Email: mpeterson@unomaha.edu

Dr. Vasilis Delis
Research-Academic Computer
Technology Institute
Kolokotroni 3
GR 262 21 Patras, Hellas, Greece
Email: delis@cti.gr

ISBN 978-3-642-07063-1 e-ISBN 978-3-540-34238-0

Springer is a part of Springer Science+Business Media
springeronline.com

Softcover reprint of the hardcover 1st edition 2006

Cover design: E. Kirchner, Heidelberg

Printed on acid-free paper 30/2132/AO 54321

In memory of

Professor Y.C. Lee (1948-2004)

Preface

This book introduces a new paradigm, *Geographic Hypermedia*, which emerges from the convergence of Geographic Information Science and hypermedia technology. Both GI Science and hypermedia have been rapidly evolving fields. The initial idea of *Geographic Hypermedia* was born in 2004 when the editors had been invited to organize a workshop in conjunction with the 'Hypertext' conference organized annually by the Special Interest Group of the Association for Computing Machinery.

The purpose of the workshop was to examine how hypermedia concepts and tools may be applied in geographical domains. The workshop was eventually held in conjunction with the Maps and the Internet Commission of the International Cartographic Association at the annual meeting of the Association of American Geographers in Denver, Colorado, in April 2005.

The Denver workshop was a successful event, bringing together multidisciplinary researchers and professionals in the area of Geographic Hypermedia. Researchers from four continents and well recognized institutions presented their work and exchanged opinions about the new paradigm, its content and distinct characteristics from other paradigms. Extended versions of the papers presented at the workshop along with some invited chapters from experts in the field led to the compendium of the twenty-five chapters in this book volume.

Geographic Hypermedia is not yet a mature paradigm but we aspire to provide the scientific community with a contemporary view of Geographic Hypermedia, present an overview of its progress and current status, while also promoting further discussion and research.

Emmanuel Stefanakis
Michael P. Peterson
Costas Armenakis
Vasilis Delis

Contents

PART V: Geographic Hypermedia: Applications and Services

List of Contributors

Andrienko, Gennady, Dr.
Fraunhofer Institute AIS
Schloss Birlinghoven, 53754 Sankt Augustin, Germany
gennady.andrienko@ais.fraunhofer.de
http://www.ais.fraunhofer.de/and

Andrienko, Natalia, Dr.
Fraunhofer Institute AIS
Schloss Birlinghoven, 53754 Sankt Augustin, Germany
natalia.andrienko@ais.fraunhofer.de
http://www.ais.fraunhofer.de/and

Armenakis, Costas, Dr.
Centre for Topographic Information, Natural Resources Canada
615 Booth Street, Ottawa, Ontario, Canada, K1A 0E9
armenaki@NRCan.gc.ca
http://www.cits.rncan.gc.ca/

Armstrong, Marc P., Dr.
Department of Geography, University of Iowa
316 Jessup Hall, Iowa City, Iowa 52242, USA
marc-armstrong@uiowa.edu
http://www.uiowa.edu/~geog/faculty/armstrong.htm

Badard, Thierry, Dr.
Laval University Center for Research in Geomatics
Dept. Geomatics Sciences, Casault Hall, Laval University Campus,
Quebec City, Qc, Canada G1K7P4
thierry.badard@scg.ulaval.ca
http://geosoa.scg.ulaval.ca/en/index.php

Bédard, Yvan , Dr.
Laval University Center for Research in Geomatics
Dept. Geomatics Sciences, Casault Hall, Laval University Campus,
Quebec City, Qc, Canada G1K7P4
yvan.bedard@scg.ulaval.ca
http://sirs.scg.ulaval.ca/yvanbedard

Bennett, David, Dr.
Department of Geography, University of Iowa
316 Jessup Hall, Iowa City, Iowa 52242, USA
david-bennett@uiowa.edu
http://www.uiowa.edu/~geog/faculty/bennett.htm

Buchroithner, Manfred F., Dr.
Institute for Cartography, Dresden University of Technology
Helmholtzstrasse 10, 01069 Dresden, Germany
manfred.buchroithner@tu-dresden.de
http://web.tu-dresden.de/kartographie

Bunningen, van Arthur
University of Twente
PO box 217, 7500AE Enschede, The Netherlands
a.h.vanbunningen@utwente.nl
http://www.utwente.nl

Cao, Yiwei
Lehrstuhl Informatik V, RWTH Aachen
Ahornstr. 55, 52056 Aachen, Germany
cao@cs.rwth-aachen.de
http://www-i5.informatik.rwth-aachen.de/lehrstuhl/staff/cao

Cartwright, William, Dr.
School of Mathematical and Geospatial Sciences, RMIT University
GPO Box 2476V, Melbourne, Victoria 3001, Australia
william.cartwright@rmit.edu.au
http://user.gs.rmit.edu.au/cartwright/index.htm

Delis, Vasilis, Dr.
Research-Academic Computer Technology Institute
Kolokotroni 3, GR 262 21, Patras, Hellas
delis@cti.gr
http://dke.cti.gr/

Dolezal, John
Center for Remote Sensing and Mapping Science (CRMS)
Department of Geography, The University of Georgia
Athens, Georgia 30602 USA
johndolezal@yahoo.com
http://www.crms.uga.edu/

Gardiner, Ned, Dr.
American Museum of Natural History
Science Bulletins, Central Park West at 79th; New York,
NY 10024, USA
ned@amnh.org
http://sciencebulletins.amnh.org

Gartner, Georg, Dr.
Institute for Geoinformation and Cartography
Technical University of Vienna
Gusshausstrasse 30, 1040 Wien, Austria
georg.gartner@tuwien.ac.at
http://cartography.tuwien.ac.at

George, Randy
Micro Map & CAD
17715 Canterbury Dr Monument, CO 80132, USA
rkgeorge@cadmaps.com
http://www.cadmaps.com

Gryllakis, Augustine
Talent Information Systems SA
Karitsi Sq. 4, 10561, Athens, Greece
augril@talent.gr
http://www.talent.gr

Hardin, Danny
The National Space Science and Technology Center
320 Sparkman Drive Huntsville AL 35805 USA
danny.hardin@msfc.nasa.gov
http://www.nsstc.org/

Hu, Shunfu, Dr.
Department of Geography, Southern Illinois University Edwardsville
Box 1459, Edwardsville, IL 62026, USA
shu@siue.edu
http://www.siue.edu/~shu

Ingensand, Jens
Institute of Urban and Regional Planning & Design – Geomatics,
Swiss Federal Institute of Technology (EPFL)
CH-1015 Lausanne, Switzerland
jens.ingensand@epfl.ch
http://lasig.epfl.ch/

Jansen, Michael, Dr.
Lehr- und Forschungsgebiet Stadtbaugeschichte, RWTH Aachen
Schinkelstr. 1, 52062 Aachen, Germany
jansen@sbg.rwth-aachen.de
http://www.sbg.rwth-aachen.de/

Jarke, Matthias, Dr.
Lehrstuhl Informatik V, RWTH Aachen
Ahornstr. 55, 52056 Aachen, Germany
jarke@cs.rwth-aaachen.de
http://www-i5.informatik.rwth-aachen.de/lehrstuhl/staff/jarke

Jordan, Thomas
Center for Remote Sensing and Mapping Science (CRMS)
Department of Geography, The University of Georgia
Athens, Georgia 30602 USA
tombob@uga.edu
http://www.crms.uga.edu/

Kavouras, Marinos, Dr.
National Technical University of Athens
9, H. Polytechniou Str., 157 80 Zografos Campus, Athens, Greece
mkav@survey.ntua.gr
http://ontogeo.ntua.gr/people/m_kavouras.htm

Klamma, Ralf, Dr.
Lehrstuhl Informatik V, RWTH Aachen
Ahornstr. 55, 52056 Aachen, Germany
klamma@cs.rwth-aachen.de
http://www-i5.informatik.rwth-aachen.de/lehrstuhl/staff/klamma

Köbben, Barend, Dr.
International Institute for Geo-Information Science and Earth Observation (ITC)
PO Box 6, 7500AA, Enschede, The Netherlands
kobben@itc.nl
http://www.itc.nl/about_itc/resumes/kobben.aspx

Kokla, Margarita, Dr.
National Technical University of Athens
9, H. Polytechniou Str., 157 80 Zografos Campus, Athens, Greece
mkokla@survey.ntua.gr
http://ontogeo.ntua.gr/

Koutlis, Manolis, Dr.
Talent Information Systems SA
Karitsi Sq. 4, 10561, Athens, Greece
koutlis@talent.gr
http://www.talent.gr

Kyrimis, Kriton, Dr.
Talent Information Systems SA
Karitsi Sq. 4, 10561, Athens, Greece
kyrimis@talent.gr
http://www.talent.gr

Madden, Marguerite, Dr.
Center for Remote Sensing and Mapping Science (CRMS)
Department of Geography, The University of Georgia
Athens, Georgia 30602 USA
mmadden@uga.edu
http://www.crms.uga.edu/staff/mmadden.htm

Mantes, Thanasis
Talent Information Systems SA
Karitsi Sq. 4, 10561, Athens, Greece
mantes@talent.gr
http://www.talent.gr

Mount, Jerry
Department of Geography, University of Iowa
316 Jessup Hall, Iowa City, Iowa 52242, USA
jerry-mount@uiowa.edu
http://www.uiowa.edu/~geog/

Müller, Anita
GeoAccess, Canada Centre for Remote Sensing, Geomatics,
Natural Resources Canada
615 Booth Street, Ottawa, Ontario, Canada, K1A 0E9
amuller@NRCan.gc.ca
http://ccrs.nrcan.gc.ca/

Muthukrishnan, Kavitha
University of Twente
PO box 217, 7500AE Enschede, The Netherlands
k.muthukrishnan@utwente.nl
http://www.utwente.nl

Persson, Donata
Institute for Geoinformation and Cartography
Technical University of Vienna
Gusshausstrasse 30, 1040 Wien, Austria
donata.persson@gmx.net
http://cartography.tuwien.ac.at

Peterson, Michael P, Dr.
Department of Geography/Geology, University of Nebraska at Omaha
Omaha, NE 68182 USA
mpeterson@unomaha.edu
http://maps.unomaha.edu

Proulx, Marie-Josée
Laval University Center for Research in Geomatics
Dept. Geomatics Sciences, Casault Hall, Laval University Campus,
Quebec City, Qc, Canada G1K7P4
marie-josee.proulx@scg.ulaval.ca
http://sirs.scg.ulaval.ca/yvanbedard

Radoczky, Verena
Department of Geoinformation and Cartography
Technical University of Vienna
Gusshausstr. 30, 1040 Wien
radoczky@cartography.tuwien.ac.at
http://cartography.tuwien.ac.at

Ramos, Cristhiane da Silva
School of Mathematical and Geospatial Sciences, RMIT University
GPO Box 2476V, Melbourne, Victoria 3001, Australia
christhiane.ramos@rmit.edu.au
http://user.gs.rmit.edu.au/

Rivest, Sonia
Laval University Center for Research in Geomatics
Dept. Geomatics Sciences, Casault Hall, Laval University Campus,
Quebec City, Qc, Canada G1K7P4
sonia.rivest@scg.ulaval.ca
http://sirs.scg.ulaval.ca/yvanbedard

Savopol, Florin
Centre for Topographic Information, Natural Resources Canada
615 Booth Street, Ottawa, Ontario, Canada, K1A 0E9
fsavopol@NRCan.gc.ca
http://www.cits.rncan.gc.ca/

Siegel, Charles
TGIS Technologies Inc.
16, chemin Pelletier, Chelsea, Quebec, Canada
charles.siegel@tgis.ca
http://www.tgis.ca

Siekierska, Eva, Dr.
Centre for Topographic Information, Natural Resources Canada
615 Booth Street, Ottawa, Ontario, Canada, K1A 0E9
siekiers@NRCan.gc.ca
http://www.cits.rncan.gc.ca/

Spaniol, Marc
Lehrstuhl Informatik V, RWTH Aachen
Ahornstr. 55, 52056 Aachen, Germany
mspaniol@cs.rwth-aachen.de
http://www-i5.informatik.rwth-aachen.de/lehrstuhl/staff/spaniol

Stefanakis, Emmanuel, Dr.
Department of Geography, Harokopio University of Athens
70 El. Venizelou Ave, 17671 Kallithea, Athens, Greece
estef@hua.gr
http://www.dbnet.ece.ntua.gr/~stefanak

Tomai, Eleni, Dr.
National Technical University of Athens
9, H. Polytechniou Str., 157 80 Zografos Campus, Athens, Greece
etomai@central.ntua.gr
http:ontogeo.ntua.gr/people/e_tomai_en.htm

Toubekis, Georgios
Lehr- und Forschungsgebiet Stadtbaugeschichte, RWTH Aachen
Schinkelstr. 1, 52062 Aachen, Germany
toubekis@sbg.rwth-aachen.de
http://www.sbg.rwth-aachen.de/

Tryfona, Nectaria, Dr.
Talent Information Systems SA
Karitsi Sq. 4, 10561, Athens, Greece
tryfona@talent.gr
http://www.talent.gr

Tsironis, Yiorgos
Talent Information Systems SA
Karitsi Sq. 4, 10561, Athens, Greece
yiorgos@talent.gr
http://www.talent.gr

Tzagarakis, Manolis, Dr.
Research-Academic Computer Technology Institute
N. Kazantzaki str., University of Patras Campus, GR-26500
Patras, Greece
tzagara@cti.gr
http://elearning.cti.gr/tzag/

Vaitis, Michail, Dr.
Department of Geography, University of the Aegean
University Hill, GR-811 00 Mytilene, Greece
vaitis@aegean.gr
http://www.aegean.gr/geography

Vasiliou, George
Talent Information Systems SA
Karitsi Sq. 4, 10561, Athens, Greece
vasiliou@talent.gr
http://www.talent.gr

Wealands, Karen
Geospatial Science, RMIT University
GPO Box 2476V, Melbourne Victoria 3001, Australia
karen.wealands@rmit.edu.au
http://www.gs.rmit.edu.au/research/projects/rp1_karen.htm

Webster, Jessica
Sustainable Buildings and Communities, Natural Resources Canada
1 Haanel Drive, Ottawa, Ontario, Canada K1A 1M1
Jessica.Webster@nrcan.gc.ca
http://www.sbc.nrcan.gc.ca

Williams, Peter
Centre for Topographic Information, Natural Resources Canada
615 Booth Street, Ottawa, Ontario, Canada, K1A 0E9
pewillia@NRCan.gc.ca
http://www.cits.rncan.gc.ca/

PART I

FOUNDATIONS OF GEOGRAPHIC HYPERMEDIA

1 Geographic Hypermedia

Emmanuel Stefanakis, and Michael P. Peterson

Abstract. The convergence between Geographic Information Science and Hypermedia technology leads to the emergence of a new paradigm, named *Geographic Hypermedia* (GH). This chapter introduces GH by presenting the underlying concepts and tools; and highlighting the content and types of services that should be provided by a GH system. The chapter also explains the structure of the book and presents an overview of the twenty-four contributions.

1.1 Introduction

The integration of the web and hypermedia technologies with Geographic Information Science (GIScience) has recently led to the development of new forms of geo-representations. Currently, many geospatial solutions are web-based and provide access to distributed multimedia elements in order to support specific application domains.

Geographic Hypermedia (GH) is an emerging paradigm, which adopts and extends the concepts and tools developed in hypermedia. *Geographic Hypermedia Systems* (GHS) are software systems that allow distributed geographic content (data and services) with various forms of media being interlinked and exploited in different ways. GH supports a wide number of research and professional activities in geographical domains.

This chapter introduces the new paradigm by first defining the underlying technology (Section 1.2). Then, the related paradigms are presented (Section 1.3). Following this, the content and the types of services that should be offered by GH are described (Section 1.4). A short discussion is provided next that attempts to place GH in relation to multimedia cartography and distributed GIS (Section 1.5). The last section explains the structure of the book and introduces the twenty-four contributions (Section 1.6).

1.2 Defining Geographic Hypermedia

The integration of *Hypermedia* technology with *Geographic Information Science* (GI Science; Goodchild 1997) creates a new paradigm in *Geo-*

graphic Hypermedia (GH). GI Science focuses on the collection and manipulation of *geographic information* (GI); i.e., information about places on the Earth's surface. Currently, these technologies make wide use of hypermedia methods and tools to enrich their functionality, effectiveness and usability. After discussing hypermedia technology (Section 1.2.1), the technological aspects of the new paradigm are presented (Section 1.2.2).

1.2.1 Hypermedia and Related Technologies

Hypermedia is a technology that is based on a series of other technologies, including media, hypertext and multimedia. Hence, a short description of these elements is provided here.

According to Webopedia (Webopedia 2006), *media* is the plural of medium and defined as: (a) objects on which data can be stored (these include hard disks, floppy disks, CD-ROMs, and tapes); (b) the cables – in computer networks – linking workstations together (e.g., the normal electrical wire, the type of cable used for cable television, and the fiber optic cable); and (c) the form and technology used to communicate information (e.g., a multimedia presentation combines sound, pictures, and videos, all of which are different types of media). Based on this definition, maps themselves are media, provided that a map can store and communicate cartographic information. Additionally, all output products of any information system or data repository, a GI System, the web, even the computer itself, can be seen as media, provided that they all store and communicate information.

Hypertext (Nelson 1965, FOLDOC 2006) refers to a collection of documents (or "nodes") containing cross-references (or "links") that with the aid of an interactive browser program, allow the reader to move easily from one document to another (Fig. 1.1a). In other words, hypertext is the extension of the traditional form of "flat" or linear text to the nonlinear form, in which the paths through a document can branch-off to other documents via appropriate references. *Hypertext systems* are software systems that allow users to author, edit and follow links between different bodies of text.

Hypermedia is the extension of hypertext to include other media, such as sound, graphics and animation (FOLDOC 2006; Fig. 1.1b). In practice, many hypertext documents have at least some graphical content (just as a text often includes illustrations). Hence, the distinction between hypertext and hypermedia is not clear. For instance, the Special Interest Group of the Association for Computing Machinery (SIGWEB 2006) organizes the

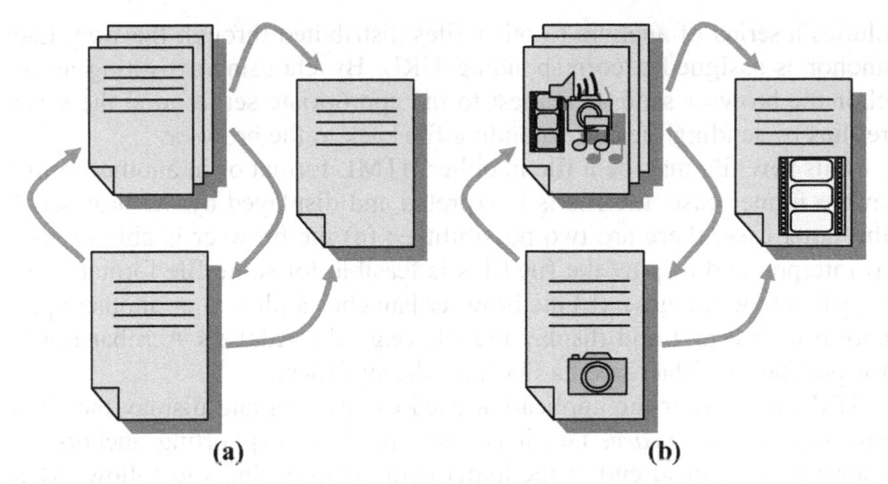

Fig. 1.1. Hypertext (a) versus hypermedia (b). A strict distinction.

foremost international conference annually, named "Hypertext", that is dedicated to both hypertext and hypermedia.

Hypermedia is an extension to hypertext. In hypermedia, other forms of media are available and it is possible to create a link between an audio file and a body of text. A myriad of hypermedia systems are available today. A list of some major systems can be found at Blustein (2006) and under SIGWEB site (SIGWEB 2006). It is worth noting that the World Wide Web (W3C 2006) is itself a hypermedia system. The web supports hyperlinks to text, graphics, sound, and video files.

An example of how hypermedia content is interlinked is the web (Wiil 1997). The World Wide Web is a subset of the Internet that establishes the communication between two Internet machines, the web browser and the web server, based on the HTTP protocol (W3C 2006) (Fig. 1.2). The browser sends a request (GET) to a specific server though the corresponding URL (i.e., a short string that identifies resources in the web) and the server responds by sending back (PUT) a file in HTML format. Hyper Text Markup Language (HTML) is the basic standard language for publishing hypertext on the web (W3C 2006).

The HTML file is interpreted and displayed by the browser. This file in-

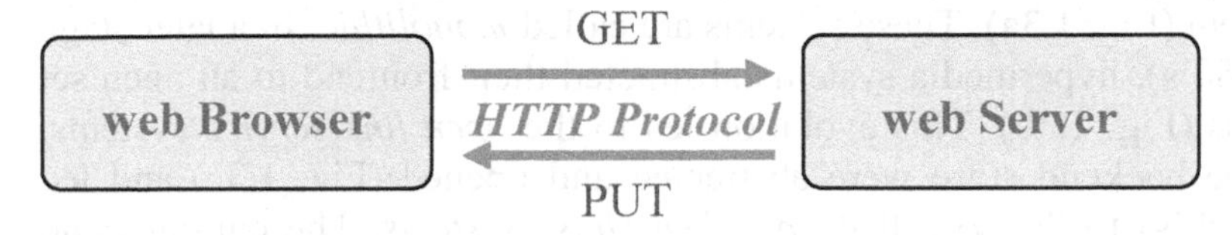

Fig. 1.2. Basic web communication.

cludes a series of anchors to other files distributed through the web. Each anchor is assigned a corresponding URL. By choosing/activating an anchor the browser sends a request to the appropriate server; and the server replies by sending the corresponding file back to the browser.

This new file may be a file in either HTML format or in another format. In the former case, the file is interpreted and displayed by the browser. In the latter case, there are two possibilities: (a) the browser is able by itself to interpret and display the file (this is feasible for some file formats, such as gif and jpg images); (b) the browser launches a plug-in or another application to interpret and display the file (e.g., the Adobe's Acrobat Reader for pdf files; or Macromedia Flash for the swf files).

If the plug-in or the application used to interpret and display the file is *non-hypermedia-aware*, i.e., it is not capable of supporting anchors, the user reaches a dead-end in the hypermedia with no links to follow. Alternatively, if the application above is *hypermedia-aware*; i.e., capable to recognize and display the anchors and the corresponding URLs; the user may activate them and get additional resources from the web. The second alternative is aligned with the framework of *Open Hypermedia Systems (OHS)*.

OHS focus on the architectural, modeling and rendering issues regarding the standardization of hypermedia services (OHSWG 2006). An OHS provides a general set of hypermedia services that can be used by other applications, programs and services in the computing environment. The minimal requirement for an OHS is the provision of basic hypermedia linking services to an open set of applications; i.e., the ability to create, delete and modify anchors and links and the ability to traverse links (Wiil 1997, Nürnberg *et al.* 1998).

The conceptual architecture of hypermedia systems that demonstrate their evolution (Nürnerg and Leggett 1997, Nürnberg *et al.* 1998) are shown in Fig. 1.3. *Client* is a frontend unit that interacts with the hypermedia system. *Storage engine* is a backend repository of hypermedia content. *Structure processor* is a server that provides structural abstractions to clients. *Link server* implements the linkage between the client and the hypermedia content.

In early hypermedia systems, a single process implements all aspects of the system (Fig. 1.3a). These systems are called *monolithic*. In a later stage (late 1980's), hypermedia systems abstracted their frontend to an open set of clients (Fig. 1.3b). This evolution led to the *open link service* systems. Next, the backend store were abstracted and opened (Fig. 1.3c) and led (mid 1990's) to the so-called *open hyperbase systems*. The current trend focuses on the abstractions at the middleware units (i.e., the structure proc-

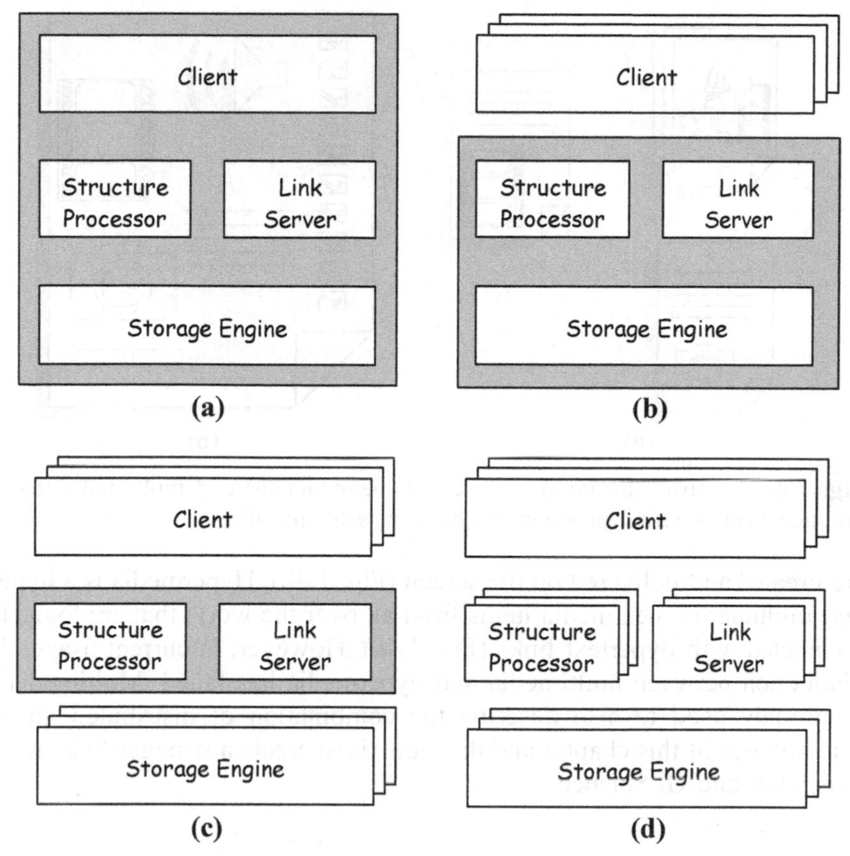

Fig. 1.3. The conceptual architectures of hypermedia systems: (a) monolithic systems; (b) open link service systems; (c) open hyperbase systems; (d) open hypermedia systems. The shading encloses the "closed" components.

essor and the link server) and leads to the development of *open hypermedia systems* (Fig. 1.3d).

A technology, which has been developed in parallel with hypertext and hypermedia is multimedia. *Multimedia* refers to "multiple media" or "a combination of media". According to FOLDOC (FOLDOC 2006), multimedia often includes concepts from hypertext and is defined as the human-computer interaction involving text, graphics, voice and video. The term "multimedia" became synonymous with CD-ROM in the personal computer world, because of the large amounts of data involved that are currently best supplied on CD-ROM, and more recently, the DVD.

In attempting to distinguish multimedia from hypermedia, it can be argued that hypermedia is a wider term, which includes multimedia. Multimedia is the combination of text, graphics, audio, video and animation that

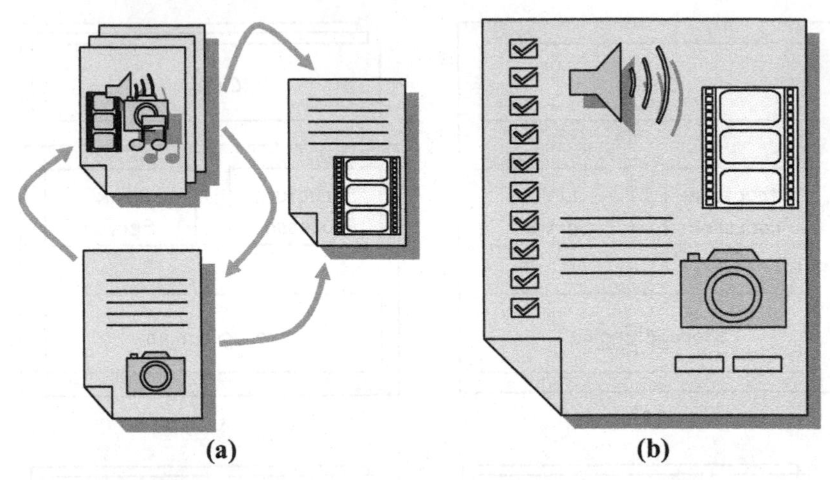

Fig. 1.4. The strict distinction between hypermedia (a) and multimedia (b). In practice, the two terms have come to have the same meaning.

are created and delivered on the screen (Fig. 1.4b). Hypermedia is a hypertext multimedia, i.e., media items from all over the world that are logically connected with hypertext links (Fig. 1.4a). However, in current usage, the distinction between multimedia and hypermedia has faded. Multimedia is commonly used to also describe the combination of distributed media. The purpose of this chapter and this book is to create a stronger linkage between GH and GI Science.

1.2.2 Geographic Hypermedia

GI Science is focused on the collection and manipulation of geographic information. Goodchild (1997) lists three main types of technologies: (a) positioning (e.g., GPS); (b) remote sensing; and (c) Geographic Information Systems (GIS).

The rapid development of the Internet has changed dramatically the collection and processing of geographic information. The Internet is already the most widely used system for the dissemination of information and services. As a consequence, it affects GIScience technology in at least three aspects (Peng and Tsou 2003): (a) the accessibility to data and information, (b) the dissemination of geographic information, and (c) the modeling and analysis of geographic information.

This situation forces a dramatic change on the architecture of systems used for the management and handling of geographic information. GI desktop systems (e.g., GIS, and remote sensing software packages) are be-

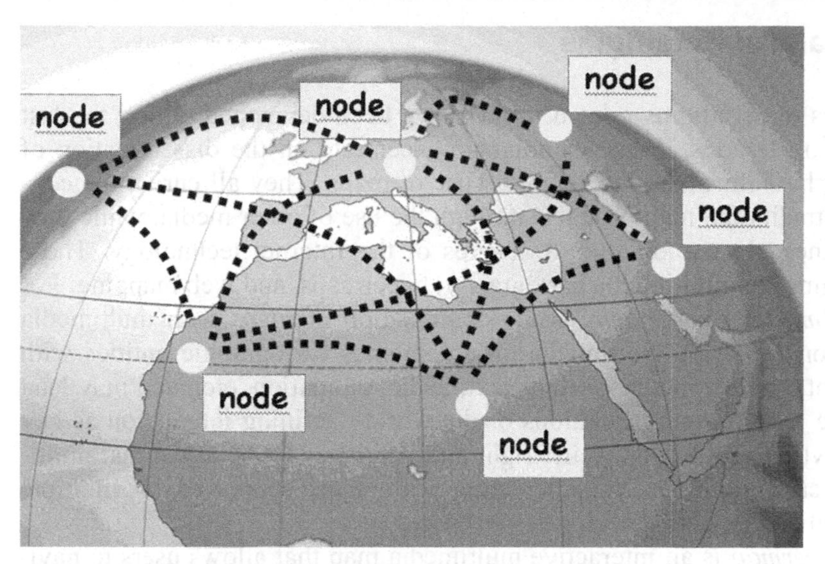

Fig. 1.5. The architecture of distributed GI Services.

ing replaced by distributed systems, called *Distributed GI Services* (Peng and Tsou 2003).

Distributed GI Services are based on an advanced network scheme (Fig. 1.5), where the individual modules of traditional GI desktop Systems as well as geographic information are distributed on numerous network nodes and are offered to other nodes upon request. A node acts either as a client or a server in this architecture; and, hence, there is no distinction between nodes.

Obviously, the individual nodes comprising the whole network adopt heterogeneous platforms and software systems. Therefore, the whole network must support the connection and interaction between heterogeneous components. Additionally, it must support the simultaneous many-to-many communication between the individual nodes.

Hypermedia technology offers a set of concepts and tools that are applicable to this new era of distributed GI Services. Each node in the architecture above is a repository of GI elements. Each GI element may be either a data/information set or an analytical/processing module. These elements are encoded in various formats and media, and are interlinked with other elements that reside in the same or other network nodes. Hence, they all together compose a hypermedia system. Provided that they focus on GI, this is a *Geographic Hypermedia System* (GHS). The associated paradigm is named *Geographic Hypermedia* (GH).

1.3. Parallel Paradigms

A series of paradigms related to GH have been developed during the last couple of decades. These paradigms concentrate on the dissemination of geographic information through the use of maps. They all enrich the content of traditional map products through the use of other media, while they simultaneously exploit the advantages of the Internet technology. These paradigms are: multimedia cartography, hypermaps, and web mapping.

Multimedia cartography is the combination of maps with multimedia (Peterson 1999). A multimedia map combines cartographic entities with different types of media, such as text, audio, animation, etc., and may lead to more realistic representations of the world. Defining interaction as key to knowledge formation, Cartwright and Peterson (1999) argue that multimedia cartography is the interaction with maps supported by multiple forms of media.

A *hypermap* is an interactive multimedia map that allows users to navigate through the map, zoom and find locations. This functionality assists the interpretation of maps and queries. The term has been first introduced by Laurini and Millert-Raffort (1990). Later, Kraak and Driel (1997) have proposed the basic functionality of hypermaps.

The web has become a widespread means for the dissemination of mapping products. The maps available through the web can be distinguished based on varying terms of interaction, with static maps incorporating no interaction at all. The dissemination of geographic data on the web in the form of maps is accompanied with special requirements for map design and a new paradigm has emerged, that of *web mapping* (Kraak and Brown 2001, Peterson 2003).

1.4. Geographic Hypermedia Systems: Content and Services

The scope of Geographic Hypermedia Systems (GHS) is twofold. Firstly, to interlink geographic content, that is available in various forms and types of media and distributed in pieces. Secondly, to support a wide number of research and professional activities in geographical domains, such as Spatial Decision Support, Geographic Database Management and Exploration, Geovisualization, Virtual and Augmented Reality, Location Based Services, Geographic Messaging Services, Hypermedia Cartography, Hypermaps and Hypermedia Atlases, and web-based and Virtual Wayfinding Services, to name a few.

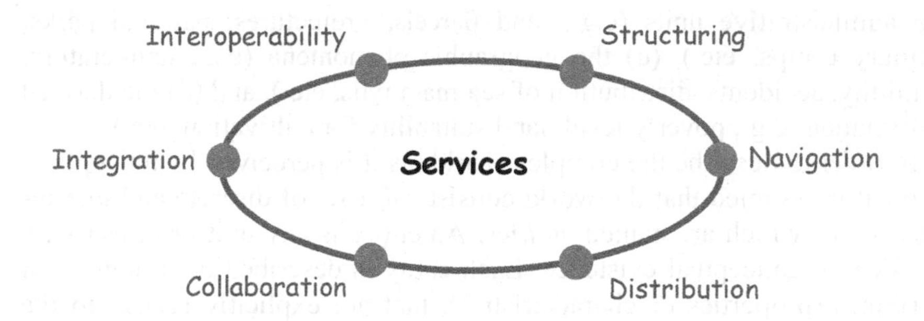

Fig. 1.6. The types of services in an Open Hypermedia System.

The development of advanced GHS should exploit the recent outcomes of the Open Hypermedia Systems (OHS) research community (OHSWG 2006). In an ideal situation, OHS constitute a central paradigm for structuring and navigating among pieces of information. This would be accomplished if all operating systems support six types of services (Grønbæk and Wiil 1997; Fig. 1.6): *integration, interoperability, structuring, navigation, distribution* and *collaboration*. Geographic Hypermedia Systems have special requirements including that these services must be able to handle geographic content.

The following subsections describe the geographic content (Section 1.4.1), the types of services that a GHS must provide (Section 1.4.2), and the efforts towards the development of geospatial standards and specifications (Section 1.4.3).

1.4.1 Geographic Content: Data and Services

GHS handle geographic content, which consists of: (a) geographic *data/information*, and (b) geographic *services*. The following paragraphs provide a brief description of these.

Geographic data constitutes a special category of data that is distributed in *space* and changes over *time*. As regards to its distribution in space, geographic data is accompanied with references to the locations in space where it occurs or the places it describes. On the other hand, as regards to its change over time, this change can be so slow, that it can be ignored (e.g., change of the seashore, climate, or distribution of ages in a country, etc.); or so fast that the rhythm of changes is of high importance (e.g., the traffic load of a highway, the temperature, the fire front in a forest fire, etc.).

Geographic data is usually classified into four categories (Maguire *et al.* 1991): (a) the physical objects (e.g., houses, roads, lakes, forests, etc.), (b) the administrative units (e.g., land parcels, prefectures, national parks, military camps, etc.), (c) the geographic phenomena (e.g., temperature, humidity, accidents, distribution of sea mammals, etc.), and (d) the derived information (e.g., poverty level, land suitability for cultivation, etc.).

In order to describe the complex world (as it is perceived by an application), it is assumed that the world consists of a set of discrete and interrelated units, which are named *entities*. An entity is any unit or object with physical or conceptual existence. Each entity is described by a number of attributes (properties or characteristics), that are explicitly related to the application. The attributes associated with geographic entities are classified into three basic categories, which are named *dimensions* of geographic entities. These dimensions are (Fig. 1.7): (a) the *identifier*, (b) the *spatial* dimension, and (c) the *thematic* dimension.

The *identifier* is the dimension which provides a means to refer to (or name) geographic entities. The *spatial* dimension incorporates all attributes that describe the spatial characteristics of geographic entities. The spatial characteristics of an entity consist of: (a) its position, (b) its geometry, (c) its graphical representation, and (d) its spatial relationships with other entities. Additionally, the *thematic* dimension includes all thematic or non-spatial attributes of geographic entities. The thematic dimension accommodates all *multimedia data* that accompany the geographic entities of an application. Multimedia data consists of sound, images, video data, virtual and augmented reality data, and describes the geographic entities.

In fact, all entities are dynamic in nature, since they lie in a continuously changed world, while time is a dimension that is tightly connected to space. However, in practice some geographic entities are usually treated as *static entities*. This is related to a time interval during which the entity attributes remain unchanged.

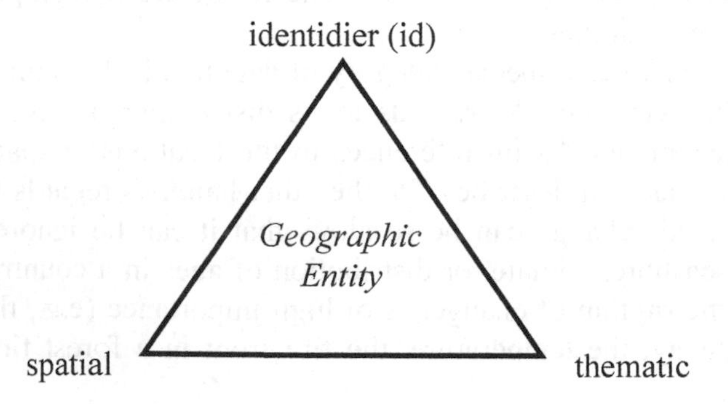

Fig. 1.7. The dimensions of a static geographic entity.

It is a common practice (which comes from traditional methodologies on handling spatio-temporal phenomena) to examine spatial and thematic properties of entities over time. This consideration highlights the behavior of an entity over time, namely *entity change*. Change takes various forms, all falling within two general categories (Frank 2001, Stefanakis 2003): (a) *life* and (b) *motion*. Life refers to change in the existence status of entities, whereas motion refers to change in the spatial and/or thematic attributes of entities.

The life and motion of individual entities, physical or conceptual, dominate our existence and perception of the environment that surrounds us. An entity may appear and disappear (e.g., a residential zone or a forest), may merge with other entities or split into two or more new entities (e.g., collection of parcels, or towns). These changes concern the life of the entity of interest. On the other hand, an entity may move with or without changing its form at the same time (e.g., pollution, a boat). Additionally, an entity may change its thematic properties without change in its spatial properties (e.g., parcel owner). These changes concern the motion of the entity of interest. Fig. 1.8 highlights some common life and motion type changes.

Geographic data is usually collected by applying error prone methodologies. Hence, the attributes of geographic entities (e.g., parcel area, street width, person age, etc.) are assigned values that decline more or less from the real ones. For a better understanding and exploitation of geographic data, a *measure of its quality* is needed. This measure includes the positional and attribute accuracy of entities, and the completeness and

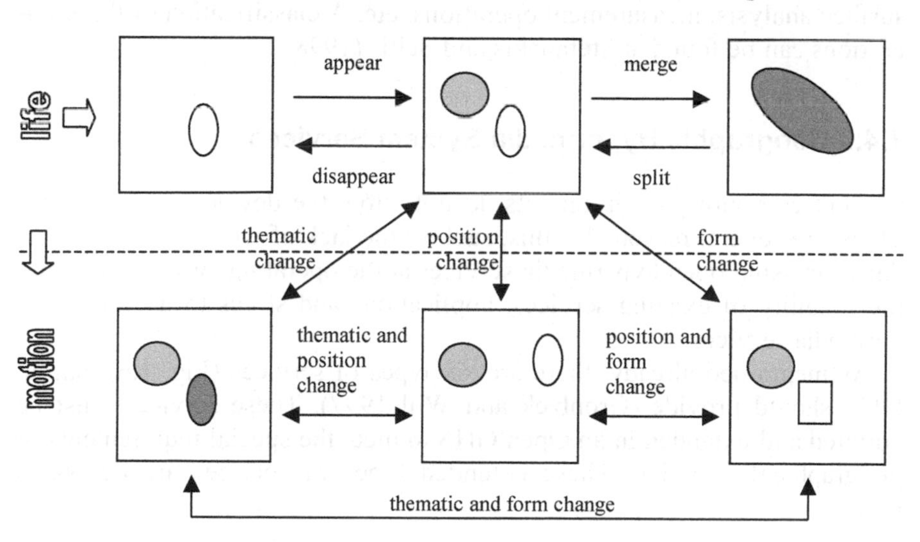

Fig. 1.8. Entity's life and motion.

logical consistency among data elements (Maguire *et al.* 1991).

Geographic services are based on the set of operations to process and analyze geographic data. The fundamental operations can be grouped into three types (Tomlin 1990): (a) data preparation, (b) data presentation and (c) data analysis operations.

Data preparation operations encompass a variety of methods for capturing data from different sources (e.g., digital or paper maps, geographic data repositories, land measurements, positioning and earth observation systems), editing and storing them appropriately in a data repository.

Data presentation operations encompass a variety of methods for the presentation of data, such as drawing maps, drafting charts, generating reports, and so on. Geographic applications have special requirements at the user interface level. This is because the appropriate visualization of geospatial data may assist people to discover or evaluate the geographic content (Andrienko and Andrienko 1999, Kraak and Brown 2001). Hence, the quality of geovisualization products must be promoted. Semantics associated to geospatial content may be exploited at this stage and lead to effective geovisualizations. Current research on the development of metaphors and graphical user interfaces (GUIs) as well as on issues related to multiple representations and interoperability of geographic data are aligned to this new trend.

Data analysis operations transform data into information and include (Maguire *et al.* 1991): classification and recording, generalization, spatial join and overlay, neighborhood analysis, connectivity, interpolation and surface analysis, measurement operations, etc. A classification of these operations can be found at Stefanakis and Sellis (1998).

1.4.2 Geographic Hypermedia System Services

Current computing environments do not allow the development of OHS. There are several reasons for this, such as the lack of widely adopted standards, missing open hypermedia services at the operating system level, and the inability of existing services, applications and stores to use open hypermedia services.

As mentioned already, there are six types of services (Fig. 1.6) that an OHS should provide (Grønbæk and Wiil 1997). These services must be adopted and extended in an Open GHS to meet the special requirements of geographical domains. These extended types of services are described next.

Integration

An OHS must be able to integrate (Grønbæk and Wiil 1997): (a) existing services, (b) third party applications and (c) third party stores. Existing services consist of scripts, programs, software agents and hardware devices.

In geographical domains, the scripts and programs involve geographic analytical and processing tools; the software agents deal with GI Systems, Mapping and/or Remote Sensing software packages; while the hardware devices involve specific devices for collecting and displaying geographical information, such as GPS receivers, geodetic stations, earth observation systems, digitizers, plotters and projectors.

Third party applications include editors, browsers, email systems, calendars, etc. Geographical third party applications (e.g., editors and browsers), should provide enhanced functionality for mapping, searching and retrieval of geographical content by spatial and/or temporal predicates.

Third party stores include databases, data and services repositories, file systems and media in general. In geographical domains, maps in digital or analog form are also present. Additionally, the repositories accommodate geographic data and services and obviously they have advanced requirements and capabilities (in modeling, processing and analysis).

Interoperability

An OHS must insure (Grønbæk and Wiil 1997) that all integrated services, applications and stores interoperate with each other; as well as with the components of other OHSs. One important issue of interoperability is how to handle the heterogeneity between data collections.

The *heterogeneity* of data collections may be considered as syntactic, schematic or semantic heterogeneity (Bishr 1998). *Syntactic* heterogeneity refers to the differences in the logical data models (e.g., relational versus object-oriented) or in spatial representations (e.g., raster versus vector). *Schematic* heterogeneity refers to the differences in the conceptual data models (the data are organized in different schemas and structures). *Semantic* heterogeneity refers to the differences in meaning, interpretation or usage of the same or related data. Semantic heterogeneity is further divided into naming (i.e., synonyms and homonyms) and cognitive (i.e., different conceptualizations) heterogeneity.

Semantic heterogeneity between data collections causes most integration problems. These problems refer to (Bernard *et al.* 2003): (a) metadata interpretation, and (b) data interpretation. As regards to metadata interpretation, different providers may use different terms. This makes it difficult for both the humans and the computers to recognize the coherence between similar terms. As regards to data interpretation, the properties of the entities involved in an application domain are expressed based on various standardized (e.g., CORINE vocabulary for land cover classification; EAA 2000) or not vocabularies. Data and metadata interpretation problems must be overcome when data from different sources have to be integrated into a single collection. This requires the adoption of common terms, which are usually provided by a standardized vocabulary with unchanged terms. All terms coming from other vocabularies will be translated in the adopted vocabulary using tools for semantic translation.

The use of *ontologies* (Fonseca and Egenhofer 1999, Guarino 1998, Wache 2001) as semantic translators is a possible approach to overcome the problem of semantic heterogeneity (Bernard *et al.* 2003). Ontologies play a critical role in associating meaning with data that computers can understand and process data automatically. Ontologies may be used (a) to assist the translation of a term (i.e., a concept) from one vocabulary (i.e., one ontology) into a term from another vocabulary; and (b) to derive super-concept and sub-concept relationships.

However, currently ontologies only define data semantics under very controlled situations. They cannot yet fully support semantic translation. They fail to cope with the semantics of services (Bernard *et al.* 2005). The GI community now addresses the following three research issues (Fonseca and Sheth 2003): (a) the creation and management of domain specific geo-ontologies for both data and services; (b) the matching geo-concepts available in the web to geo-ontologies; and (c) the integration of different ontologies as regards to both the geographic dimension and the non-geographic domain.

Structuring

An OHS must provide a rich set of structuring services (Grønbæk and Wiil 1997). Structuring services support the authoring of hypermedia scenarios. Hypermedia technology already has numerous structuring services available, such as anchors, links, presentation specifiers (pspecs), location specifiers (locspecs), file-wrapper nodes, spatial hypertext, etc.

There are special requirements for structuring services in geographic hypermedia. These services need to take into consideration the nature of geographic data involved (e.g., their spatial and temporal dimensions) as well as the individual features of the processing and analytical operations (as regards to the input parameters and the results produced). Appropriate graphical user interfaces must be available to GH scenario authors, so that they will be capable to discover, exploit and highlight the geographic content.

Navigation

An OHS should provide a rich set of navigation services (Grønbæk and Wiil 1997). These services assist the user to understand the OHS content and prevent him by being "lost in the hyperspace". Examples of typical navigation services are link traversal, querying, graphical overviews, guided tours, fish-eye views, paths, etc.

A geographic hypermedia system must provide advanced navigation services, where the location and distribution of geographic objects will be highlighted. Geovisualization techniques (Kraak and Dreil 1997) must be adopted and extended for an enhanced understanding of the distribution and behavior of spatio-temporal phenomena.

Distribution

The pieces of hypermedia content interlinked by an OHS may be running on different computing environments in local area network and across different Internet domains. The wired (e.g., web) and wireless (e.g., mobile systems) environments accommodate the hypermedia content and are interlinked with each other.

Geographic hypermedia systems must exploit and extend the outcomes of the research and professional activities towards the development of Internet and mobile systems, such as distributed GI Systems (Peng and Tsou 2003).

Collaboration

The collaborative authoring of both hypermedia contents and structure is proved very valuable in the hypermedia community (Grønbæk and Wiil 1997). Hence, an OHS should provide support for collaboration. The collaboration can take place in asynchronous or synchronous mode, depending on how the authors interact with each other in the creation of content and the structuring of the individual sessions in a hypermedia scenario.

The nature of geographic domains forces the adoption of collaborative authoring in geographic hypermedia systems. The involvement of the user at all stages of the development of a GI application has also proved valuable in the past.

1.4.3 Developing Standards and Specifications

A reference needs to be made to the geospatial standards and the proposed specifications that have been widely adopted recently. These standards and specifications affect the way geographic content is described and interlinked in the Internet; hence they affect the evolution of Geographic Hypermedia.

The Open Geospatial Consortium (OGC 2006) in parallel with World Wide Web Consortium (W3C 2006) and the International Standards Organization (ISO 2006) develop standards and specifications to support the interoperability between repositories of geospatial information. Table 1.1 provides some representative standards/specifications related to information content.

The scope of these standards and specifications is to enable an application developer to use any geospatial content (data and services) available on "the net" within a single environment and a single workflow (McKee and Buehler 1998). However, it should be noticed that standardization might not solve the problem of interoperability by itself because of the following reasons (Stoimenov and Dordevic-Kajan 2002):

- The construction and maintenance of a single and integrated model is a hard task.
- The requirement to communicate with geospatial sources that do not conform to the adopted standard will always be present.
- Existing geospatial sources have their own models which may not always be mapped to the common model without information loss.

Table 1.1. Standards and Specifications for geospatial content.

Organization	Standard/Specification name
W3C	HyperText Transfer Protocol (HTTP)
W3C	Simple Object Access Protocol (SOAP)
W3C	Web Services Description Language (WSDL)
OGC	Web Map Server (WMS)
OGC	Web Coverage Server (WCS)
OGC	Web Feature Service (WFS)
OGC	Geography Markup Language (GML)
OGC	Catalog Service (CS)
ISO-TC211	Rules for Application Schema (DIS19109)
ISO-TC211	Methodology for Feature Cataloguing (DIS19110)
ISO-TC211	Spatial Referencing by Geogr. Identifiers (CD19112)
ISO-TC211	Metadata (IS19115)
ISO-TC211	Metadata Implementation Specification (DTS19139)

– The standards/specifications are subject to continuous change, but systems will not all simultaneously change to conform.

1.5 Discussion

Geographic Hypermedia (GH) involves the interlinking of geographic content and media. The geographic content (data and services) is distributed with various forms of media being interlinked and exploited in different ways. Special attention is given on the integration of heterogeneous geographic content and consequently on interoperability issues. Additionally, the issues related to authoring and effective use of GH applications are considered.

In contrast to multimedia cartography or multimedia GIS, GH focuses on the integration of distributed geographic content, whereas multimedia cartography/GIS (Cartwright *et al.* 1999) mostly avoids the use of any integration and leans to a more "artistic" considerations that is usually not dependent on a database. Hence, GH encloses a wide research field, that of content integration, which affects the design, implementation and use of GH applications. In addition, GH subsumes multimedia cartography/GIS provided that their concepts and tools are used in GH system development.

In contrast to a distributed GIS, GH interlinks geographic content that may be available at, and offered through, any software package and device. On the other hand, a distributed-GIS (either web-GIS or Internet GIS; Peng and Tsou 2003) focuses on the integration of GI-driven data and services that are usually available at a GIS or Map Server. Hence, GH is a wider term that encompasses the specialized field of distributed-GIS.

1.6 About this Book

This Book consists of twenty-five (25) chapters, including this one. These chapters present different issues related to Geographic Hypermedia and concentrate either in the concepts or in the tools involved in this new paradigm. An attempt to group these chapters based on their content and scope leads to five groupings.

Part I, entitled "Foundations of Geographic Hypermedia" comprises of three chapters. Chapter 1 provided a definition of Geographic Hypermedia, and presented its components, services and related technologies. Chapter 2 gives an introduction to new media tools, investigates how they affect the mapping paradigm, and poses the question which media is appropriate for a specific case. Chapter 3 provides design guidelines for hypermedia GI Systems, distinguishes between discrete and distributed systems and describes the software requirements, the file formats, and the data structures involved, including a prototyping system.

Part II, entitled "Data Resources and Accessibility Issues" consists of six chapters. Chapter 4 provides a survey of geographic data resources and gateways, and shows how to search and retrieve geospatial data from the web. Chapter 5 highlights the use of metaphors to access geographic information. Chapter 6 provides a definition of atlases and digital atlases, describes their characteristics, the alternative delivery media, and proposes a classification of them. Chapter 7 highlights the accessibility issues; specifically how the accessibility of hypermedia maps can be assisted. Chapter 8 focuses on the way to provide accessibility using mobile hypermedia systems. Chapter 9 proposes a new medium/platform for publishing, searching and enhancing communication on the web.

Part III, entitled "Technologies for Content Integration" comprises of five chapters. Chapter 10 gives an introduction to Spatial On-Line Analytical Processing (SOLAP) and how it can be combined with hypermedia technology; it suggests the management of geographic hypermedia documents into a data cube and presents a prototype system. Chapter 11 introduces a conceptual approach to integrate geographic, multimedia and cultural heritage information and collaborative aspects into a single model; this approach may be used to design and implement a web-based information system on top of a database. Chapter 12 considers the development of open hypermaps to achieve the effective information and the exchange of services between them. Chapter 13 presents the main existing standards for open GI Systems, hypermedia systems and cultural heritage systems; it provides a comparison between them and discusses interoperability issues. Chapter 14 focuses on the role of semantics in a geographic hypermedia system; the issues of extraction of geosemantics, ontology engineering and ontology integration.

Part IV, entitled "Analytical Functionality and Geovisualizations" consists of six chapters. Chapter 15 provides a typology of functions for visual exploration on thematic maps. Chapter 16 proposes a comprehensive data exploration through the combination of multiple tools; it sketches a taxonomy of tool and display combinations described in terms of possible input data and tool outputs. Chapter 17 discusses the role of geovisualization and hypermedia, and in concepts in decision-making; two prototype systems are presented. Chapter 18 shows how geovisualization techniques may assist the extraction of spatial relationships and thematic/geometric inconsistencies. Chapter 19 presents hypermedia-based concepts and techniques that lead to effective visualization of time-dependent geographic information. Chapter 20 provides a gallery of Scalable Vector Graphics (SVG) based applications and interfaces for geographic hypermedia.

Part V, entitled "Geographic Hypermedia: Applications and Services" comprises of five chapters. Chapter 21 discusses the principles and communication methods to assist wayfinding in urban areas. Chapter 22 presents the development of a campus location based service (LBS) and investigates the necessary infrastructure for LBS's. Chapter 23 proposes the development of a web-GI System according to human-computer interaction (HCI) guidelines; an example context is examined. Chapter 24 provides guidelines to compile hypermedia content and educate people with various experiences/training on earth and biodiversity issues using high definition video (HDTV) technology. Chapter 25 shows how mobile GI Systems may support education, presenting an example application at a university campus.

References

Andrienko G, Andrienko N (1999) Interactive maps for visual data exploration. International Journal of Geographical Information Science, vol. 13(4), Taylor-Francis, pp 355-374

Bernard L, Einspanier U, Haubrock S, Hübner S, Kuhn W, Lessing R, Lutz M, Visser U (2003) Ontologies for intelligent search and semantic translation in spatial data infrastructures. Photogrammetrie - Fernerkundung - Geoinformation, vol 6, pp 451-462, http://ifgi.uni-muenster.de/~lutzm/pfg03_bernard_et_al.pdf

Bishr Y (1998) Overcoming the semantic and other barriers to GIS interoperability. International Journal of Geographical Information Science. Special Issue on Interoperability in GIS, vol 12(4), Taylor & Francis, pp 299-314

Blustein J (2006) alt.hypertext: frequently asked questions (FAQ), http://www.csd.uwo.ca/~jamie/hypertext-faq.html)

Cartwright W, Peterson MP, Gartner G (eds) (1999) Multimedia cartography. Springer

Cartwright W, Peterson MP (1999) Multimedia cartography. In Cartwright W, *et al*. (1999) Multimedia cartography. Springer

EAA (2000) CORINE land cover: technical guide. European Environmental Agency

FOLDOC (2005) FOLDOC: free on-line dictionary of computing. Imperial College, Department of Computing, http://foldoc.doc.ic.ac.uk/foldoc

Fonseca F, Egenhofer M (1999) Ontology-driven information systems. In: Proceedings of the 7th ACM Symposium on Advances in GIS, ACM Press, Kansas, MO, pp 14-19

Frank A, Raper J, Cheylan JP (eds) (2001) Life and motion of socio-economic units. GISDATA 8, Taylor & Francis

Garcia Molina H, Ullman JD, Widom J (2000) Database Systems Implementation. Prentice Hall.

Goodchild MF (1997) What is geographic information science?, NCGIA Core Curriculum in GIScience, http://www.ncgia.ucsb.edu/giscc/units/u002/u002.html

Grønbæk K, Wiil KU (1997) Towards a common reference architecture for open hypermedia. Journal of Digital Information, vol 1(2), http://jodi.tamu.edu/

Guarino N (1998) Formal ontology in information systems. In: Proceedings of the Formal Ontology in Information Systems (FOIS), IOS Press, Trento, Italy, pp 3-15

Kraak M-J, van Dreil R (1997) Principles of hypermaps. Computers & Geosciences: Special issue on exploratory cartographic visualization, vol 23(4), pp 457-464

Kraak M-J, Brown A (eds) (2001) Web cartography. Taylor & Francis

Laurini R, Millert-Raffort F (1990) Principles of geomatic hypermaps. In: Proceedings of the 4th Conference on Spatial Data Handling, Zurich, Switzerland, pp 642-651

Maguire DJ, Goodchild MF, Rhind, DW (1991) Geographic information systems: principles and applications. Longman.

McKee L, Buehler R (1998) The open GIS guide. Open GIS Consortium Inc, http://www.OpenGIS.org/techno/ specs.htm

Nelson TH (1965) A file structure for the complex, the changing and the indeterminate. In: Proceedings of the 20th National ACM Conference, ACM, pp 84-100

Nürnberg PJ, Leggett JJ (1997) A vision for open hypermedia systems. Journal of Digital Information, vol 1(2), http://jodi.tamu.edu/

Nürnberg PJ, Leggett JJ, Wiil UK (1998) An agenda for open hypermedia research. In: Proceedings of the 9th ACM conference on hypertext and hypermedia, Pittsburgh, Pennsylvania, pp 198-206

OHSWG (2006) Open Hypermedia Systems Working Group. http://www.csdl.tamu.edu/ohs

Peng Z-R, Tsou M-H (2003) Internet GIS: distributed geographic information services for the Internet and wireless network. Wiley

Peterson MP (1999) Elements of multimedia cartography. In: Cartwright W, *et al.* (1999) Multimedia cartography. Springer

Peterson MP (2003) Maps and the Internet. Elsevier

SIGWEB (2006) SIGWEB: Special Interest Group on Hypertext, Hypermedia, and the web. ACM, http://www.sigweb.org

Stefanakis E, Sellis T (1998) Enhancing operations with spatial access methods in a database management system for GIS. Cartography and GIS (Journal), ACSM, vol 25(1), pp 16-32

Stefanakis E (2003) Modeling the history of semi-structured geographic entities. International Journal of Geographical Information Science, Taylor & Francis, vol 17(6), pp 517-546

Stoimenov L, Derdevic-Kajan S (2002) Framework for semantic GIS interoperability. Facta Universitatis, Ser. Math. Inform, vol 17, pp 107-125

Tomlin CD (1990) Geographic information systems and cartographic modeling. Prentice Hall.

Wache H, Voegele T, Visser U, Stuckensmhmidt H, Scuster G, Neumann H, Huebner S (2001) Ontology-based integration of information, a survey of existing approaches. Workshop on ontologies and information sharing, IJCAI01, pp 108-117

webopedia (2005) webopedia: the only online dictionary and search engine you need for computer and Internet technology definitions. http://www.webopedia.com

Wiil KU (1997) Open hypermedia: systems, interoperability and standards. Journal of Digital Information, vol 1(2), http://jodi.tamu.edu/

WWW (2006) The World Wide web Consortium, http://www.w3.org

2 New Media: From Discrete, to Distributed, to Mobile, to Ubiquitous

William Cartwright, Michael Peterson, and Georg Gartner

Abstract. *New Media* includes a range of new delivery and display platforms; among them are the World Wide Web, interactive digital television, mobile technologies, interactive hyperlinked mapping services, and enhanced mapping packages that are “linked” to large databases—national or global (Cartwright *et al*. 2001). New Media now provides a unique conglomerate media form for representing geospatial information in innovative ways. The many cartographic products developed and published using New Media illustrate the enthusiasm with which the geospatial science community has embraced it as a tool for representing geography. It is argued that this 'new' method of access to and representation of geospatial information is different to aforeused methods and therefore, whilst New Media applications can be considered to be at a fairly immature stage of development (compared to paper maps – here paper maps have a 500 years or so ‘start’ on electronic counterparts), much research has been undertaken to develop strategies for ‘best practice’ so as to overcome any deficiencies. This chapter gives an overview of how multimedia / hypermedia mapping has developed using discrete, distributed, mobile and ubiquitous media and devices. It provides an overview of the applications of New Media tools and communications systems that cartography has adopted / adapted to deliver timely and appropriate geoinformation. It provides examples of the use of discrete interactive multimedia, distributed interactive multimedia, and delivered via the World Wide Web (Web) or via intranets and mobile information delivery applications. Finally, it looks at how the concept of ubiquitous computing might be used to facilitate a ‘different’ paradigm for geographical information delivery.

2.1 Introduction

Cartography has always used / developed New Media mapping tools. When map cartographers/publishers applied printing to map production they used this 'New Media' to facilitate quicker, more accurate and cheaper versions of their works. The quest for more speed, lower compilation and production costs and an efficient communication system has led cartographers to embrace new technology. Developments in printing, were employed by cartography in its various guises - woodblock printing, letterpress, gravure, lithography and offset lithography, it was the norm for over 500 years. And, different 'manuscript to plate' methods were developed and used, like the application of photography, computer-driven exposure devices and direct computer-to-plate systems. Printing was complemented and enhanced by precision machines, computers (at first for computational mapping applications and later as complete interactive publishing systems). Later, optical storage media like videodisc, CD-ROM, DVD and their many configurations, communications systems – intranets and the Internet, and interactive installations in the form of hypermedia provided alternative publishing vehicles. Then visualization systems became mobile, providing geographical information 'where needed' and delivered on different, smaller platforms. Finally, the provision of this information became ubiquitous, delivering information in a manner that is virtually invisible to the final information consumer. Different and innovative mapping systems have developed and products have been produced to show 2D, 2½D, 3D and 3D+time (4D), plus n-dimensional data elements. There has been a digital convergence, and relatively inexpensive tools exist to develop and provide a plethora of (geo) information exploration devices. Tools and techniques are readily available, and the best methods for delivering effective geoinformation tools are being explored.

The following sections of this chapter provide information about how the tools for delivering geohypermedia have developed: from systems 'sitting' on a multimedia computer to systems that are now so commonplace that they have become ubiquitous.

2.2 Discrete

Here, discrete cartographic media refers to products made available through the use of isolated computers regardless of whether they are desktop, notebook or personal assistant. Nevertheless, it is recognised that all cartographic media can be discrete. However, the focus here is on discrete publishing media, and not discrete cartographic products *per se*. The packages are stored in digital form on floppy disk, hard disk drive, optical disk, videodisc or computer tape. They were applied to provide innovative access to image collections, data, information sites brimming with maps, 'do-it-yourself' map generation, map collections, atlases and hybrid products.

The analog videodisc was the storage medium used on the first product where the term 'multimedia' was used. In fact it was a mapping project:

the *Aspen Movie Map*, produced by MIT's Machine Architecture Group (which later led to the creation of the Media Lab) in 1978 (Negroponte 1995b). Videodiscs stored analogue video signals and were controlled by programs executed on a computer to which the videodisc was attached. This storage medium provided 36 minutes of PAL or NTSC read-only video from a 12" standard disc or 60 minutes from an extended play disc. They could store the equivalent of 52,000 slides and provided two audio tracks. As well as the MIT product, other examples of the application of videodiscs to the visualization of geographical information were The *Domesday* videodisc system (British Broadcasting Commission 1985), the Canadian Energy, Mines and Resources prototype, *Canada on Video Disk*, produced in 1987 (Duncan 1992) and the *Queenscliff* prototype videodisc, 1987 (Cartwright 1990).

Limited as they were, and constrained by underdeveloped user interfaces and interrogation routines, interactive videodisc products heralded the future of the application of hypermedia to geography. Products published on the 'Laservision' videodisc standard proved to be the forerunner of later products developed using CD-ROM.

CD-ROM, jointly developed by Philips and Sony in 1982, proved to be a most popular medium due to its (relatively) large, robust storage capacity. The media stored a minimum of 540 MB of data made it attractive to map producers looking for a robust storage medium that could store the large amounts of data associated with maps and other geographical visualization artefacts. CD-ROM was overtaken by DVD-ROM during late 1995 and by early 1996 some titles previously published on CD-ROM were reissued on DVD (Hamit 1996). By around 1998 DVD generally replaced CD-ROMs in most machines that required large storage capacities. Maps and other related products were published using this medium, and a wide range of innovative products resulted. Typical of these products were early publications that consisted of scanned maps from existing paper products.

Also around the same time, the introduction of the World Wide Web focussed interest away from discrete media to distributed applications. Writers like Negroponte (1995a) saw all kinds of package media slowly dying out. This was predicted for two reasons: the approaching 'costless' bandwidth of the Internet, allowing almost a limitless distribution system; and solid-state memory catching-up to the capacity of the CD, giving the prospect of massive data storage at minimal cost. The future demise of discrete multimedia was predicted by Louis Rosetto, the founder of *Wired*, and he called CD-ROMs the 'Beta of the '90s', referring to the quickly-defunct 'Betamax' video format (Negroponte 1995a). Negroponte (1995a, p. 68) agreed with him and said that: "*It is certainly correct that, in the long term, multimedia will be predominantly an on-line phenomenon.*"

By the mid to late 1990s the Internet, and more particularly the use of the World Wide Web (web), became the focus of interactive multimedia developers. Discrete media was pushed aside somewhat in the move towards the communication system that changed forever how we access information, including geoinformation.

2.3 Online / Distributed

From its beginnings in 1969 until about 1983, the Internet was under the control of the U.S. military. In that year, the U.S. National Science Foundation (NSF) began a major investment in the network through funding for a series of supercomputer centers. The purpose of these centers was to allow access to high speed computers by researchers from different locations. The NSF investment led to increased data communication speeds but the system was primarily still relegated to academic research scientists and the U.S. military. Email began to be used in the 1980s and this slowly increased the number of Internet users but the system was still difficult to use. Sending and receiving files required memorizing text commands. The initial World Wide Web, as conceived and implemented by Tim Berners-Lee at the beginning of the 1990s in a research laboratory in Switzerland, was still based on text. His hypertext WWW system allowed text elements to be linked to other text files.

Bringing the Internet to the masses required adding graphic elements to Internet information delivery. The first graphical WWW browser was a program called Mosaic introduced by the National Center for Supercomputer Applications at the University of Illinois in the US. Marc Andreesen and Eric Bina designed the program as a “consistent and easy-to-use hypermedia-based interface into a wide variety of information sources.” The concept, Andreessen says, "was just there, waiting for somebody to actually do it."

The Internet facilitated the wide-spread adoption of the Mosaic program in a matter of months. Never before had a program enjoyed almost immediate success – and distribution via the Internet made it possible. The appeal of Mosaic was primarily the incorporation of graphics. Not everyone was pleased with this new development. Tim Berners-Lee, who designed the Web only a few years before, admonished Andreesen and Bina by telling them that adding images to the Web was going to bring in a flood of new users who would do things like post photos of nude women. Andreesen later admitted that Berners-Lee was right on both counts.

Table 2.1. Browser market share first two months of 2006 and the last two months of 2005. Explorer and Netscape both lost market share to Firefox.
[IE: Internet Explorer; Ffox: Firefox; Moz: Mozilla; O: Opera; N: Netscape]

Month	IE6	IE5	Ffox	Moz	N7	O8	O7
February 2006	60.5%	5.8%	25.4%	2.9%	0.4%	1.4%	0.1%
January 2006	61.3%	5.5%	25.0%	3.1%	0.4%	1.4%	0.2%
December 2005	61.5%	6.5%	24.0%	2.7%	0.4%	1.3%	0.2%
November 2005	62.7%	6.2%	23.6%	2.8%	0.4%	1.3%	0.2%

Source: http://www.w3schools.com/browsers/browsers_stats.asp March 2006.

The commercialization of the Internet in the decade following 1995 is marked both by a large increase in the number of Internet users and a "browser war" with companies attempting to gain control of commerce through a standard browser. Mosaic had morphed into a commercial product called Netscape by October of 1994. Microsoft's Explorer was released in 1996. Toward the end of the decade, an open source program called Firefox began to break the hold of commercial interests (see Table 2.1). *Firefox* is viewed as a faster, trimmer Web browser that isn't subject to the crashes and security gaps that afflict the market-leading Internet Explorer. A general trend toward open source software, including a hugely popular web server program called Apache, began to take hold as dissatisfaction increased with commercial control of the Internet.

Tracking the number of Internet users is an inexact science. According to the Computer Industry Almanac, there are currently one billion Internet users or nearly 16% of the world's population. This is up from 533 million Internet users worldwide at year-end 2001 which at that time represented only 8.7% of the world's population. There were only 200 million Internet users at year-end 1998. It is expected that the number of Internet users will reach 1.46 billion by 2007. Most of the current Internet users are located in the top 15 countries (see Table 2.2). The major growth in the use of the Internet is coming from the East and South Asia, Latin America, and Eastern Europe. India is now ranked 5th in terms of the share of world Internet users. In 2001, India was not even in the top 15 (see Table 2.3).

It has also been observed that there is a disparity in the number of male and female Internet users, particularly in certain countries. In 2000, male-female ratio ranged from 94:6 in Middle East to 78:22 in Asia, 75:25 in Western Europe, 62:38 in Latin America, and finally 50:50 in USA (Dholakia, et.al, 2003). Updated data are presented in Table 2.4.

There has been great interest in the Internet by cartographers. Research has examined a variety of issues associated with the distribution of maps through the Web. The 2003 volume entitled Maps and the Internet (Peter-

Table 2.2. Top 15 nations in Internet use at year-end 2004. The last column indicates the percent of the world total. Data for some countries are not available.

Rank	Nation	Internet Users 2004 (millions)	Share of World Users
1	United States	186	19.86%
2	China	100	10.68%
3	Japan	78	8.35%
4	Germany	42	4.48%
5	India	37	3.96%
6	UK	33	3.54%
7	South Korea	32	3.39%
8	Italy	26	2.73%
9	France	25	2.72%
10	Brazil	22	2.39%
11	Russia	21	2.27%
12	Canada	20	2.19%
13	Mexico	14	1.49%
14	Spain	13	1.44%
15	Australia	13	1.39%

Source: Computer Industry Almanac (2004)

son 2003) identified five areas of study: Internet map use, Internet map delivery, Internet multimedia mapping, Internet mobile mapping, and Internet cartography theory. Research in Internet map use has found that usage grew rapidly after 1997, particularly at commercial sites. It has been shown that the growth in the use of maps through the Internet expanded at an exponential rate. Users were quick to adapt to map delivery through this new medium.

In 2005, Google.com introduced two online map services. The first, Google Earth, allowed the user to fly in a virtual sense between locations using a combination of satellite images, air photos, and maps. The second, Google Local, made it possible to add features to a map based on an address. Both services allow users to interact with representations of the earth in different ways than was possible in the past and many see these services as the beginning of a new era in Internet cartography. While Google Earth has gained considerable attention, the major development may be Google Local because it allows users to add information to maps for everyone else to see. Analogous to a Wiki, these maps represent a type of collective online map that a community of users can help construct. Rather than a cartographer deciding what to include on a map, this Google map will be made by users to show what they think is important.

Table 2.3. Top 15 nations in Internet use at year-end 2001. The last column indicates the percent of the world total. Data for some countries are not available.

Rank	Nation	Internet Users 2001 (millions)	Share of World Users
1	United States	149	41.92%
2	China	33.7	9.48%
3	UK	33	9.29%
4	Germany	26	7.32%
5	Japan	22	6.19%
6	South Korea	16.7	4.70%
7	Canada	14.2	4.00%
8	Italy	11	3.10%
9	France	11	3.10%
10	Russia	7.5	2.11%
11	Spain	7	1.97%
12	Netherlands	6.8	1.91%
13	Taiwan	6.4	1.80%
14	Brazil	6.1	1.72%
15	Australia	5	1.41%

Source: Computer Industry Almanac (2001)

Whether the change brought by the Internet is an evolution or a revolution for cartography has yet to be determined. Cartography has always been subject to developments in technology. Printing may represent the best analogy to the Internet because it also increased the distribution of maps. But, the medium of paper also limited their distribution. In addition to increasing the distribution of maps, the Internet also has the potential of changing the way spatial information is communicated to people. It is clear that maps will continue to evolve to take advantage of this new medium.

2.4 Mobile

In Europe, the first generation of mobile telephones appeared in the mid 1970's in Scandinavia and was based on analogue techniques. The second generation of mobile handheld devices brought digital transfer technologies as the "Global System for Mobile Communications" (GSM) and made the wireless phones a mass market phenomenon. Today, multiple standards are used in worldwide mobile communications. Different standards serve different applications with different levels of mobility, capability, and ser-

Table 2.4. Gender differences in Internet use by country ranked by disparity in male usage.

	Internet Users by Gender	
Country	*Male %*	*Female %*
Germany	63.4	36.6
France	61.9	38.1
Italy	60.9	39.1
Spain	60.9	39.1
Belgium	60.6	39.4
Netherlands	59.8	40.2
Brazil	59.7	40.3
Switzerland	58.7	41.3
Japan	58.4	41.4
Austria	58.1	41.9
Norway	58.0	42.1
UK	57.2	42.8
Israel	57.1	42.9
Hong Kong	56.6	43.4
Singapore	56.5	43.5
Denmark	55.9	44.1
Taiwan	55.8	44.2
Ireland	54.8	45.2
Sweden	54.8	45.2
South Korea	54.4	45.7
Mexico	54.0	46.0
Finland	53.9	46.1
New Zealand	52.5	47.5
Australia	51.6	48.4
Canada	49.0	51.0
United States	47.3	52.2

Source: Nielsen/NetRatings, 2003

vice area (paging systems, cordless telephone, wireless local loop, private mobile radio, cellular systems, and mobile satellite systems). Many standards are used only in one country or region, and most are incompatible. GSM is the most successful family of cellular standards, supporting some 250 million of the world's 450 million cellular subscribers with international roaming in approximately 140 countries and 400 networks.

When “Wireless application protocol” (WAP) started some years ago it was for the first time ever that mobile devices had restricted access to the Internet and content that was prepared especially for the use on mobile clients with small displays. Although it does not allow the provision of graphics other than in a very basic presentation, it has been used for first attempts. With 3rd generation technology UMTS it is possible to give continuous access to most of the internet sites, graphical presentations included.

The new so called "3rd Generation" (3G) of mobile phones features not only an IP-based technology but allows also for the first time so called “rich calls” transferring several user data streams simultaneously. This is also often referred to as “multimedia calls”. It was a question of data transfer rates which did not allow other than voice calls up to now. But users and developers of wireless devices always had the idea not only to transmit “simple” voice calls but also all other forms of digital data. The new technologies as "Global Packet Radio Switch" (GPRS) and the latest, on air since 2001, "Universal Mobile Transmission System" (UMTS) seem to make this idea become true for the first time in mobile communication. This will be possible only with the transmission rates proposed for the third generation of mobile devices as UMTS will be. The difference in speed between GSM and UMTS can be given by factor 50, in rare cases up to a factor of 200. This is a factor of 6 compared to ISDN and enables video transmission as well as audio files. Because UMTS technology enables the transfer of many different data formats in fast growing transmission rates, the development of complete new and attractive applications is initiated. Still, there are only very few ideas, prototypes and even less running applications trying to take advantage of the UMTS possibilities. But due to telecommunication companies this market will grow up and is currently highly focused in research and development.

There has been a massive growth in the use of the wireless Internet via cell phones. The wireless Internet share is increasing and it is predicted that that most of the growth in the use of the Internet will come from the wireless sector. However, it is likely that a wireless Internet user will also use a wired network.

Telecommunication infrastructure (mobile network), positioning methods, mobile in- and output devices and multimedia cartographic information systems are prerequisites for developing applications, which incorporate the user’s position as a variable of an information system. Integrating geospatial information into such a system, normally cartographic presentation forms are involved. Thus, the resulting system can be called a "map-based location based service" (LBS).

Different levels of solutions for presenting information within map-based LBS have developed:

- Cartographic presentation forms without specific adaptations
- Cartographic presentation forms adapted to specific requirements of screen display
- New and adapted cartographic presentation forms
- Multimedia add-ons, replacements and alternative presentation forms

Rules and guidelines have been developed during the last years to adapt cartographic presentations to the specific requirements of screen displays (Neudeck 2001). A lively discussion about new and special guidelines for map graphics regarding the very restrictive conditions of TeleCartography and mobile internet has brought up various suggestions and proposals (see Reichenbacher 2003, Gartner and Uhlirz 2001).

Common rules or standards for cartographic presentations on screen displays are not defined yet, due to the permanently changing determining factors. Display size and resolution of state-of-the-art devices are permanently increasing and colour depth is no longer a restricting factor. Parameters of external conditions during the use of the application (weather, daylight) are hard to model. The needs of an interactive system have to be incorporated into the conception of the user interface, which includes soft keys as well as functionalities for various multimedia elements. As a general approach for including the various parameters within a model of map-based LBS the concept of "adaptation" (in terms of user-dependent adaptation of a cartographic communication process) has been brought up (Reichenbacher 2003). The concept is to describe links or mutual dependencies between various parameters and the results are connected to impacts to the data modelling and cartographic visualization. Furthermore, new cartographic presentation forms especially designed for restricted and small screen displays have been developed (see e.g. “focus-map” by Klippel 2003).

For the presentation of geospatial information within LBS and on small displays additional multimedia elements and alternative presentation forms may become potential improvements. Methods of "Augmented Reality" (AR) link cartographic presentation forms (e.g. 3D graphic) to a user’s view of reality, e.g. at applications like navigation systems. Cartographic AR-applications try to create a more intuitive user interface (Reitmayr and Schmalstieg 2003). Kolbe (2003) proposes a combined concept of augmented videos, which realises positioning and information transfer by means of video.

2.5 Ubiquitous

Ubiquitous computing has been named as the 'third wave' in computing, or "...the age of calm technology, when technology recedes into the background of our lives" by the father of ubiquitous computing, the late Mark Weiser (1996). In his disquisition on "The Computer for the 21st Century" he assumed that in the near future a great number of computers will be omnipresent in our everyday life and that they will soon be interconnected in a ubiquitous network. We now see this type of computing in the form of handheld PCs, mobile phones, wireless sensors, radio tags and Wi-Fi (Baard 2003). Designers of ubiquitous systems envision seeding private and public places with sensors and transmitters that are embedded into objects and hidden from view, providing for the deployment of things like 'Audio Tags', which plays an infrared sensor-triggered message once a person is within a pre-determined proximity (Wired News 2003). In the mapping world, the interest in ubiquitous cartography has been formalised with the International Cartographic Association's Commission on Ubiquitous Mapping (http://ubimap.net/). It has as a goal to explore the potential that ubiquitous computing has for mapping.

Currently ubiquitous mapping is being delivered via cellular telephone systems, through the use of wireless Internet 'zones and sensors that upload current data like train timetables etc. to users who have subscribed to a service. Such services are deemed to be ubiquitous when a user does not need to 'log-in' or actively connect to a service. According to Morita (2004), ubiquitous mapping provides the ability to create and use maps any place at any time to resolve spatial problems.

Most recently geo-scientists started to discover the possibility to use the omnipresent computer landscape for exploring our spatial environment. Fairbairn (2005) explains the term 'Ubiquitous Cartography' as a technological and social development, made possible by mobile and wireless technologies, that receives, presents, analyses and acts upon map data which is distributed to a user in a remote location. Furthermore he predicts that this new approach to maps will revolutionize the way many people interact with maps. To Ota (2004) "the definition of ubiquitous mapping is that people can access any map at (sic) anywhere and anytime through the information network" (pp. 167).

A prominent field of application in the context of ubiquitous cartography is the support of orientation and navigation functions. Within the last few years a lot of research and development has taken place concerning Location Based Services, which could now be supplemented and expanded with the help of ubiquitous methods, and maybe in the future they could even be replaced. Yet research is still in the early development stage that still requires many new challenges. The improper usage of ubiquitous systems could easily lead to an overload of impressions. A lot of information that might even be completely independent from each other could overstrain the user and hinder effective information extraction. To avoid this effect the aim of such a system should concentrate on providing information about the environment without overstraining the user. At decision points the information should be unmistakably clarified but everywhere else, where guidance is not implicitly necessary, additional information should be provided in an unobtrusive way. User friendliness is therefore the main ambition of ubiquitous cartography applications.

The concept of ubiquity requires an intensive analysis of appropriate presentation forms for particular contexts. Beside the yet unspecified visualisation of the basic data material, namely the depiction of three-dimensional space on a two-dimensional display, additional visualisation techniques need to be considered that evolve from the possibility of interconnected data exchange. The basic assumption in this context is that a harmonized combination of active and passive systems with various presentation forms supports the wayfinding process best. Radoczky (2003) has shown that various presentation forms used for navigating a mixed indoor/outdoor environment lead to different mental representations and subjective acceptance.

The development of ubiquitous cartography so far gives an indication, that although maps will play a prominent role in ubiquitous environments, the nature of quick and individually tailored presentations of location-based and time-dependent information will lead to a wide variety of different cartographic presentation forms, from schematic 2d-graphics to interactive 4D-presentations. In terms of the contents, which are presented by these various forms, the concept of ubiquitous cartography implicates individually tailored contents, which remains a major challenge for contemporary cartography.

2.6 Conclusion

We now have access to information online, wired and ubiquitously. We can receive information about geography, our geography and related services. Access to information has changed forever and we are living in an information era that provisions information to us at home, in the workplace, in school and whilst we are 'on-the-move'. For designers and provisioners of geographical information artefacts this wealth of information-provision tools and communications systems offers challenges that ask us to question which delivery mode / media composite is appropriate for consumers, sometimes unknown, and geographical information delivery, chosen from a plethora of continually-updated data sources that are available and usually accessible. GeoHypermedia installations and products, built from data provisioned via discrete resources, on-line, wirelessly and ubiquitously provide users with a choice of information resources hitherto unimagined.

References

Baard M (2003) A connection in every spot. Wired News, 16 October, http://www.wired.com/news/print/0,1294,60831,00.html, web page accessed 19 November 2003

British Broadcasting Commission (1985) Domesday project, promotional booklet.

Camielon Project (2005) BBC Domesday, http://www.si.umich.edu/CAMILEON/domesday/pictures.html, web page accessed April 11, 2005

Cartwright WE (1990) Mapping and videodiscs: some observations on the design, production and program assembly of atlases on videodisc based on research results from pilot VideoAtlas of Queenscliff, Victoria, Australia. In: Proceedings of the British Cartographic Society Annual Symposium, Newcastle, United Kingdom

Cartwright W, Crampton J, Gartner G, Miller S, Mitchell K, Siekierska E, Wood J (2001) User interface issues for spatial information visualization. CaGIS, vol 28 (1), pp 45 – 60

Computer Industry Almanac (2004) Worldwide Internet Users will Top 1 Billion in 2005. http://www.c-i-a.com/200010iuc.htm

Cyberatlas (2002) European Women Surf to a Different Drum, http://cyberatlas.internet.com/big_picture/demographics/

Dholakia RR, Dholakia N, Kshetri N (2003) Gender and Internet usage in Bidgoli. In: Hossein (ed.) The Internet Encyclopedia, Wiley, New York

Duncan DJ (1992) Videodisk mapping for commercial use. In: Proceedings of the Canadian Conference on GIS, Ottawa, vol 1, pp 833 - 844

Fairbairn D (2005) Lecturer's appointment helps map future of Geomatics. http://www.ceg.ncl.ac.uk/news/news.htm, web page accessed June 2005

Gartner G, Uhlirz S (2001) Cartographic concepts for realizing a location based UMTS service: Vienna City Guide. In: Mapping the 21st Century - Proceedings of the 20th ICC, Beijing, pp 3229-3238

Hamit F (1996) DVD technology: exploding room for images in on-disk multimedia. Advanced Imaging, pp. 42 - 44

Klippel A (2003) Wayfinding Choremes: conceptualizing wayfinding and route direction elements. Doctoral dissertation, University of Bremen

Kolbe T (2003) Augmented videos and panoramas for pedestrian navigation. In: LBS & TeleCartography, Geowissenschaftliche Mitteilungen, vol 66, pp 45-52

Morita T (2004) Ubiquitous mapping in Tokyo. In: Proceedings of the 1st International Joint Workshop on Ubiquitous, Pervasive and Internet Mapping, pp 7–16

Negroponte N (1995a) Digital videodiscs: either format is wrong, Wired, pp 222

Negroponte N (1995b) Being Digital. Hodder and Stoughton, Rydalmere

Neudeck S (2001) Gestaltung topographischer Karten für die Bildschirmvisualisierung (Design of topographic map for screen display). Schriftenreihe des Studienganges Geodäsie u. Geoinformation der Univ. der Bundeswehr München, vol 74

Ota M (2004) Ubiquitous path representation by the geographic data integration. In: Proceedings of the 1st International Joint Workshop on Ubiquitous, Pervasive and Internet Mapping, Tokyo, pp. 166 – 172

Peterson M (ed) (2003) Maps and the Internet. Elsevier Press, Amsterdam, Cambridge

Radoczky V (2003) Kartographische Unterstützungsmöglichkeiten zur Routenbeschreibung von Fußgängernavigationssystemen im In- und Outdoorbereich. Diplomarbeit am Institut für Kartographie und Geo-Medientechnik, TU-Wien

Reichenbachter T (2003) Adaptive methods for mobile cartography. In: Proceedings of the 21st ICC, Durban

Reitmayr G, Schmalsteig D (2003) Collaborative augmented reality for outdoor navigation and information browsing. In: LBS & TeleCartography, Geowissenschaftliche Mitteilungen, vol 66, pp 53-59

Weiser M (1996) Ubiquitous Computing. http://sandbox.xerox.com/hypertext/weiser/UbiHome.html, web page accessed 19 November 2003

Wired News (2003) Balancing utility with privacy, http://www.wired.com/news/print/0,1294,60871,00.html, web page accessed 19 November 2003

3 Design Issues Associated with Discrete and Distributed Hypermedia GIS*

Shunfu Hu

Abstract. The advancement of computer technology enables the integration of geographic information system (GIS) and multimedia technologies that allow to incorporate not only spatial-temporal geographic information in image/vector format, but also multimedia geographic information in descriptive text, scanned ground photographs, graphics, digital video and sound. The concept of hypermedia GIS is defined in this chapter. Design issues on the development of hypermedia GIS for use on individual personal computers (PC) (i.e., discrete system) and on the Internet (i.e., distributed system) are discussed. Software requirement, file format and data structure used in each system are described. The discrete and distributed hypermedia GIS provide the essential concepts and techniques for many new GIS applications such as visualization, spatial decision support systems and spatial database management and exploration.

3.1 Introduction

The advancement of computer technology enables the integration of digital geographic information systems (GIS) and multimedia technologies to incorporate not only spatial geographic information in image/vector format, but also multimedia information. The term "multimedia" implies the use of a personal computer (PC) with information presented through the following media: 1) text (descriptive text, narrative and labels); 2) graphics (drawings, diagrams, charts or photographs); 3) digital video (television-style material in digital format); 4) digital audio sound (music and oral narration); and 5) computer animation (changing maps, objects and images) (Bill 1994). Multimedia technology has been extensively utilized by commercial encyclopedia CD-ROMs (e.g., *Microsoft Encyclopedia*) to provide a multi-sensory learning environment and the opportunity to improve concept understanding.

---

* Adapted with permission from GIScience and Remote Sensing, Vol. 41, No. 4, 371-383.

The integration of multimedia technology and computer-assisted mapping systems has gone through several stages. The first stage was the development of interactive maps and electronic atlases during the 1980s. The interactive map is a computer-assisted form of map presentation and is characterized by an intuitive graphical user interface that allows the user to manipulate the map features (points, lines or polygons) through a computer mouse (Peterson 1995). The link to multimedia information is achieved through superimposing "hotspots" on the cartographic features of the map or on digital remote sensor data (e.g., digitized aerial photographs). Interactivity becomes a key feature of the interactive maps, which allow the user to explore more detailed information in the area predefined by the map developer. Examples of early electronic atlases include the *Domesday Project* and *Goode's World Atlas* (Openshaw and Monnsey 1987, Rhind *et al.* 1988, Espenshade 1990).

The second stage was the development of the "hypermap" in the early 1990s. Coined by Laurini and Milleret-Raffort (1990), the term "hypermap" was described as multimedia hypertext documents with geographical access. In other words, the hypermap is an interactive, digital multimedia map that allows users to zoom and find locations using a hyperlinked gazetter (Cotton and Oliver 1994). The underlining principle of the hypermap is the concept of hypertext. Hypertext represents a single concept or idea. By activating pre-defined hyperlinks, it is possible for the user to connect a hypertext to other non-linear text information (Nielson 1990). If the hypertext is linked to multimedia information, the term "multimedia hypertext" or "hypermedia" is used. Therefore, hypermap is also called "cartographic application of hypertext" or "hypermedia mapping" (Cartwright 1999). The development of hypermaps was made possible with Apple Corporation's Hypercard software developed for the Macintosh computer released in 1987 (Ravenau *et al.* 1991). Examples of hypermaps include the *Glasgow Online Digital Atlas* (Raper 1991) and HYPERSNIGE (Camara and Gomes 1991).

The third stage was the integration of hypermedia systems (which feature hypertext, hyperlinks and multimedia) and geographic information system (GIS), referred to as hypermedia GIS. A GIS is used to capture, retrieve, manipulate, and display geographic information. It is used to link cartographic features and their alpha-numeric attributes to perform spatial analysis (Burrough 1986, Star and Estes 1989, Clarke 1995). However, GIS is limited in handling multimedia information (Shepperd 1991). Recently, there has been increasing interest in integrating multimedia information in GIS (Bill 1994, Hu 1999, Hu *et al.* 2000, Hu *et al.* 2003). GIS development in recent years has seen fast advancement in both the desktop computing environment and on the Internet. The former is referred to "discrete GIS" and the latter is "distributed GIS" (Peng and Tsou 2003). "Discrete GIS" is a standalone GIS software package such as ArcView 3.3. or ArcGIS 9.0 that is installed typically on a PC or Unix machine. This traditional GIS is a closed, centralized system that incorporates a graphical user interface, programs, and data. Each system is platform dependent (e.g., Unix, Windows, Macintosh) and application dependent (e.g., ArcView, Arc/Info, or MapInfo). Data to be used in the discrete GIS application can be stored on computer hard drives or compact discs. Access to the data by the GIS program is usually fast. The term "Distributed GIS", on the other hand, is defined as GIS software and related data that are distributed via different computers located in different physical locations through a computer network such as a local area network (LAN) or the Internet. A distributed GIS applies the dynamic client/server concept in performing GIS analysis tasks through standard interfaces of the World Wide Web (WWW). Web-based GIS applications provide GIS database query or interactive map exploration on the Internet (Peng and Nebert 1997, Abel *et al.* 1998, Peng 1999, Dragicevic *et al.* 2000, Myer *et al.* 2001, Zhang and Wang 2001). Unfortunately, access to the data (stored on a computer server) by the Internet user (client) through the computer network is usually relatively slow when compared to processing in a discrete GIS environment.

Since discrete GIS and distributed GIS are different computing environments, the integration of a hypermedia system within a discrete or distributed GIS needs to be addressed in an appropriate manner. There is currently no single literature available that discusses the design issues related to the development of discrete hypermedia GIS and distributed hypermedia GIS. The objective of this chapter is to describe software requirements, file formats and data structures for the development of discrete and distributed hypermedia GIS, respectively. In order to explain how the two systems are developed, the author will use a portion of the vegetation database generated for the Everglades National Park (Welch *et al.* 1999). In both cases, emphasis is placed on the design and development of the hypermedia GIS system that includes: 1) a hypermedia vegetation plant community database; 2) a hypermedia browser; and 3) the integration of the hypermedia system and GIS application.

3.2 Design of a Discrete Hypermedia GIS

The discrete hypermedia GIS can be seen as an integrated system between a standalone GIS that provides tools to manipulate map features and their attributes stored in a GIS database and a hypermedia system that includes a hypermedia database (including hypertext and hyperlinks) and a browser. The linkage between the two different structured databases can be established through a common identification number (ID) as seen in a relational database management system. The map features can be manipulated through commercial GIS software (e.g., ArcView) or customized GIS program developed using Visual Basic (Microsoft 1994) and MapObjects (ESRI 1996). In the design of the hypermedia system, hypertext is employed to highlight key features (e.g., plant species) and to facilitate access to further detailed information about these features. The development of a hypertext document on the WWW requires the use of the Hypertext Markup Language (HTML - a programming language for publishing text-based documents on the Internet), and an Internet browser (e.g., Microsoft Internet Explorer or Netscape Navigator). The following sections discuss the design of a discrete hypermedia GIS, including: 1) the data structure of a hypermedia database; 2) the development of a standalone browser to visualize and explore the hypermedia database; and 3) the integration between the hypermedia system and a GIS application.

3.2.1 Hypermedia Database

Descriptive text information about the plant community is fundamental to the preparation of the Everglades hypermedia vegetation plant community database. Hypertext is employed to highlight key features (e.g., plant species) and to facilitate access to additional detailed information about these features (e.g., text, ground photographs, digital video and sound). The construction of hypertext involves the use of three elements: 1) a hyperlink; 2) the text information associated with the hypertext, and 3) the media (i.e., scanned photographs or digital pictures, video or audio) associated with the hypertext. Accordingly, the data structure of the hypermedia database will contain three fields, namely, Hyperlink, text and Media (Fig 3.1a).

Visual Basic 6.0 comes with a utility called Visual Data Manager that allows the development of databases in various formats such as Microsoft Access 7.0 and dBASE 5.0. Fig. 3.1b displays a Hypermedia Database Editor, generated in Visual Basic to create, update, and save a hypermedia database in Microsoft Access Database format (.mdb). The Hyperlink field stores the name of the hyperlink for each hypertext. The Text field stores the descriptive text information associated with the hypertext. The syntax of HTML was used to organize text information, set up hypertext and establish hyperlinks. The Media field of the database stores only the file names of photographs (.bmp), digital video (.avi) or audio sound (.wav)

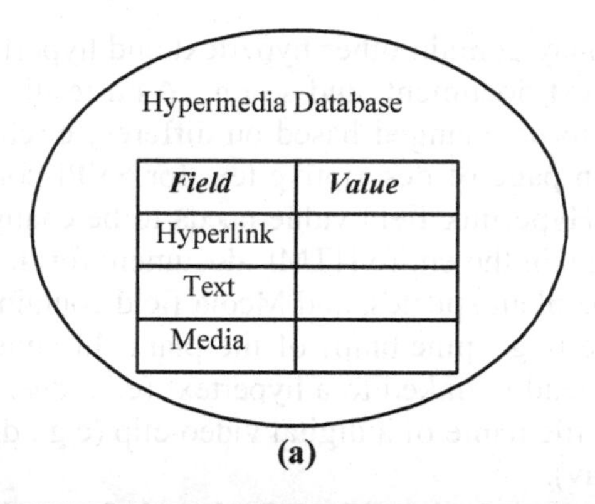

(a)

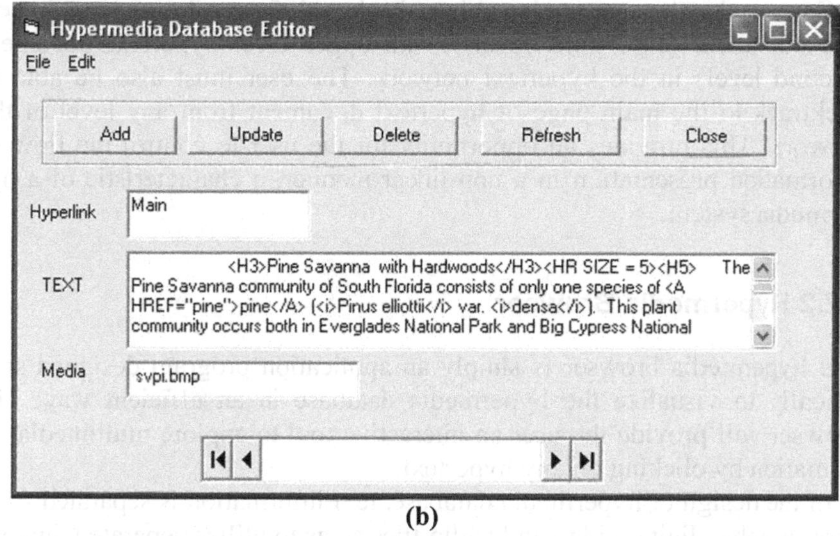

(b)

Fig. 3.1. Hypermedia database structure (a). Hypermedia database editor (b).

associated with the hypertext. In Fig. 3.1b, the Everglades SVPI plant community (i.e., Pine Savanna with Hardwoods) is used as an example for the field value. The Hyperlink field contains the value of "Main", indicating that it is the main page of the database, the Text field contains the entire HTML document, and the Media field contains the file name (e.g., svpi.bmp) of the scanned ground photograph for the panoramic view of the SVPI plant community.

There are multiple levels of hypertext within a hypermedia database. The first level is the main page of hypertext document for a plant community, which may contain hypertext and hyperlinks. The user is able to click on any hypertext in the first level and reach the second level hypertext document which may contain other hypertext and hyperlinks, and then the third level hypertext document, and so on. As a result, the values of the three fields need to be changed based on different circumstances. For instance, if the main page of descriptive text for SVPI contains a hypertext "pine," then, the Hyperlink field value needs to be changed to "pine," the Text filed will contain the entire HTML document for descriptive information about the pine plant species, and Media field contains the file name of a close-up picture (e.g., pine.bmp) of the plant. In another instance, if a digital video or sound is linked to a hypertext (e.g., deer), then the Media field contains the file name of a digital video clip (e.g., deer.avi) or an oral narrative (deer.wav).

Conversely, the user is also able to backtrack from a lower level hypertext document (e.g., third level) to an upper level hypertext document (second level) in the hypertext network. The user must also be able to backtrack to the main page of hypertext document from any level of the network. This provides an opportunity for the user to control the flow of information presentation in a non-linear manner, a characteristic of a hypermedia system.

3.2.2 Hypermedia Browser

The hypermedia browser is simply an application program designed specifically to visualize the hypermedia database in an efficient way. The browser will provide the user an interactive tool to explore multimedia information by clicking on any hypertext.

In the design of hypermedia database, text information is separated from photographs, digital video and audio files so as to utilize separate windows to convey information associated with the same text. The utilization of both descriptive text and a photograph for the same hypertext typically improves understanding. Consequently, two display windows, namely a Text window and a Graphic window, may be required to display descriptive text and photographs, respectively. No display window is required for the digital video and sound because the digital video can be played back in a window provided by the Windows operating system, and audio sound can be played back as background.

A few procedures are required in order to display text information (including hypertext and hyperlinks) in the Text window, display photographs in the Graphic window and play back digital video or sound (Fig. 3.2). The procedure, OpenDatabase( ), is required to open the hypermedia database in Access format and then split it into two streams: Text and Media, corresponding to the Text and Media fields of the database. For the

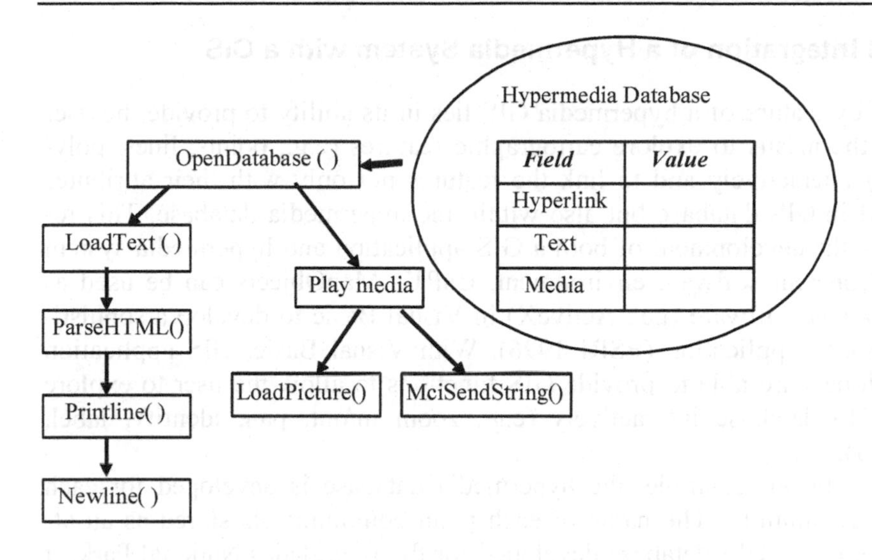

Fig. 3.2. Basic procedures to create a hypermedia browser.

Text stream, LoadText( ) is used to load text from the Text field and to check if there is any HTML syntax. If the HTML syntax exists, a third procedure, ExecuteHTML( ), is used to process the HTML syntax which includes the one used to set the hypertext in blue and underlined, and the one used to establish the hyperlink. This process continues until all HTML syntax in the Text field has been checked. Two other procedures are required to print a line of text and start a new line in the Text window.

For the Media stream, there are three cases: 1) scanned photograph (.bmp); 2) digital video (.avi); and 3) audio sound (.wav). In the case of a scanned photograph, the *LoadPicture ( )* function, available in Visual Basic, can be used to display the photograph in the Graphic window. The code in Visual Basic is:

```
Case “bmp”
GraphicWindow.Picture = LoadPicture(App.path & media)
```

In the case of digital video file (.avi) or sound file (.wav), a specialized function, mciSendString ( ), can be used (Aitken 1996, Jarol and Potts 1995). The Visual Basic code is:

```
Case “wav”, “avi”
Dummy = mciSendString (“play” & App.Path & media, 0, 0, 0)
```

3.2.3 Integration of a Hypermedia System with a GIS

The key feature of a hypermedia GIS lies in its ability to provide the user with the means to explore cartographic features (e.g., points, lines, polygons) interactively and to link the features not only with their attributes stored in GIS database but also within the hypermedia database. This requires the development of both a GIS application and hypermedia system in a coherent software environment. ESRI's MapObjects can be used as component software (i.e., ActiveX) in Visual Basic to develop a sophisticated GIS application (ESRI 1996). With Visual Basic, GIS application developers are able to provide GIS functions to allow the user to explore the GIS database interactively (e.g., zoom in/out, pan, identify, label, search).

In our example, the hypermedia database is developed for each plant community. The name of each plant community is stored as an attribute of the GIS database developed for the Everglades National Park. It is feasible to establish the linkage between the GIS database and hypermedia database through this common key (i.e., the name of the plant community, or just its abbreviation). In addition, since both GIS application and hypermedia system are all developed in the Visual Basic programming environment, they are coupled coherently. Further more, both the application software and the databases can be stored on a CD-ROM or DVD-ROM for wide distribution.

Fig. 3.3 demonstrates a prototype hypermedia GIS developed for the Everglades National Park in which basic GIS tools are provided to the user to manipulate the GIS database. At the same time, the user can identify the name of the plant community (e.g., SVPI) by clicking on a polygon feature from the digital map and a hyperlink is activated with the hypermedia system which displays the panoramic view of the plant community and the main page of the descriptive text information. The hypertext and hyperlinks in the Text window allow the user to explore more detailed information about this plant community.

3.3 Design of Distributed Hypermedia GIS

The distributed hypermedia GIS is designed for use on the Internet. Such applications are based upon the interactions between client and server computer systems through network technology. The client side allows Internet users to access remote computers on the Internet by providing requests through standard Web browser software such as Microsoft Internet

Explorer, Netscape Navigator, or other custom-generated software such as ESRI ArcExplorer.

The server side consists of at least three components, including the web server, the map server and the data server. Web server software such as Netscape's FastTrack Server or Microsoft's Internet Information Server provides the capability to manage and respond to requests from the client side.

The map server, interacting with the web server, implements data processing in a GIS application. Examples of such map servers include ESRI ArcView Internet Map Server (IMS), MapObjects IMS, and ArcIMS, MapInfo MapX and MapXtreme. The data sever provides various data sets such as ESRI ArcView shapefiles, ArcInfo coverages, remotely sensed data, and/or other statistical data. Typically, the components on the server side can be placed on more than one computer.

The network technology provides Internet software components that communicate with each other on various computers connected by the network. Those components include HTTP (i.e., hypertext transfer protocol), and TCP/IP (transmission control protocol / Internet protocol). Protocols are the languages that make Internet communication possible.

The distributed hypermedia GIS is based upon interactions between three components: 1) a web-based GIS application developed to manipulate the cartographic features and their attributes; 2) a web-based hypermedia system designed to manipulate multimedia information including hypertext, hyperlinks, graphics, photographs, digital video, and sound; and 3) a mechanism linking the web-based GIS application and the hypermedia system.

3.3.1 Web-based GIS application

In addition to the requirement of developing a standalone GIS application running on a PC, the web-based GIS requires additional software components: web server and map server. In the design of a distributed hypermedia GIS, the Microsoft Internet Information Server (MIIS) is employed as the web server and the ESRI MapObjects Internet Map Server (IMS) as the map server. MapObjects IMS is the IMS extension to MapObjects (ESRI 1998). It provides ready-to-use software components such as ESRI-Map(n).dll, IMSAdmin.exe, IMSCatalog.exe, IMSLaunch.exe, and WebLink.ocx that enable the developer to run MapObjects applications on the Internet.

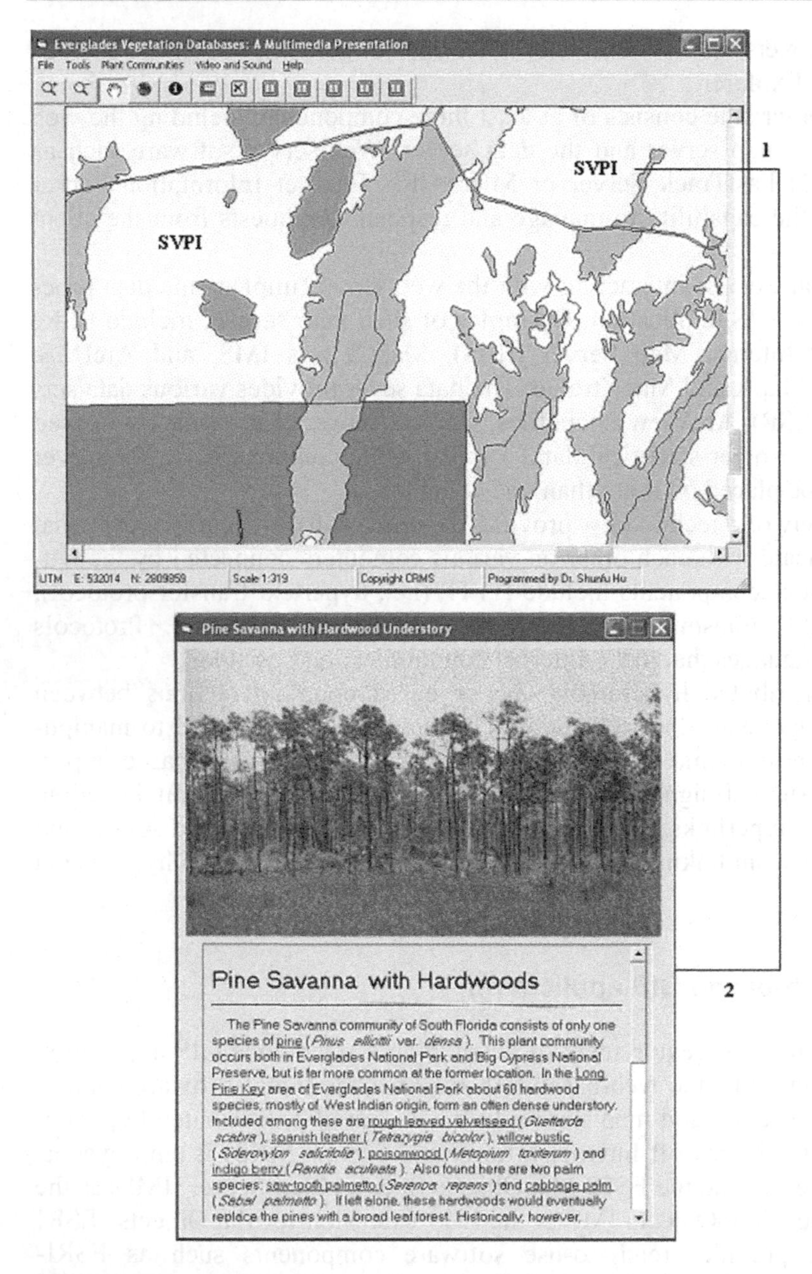

Fig. 3.3. The two basic components of a discrete hypermedia GIS for the Everglades vegetation database: (1) the GIS component used to explore the GIS database of a vegetation plant community via standard GIS operations; and (2) the Hypermedia component to access the hypermedia database for the same plant community.

3.3.2 Web-based Hypermedia system

With the essential features of a hypertext system commonly seen on the WWW, it is feasible to develop a web-based hypermedia application or web homepage. There are numerous ways to design excellent hypertext documents for use on the Internet. Software for developing interactive hypermedia presentations for use on the Internet includes HTML, Macromedia Director and Dreamweaver, and Adobe Streaming Media Collection, to name just a few. Since the hypermedia database developed for each plant community in the discrete hypermedia GIS was edited in HTML, each hypertext document is now readily employed on the WWW. All the HTML syntax in each hypertext document can be interpreted by the Internet Explorer without an additional hypermedia browser.

Due to the limitation of the current Internet speed, a digital video file that usually can be played back efficiently on an individual PC is not suitable for use on the Internet because of its large file size. The remedy for that is to change the format of the digital video file from Windows AVI (Audio Video Interleave) to Apple QuickTime movie format (.mov). The only additional requirement for the client computer is to install a QuickTime movie player (a free shareware that can be downloaded from Apple's website). Similarly, scanned pictures in Windows bitmap format (.bmp) need to be compressed to JPEG (Joint Photographic Experts Group) format (.jpeg) to allow fast delivery over the network.

Fig. 3.4 is an illustration of the web-based GIS application for visualizing the Florida Everglades vegetation database. Once the program is up running on a computer server and the user knows the web address or the server name, he/she is able to access the GIS database through standard Internet browser. GIS functions such as Zoom, Pan, and Identify are provided to allow the user to manipulate the vector GIS database. For instance, the user can select the "Zoom in" option and zoom into an area on the digital map, then select the "Identify" option, click on any polygon, and the alphabetic letters representing the plant community in that polygon will be displayed on the client's computer screen. Further, the user can select the "Hyperlink" option, click on one polygon, and be directed to the web-based hypermedia system or web site containing a ground panoramic view of that plant community and a descriptive text, including hypertext and hyperlinks, about the plant community (Fig. 3.5). Video clips can be either linked directly to a map feature or to hypertext.

3.4 Conclusions

Hypermedia GIS provides more visualization capabilities over traditional GIS: interactivity, user control of information flow, and multimedia presentation. The design of an operational hypermedia GIS can be a challenging task for the GIS community. The developers of hypermedia GIS applications need to clearly understand the difference between the one used on a PC by individual users and the one used on the Internet. The former has the flexibility of being able to deal with large file sizes associated with both the GIS database and hypermedia database, especially when digital video and high-resolution photographs are utilized. The later has the advantages of platform independence and wide accessibility.

Both discrete and distributed hypermedia GIS are generating great interest in the GIS community. With the integration of multimedia information in GIS, the hypermedia GIS is able to handle geographic information in

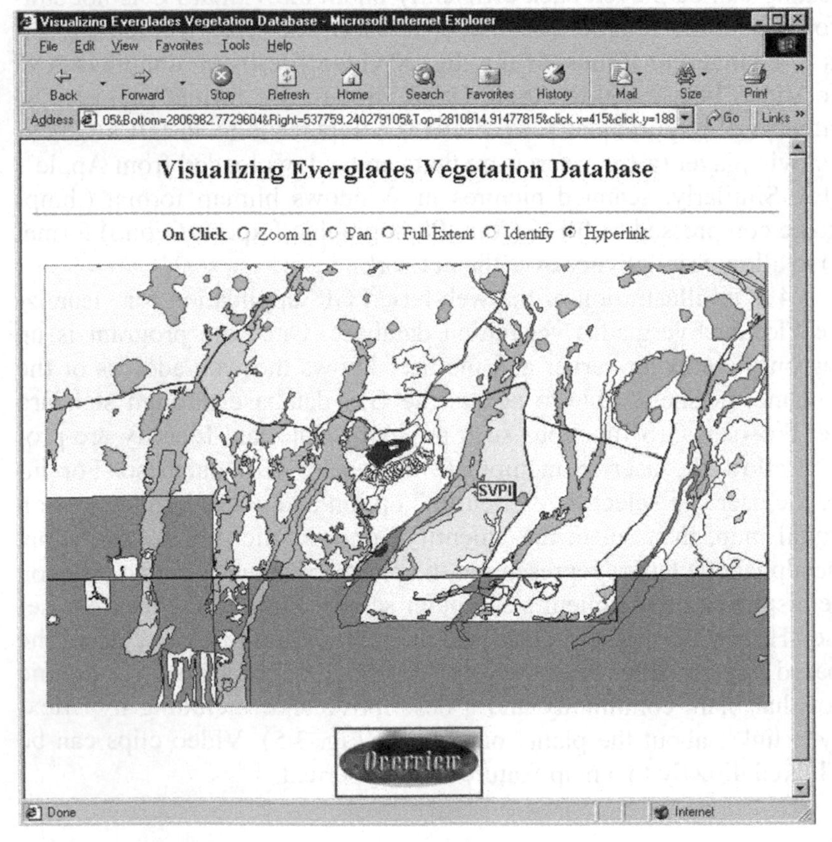

Fig. 3.4. Interactive map interface of a web-based GIS application.

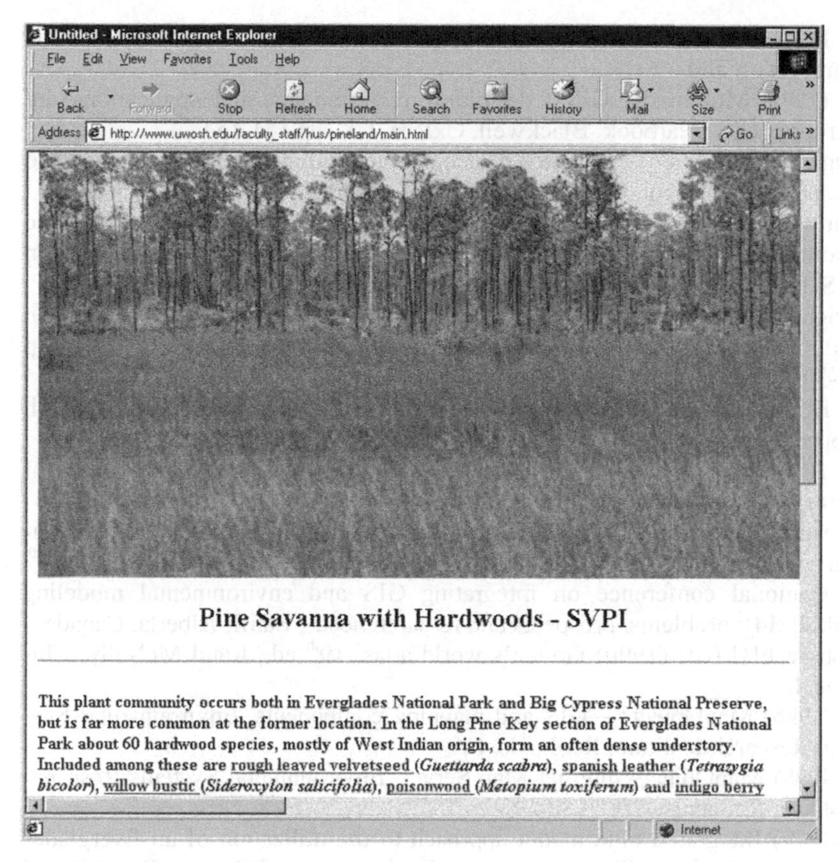

Fig. 3.5. Web-based hypermedia system (hypermedia database, hypertext, hyperlinks via Internet Explorer).

any format (e.g., spatial data in image/vector format, attribute data in alphanumeric format, and multimedia data in the form of text, graphics, photographs, and digital video). It provides the essential concepts and techniques for many new GIS applications such as visualization, spatial decision support systems and spatial database management and exploration.

References

Abel D, Kerry T, Ackland R, Hungerford S (1998) An exploration of GIS architectures for Internet environments. Computer, Environment and Urban Systems, vol 22(1), pp 7-13.

Aitken PG (1996) Visual basic for windows 95 insider. John Wiley & Sons, New York, NY, p 580

Bill R (1994) Multimedia GIS - definition, requirements, and applications. In: European GIS Yearbook, Blackwell, Oxford, UK, pp 151-154.

Burrough PA (ed) (1986) Principles of geographical information systems for land resources assessment. Oxford Science Publications, New York, NY, p 193

Camara A, Gomes AL (1991) HYPERSNIGE: a navigation system for geographic information. In: Proceedings of EGIS'91, second European conference on GIS, Brussels, Belgium, pp 175-179.

Cartwright W (1999) Development of multimedia. In: Cartwright W, Peterson MP, Gartner G. (eds) Multimedia cartography. Springer, New York, NY, pp 11-30.

Clarke KC (ed) (1995) Analytical and computer cartography. 2nd ed., Prentice Hall, Upper Saddle River, NJ, p 352

Cotton B, Oliver R (1994) The cyberspace lexicon – an illustrated dictionary of terms from multimedia to virtual reality. Phaidon Press Ltd, London, UK

Dragicevic S, Balram S, Lewis J (2000) The role of web GIS tools in the environmental modeling and decision-making process. In: Proceedings of the 4th international conference on integrating GIS and environmental modeling (GIS/EM4): problems, prospects and research needs, Banff, Alberta, Canada

Espenshade EBJ (ed) (1990) Goode's world atlas. 19th ed., Rand McNally, Chicago, IL

ESRI (1996) MapObjects – GIS and mapping components. Environmental Systems Research Institute, Redlands, CA

ESRI (1998) MapObjects Internet Map Server. Environmental Systems Research Institute, Redlands, CA

Hu S (1999) Integrated multimedia approach to the utilization of an Everglades vegetation database. Photogrammetric Engineering and Remote Sensing, vol 65(2), pp 193-198

Hu S, Gabriel AO, Bodensteiner LR (2003) Inventory and characteristics of Wetland habitat on the Winnebago Upper Pool Lakes, Wisconsin, USA: an integrated multimedia-GIS approach. Wetland, vol 23(1), pp 82-94

Hu S, Gabriel AO, Lancaster C (2000) An integrated multimedia approach for Wetland management and planning of Terrell's Island, Winnebago Pool Lakes, Wisconsin. The Wisconsin Geographer, vol 15-16, pp 34-44

Jarol S, Potts A (1995) Visual basic 4: multimedia adventure set. The Coriolis Group Inc, Scottsdale, AZ

Laurini R, Milleret-Raffort F (1990) Principles of geomatic hypermaps. In: Proceedings of the 4th international symposium on Spatial Data Handling, Zurich, Switzerland, vol 2, pp 642-655.

Meyer J, Sugumaran R, Davis J, Fulcher C (2001) Development of a web-based watershed level environmental sensitivity screen tool for local planning using multi-criteria evaluation. In: Proceedings of ASPRS Annual Conference, St. Louis, Missouri

Microsoft (1994) Microsoft visual basic: programmer's guide. Microsoft Corporation, Seattle, WA, p 857

Nielson J (1990) Hypertext and hypermedia. Academic Press Professional, Boston, MA, p 296

Openshaw S, Mounsey H (1987) Geographic information systems and the BBC's Domesday interactive videodisk. International Journal of Geographical Information Systems, vol 1(2), pp 173-179

Peng Z, Nebert D (1997) An Internet-based GIS data access system, Journal of the Urban and Regional Information Systems Association, vol 9(1), pp 20-30

Peng ZR (1999) An assessment framework for the development of Internet GIS. Environment and Planning B: Planning and Design, vol 26(1), pp 117-132

Peng ZR, Tsou MH (eds) (2003), Internet GIS – distributed geographic information services for the Internet and wireless networks, John Wiley & Sons, Hoboken, NJ, p 679

Peterson MP (ed) (1995) Interactive and animated cartography. Prentice Hall, Upper Saddle River, NJ, p 257

Raper J (1991) Spatial data exploration using hypertext techniques. In: Proceedings of EGIS'91, second European conference on GIS, Brussels, Belgium, pp 920-928

Raveneau JL, Miller M, Brousseau Y, Dufour C (1991) Micro-atlases and the diffusion of geographic information: an experiment with hypercard. In: Taylor FDR (ed) GIS: The Microcomputer and Modern Cartography. Pergamon Press, Oxford, UK, pp 263-268

Rhind DP, Armstrong P, Openshaw S (1988) The Domesday machine: a nationwide geographical information system. Geographical Journal, vol 154(1), pp 56-58

Shepherd ID (1991) Information integration and GIS. In: Magurie DJ, Goodchild, MF, Rhind DW (eds) Geographical Information Systems: Principles and Applications, vol 1, Longman Scientific and Technical Publications, Essex, UK, pp 337-357

Star J, Estes J (eds) (1989) Geographic information systems: an introduction. Prentice Hall, Englewood Cliffs, NJ, p 303

Welch R, Madden M, Doren RF (1999) Mapping the Everglades. Photogrammetric Engineering and Remote Sensing (special issue), vol 65(2), pp 163-170

Zhang X, Wang YQ (2001) Web based spatial decision support for ecosystem management. In: Proceedings of ASPRS Annual Conference, St. Louis, Missouri.

PART II

DATA RESOURCES AND ACCESSIBILITY ISSUES

4 Using Data from Earth Orbiting Satellites in Geo-Hypermedia Applications: A Survey of Data Resources

Danny Hardin

Abstract. Every day a suite of satellites in orbit about the Earth captures massive amounts of data in a vast spectrum of wavelengths. The Earth's systems are being measured more accurately, more frequently, and with higher resolution than ever before in our history. The amount of data generated is prodigious, pouring into data archive centers at over 1000 GBytes per day. This presents a challenge to those who wish to locate and use data for a specific application over a constrained geospatial area and time span. The vast majority of the data are free, or available at low cost. However, there remain barriers to its use because in many cases the data are not in a preferred format, it is difficult to locate, it is hard to extract a specific data item from the massive inventories, or simply because users are not aware that an important data set exists. In this chapter, you will learn how to find data, by using data catalog services, and how to order data, by using data search and order systems. You will also be presented with summary information on the data resources available from NASA's nine data archive centers.

4.1 Introduction

At this writing, there are18 active NASA satellite missions dedicated to measurements of the Earth's systems – atmosphere, ocean, ice, land, and life (NASA Current Missions). There are many other Earth orbiting satellite missions from US government agencies such as the National Oceanic and Atmospheric Administration (NOAA), and from other countries such as the European Space Agency and the Space Agency of Japan. Unclassified data from military satellites such as the Defense Meteorological Satellite Program (DMSP) contribute further volume. And imagery from commercial orbiters such as IKONOS and SPOT round out a comprehensive suite of instruments aimed at the Earth. Added to this are measurements from instruments at the Earth's surface (land and oceans), and from short duration orbital and atmospheric flights. The total amount of available data is enormous. In this chapter descriptions are limited to data that are generally available at little or no cost to researchers, educators, and the general public. This can be done by considering only data sets that are (or have been) collected by NASA funded instruments or are available from NASA funded data archive centers. This is not a significant limitation as there are thousands of data sets that fall into this classification. In the following sections, you will discover how to find data by using data catalog services and how to order data by using data search and order systems. Summary information pertaining to data resources available from NASA's nine data archive centers is provided.

4.2 Discovering Data

There are thousands of data sets available from a variety of sources, including satellites, ground-based instruments on land and ocean, aircraft, balloons, rockets, and special spacecraft such as the Space Shuttle and Space Station. While the vast majority of data are from the past 30 years, there are many that date back over a century. Data sets exist in a variety of forms and resolutions. And, they are stored in archive centers, research labs, universities, and desk drawers around the world. Given all this, how is it possible to determine the kind of data available and its location? Fortunately, there are data directories that make the task simple. If you want to know what data are available, a directory should be your first stop. This section describes some of the best.

4.2.1 The Global Change Master Directory (GCMD)

If you need to know what data are available there is no better starting place than the Global Change Master Directory (GCMD) (Global Change Master Directory). Do not let the "Global Change" part of their name mislead you. The GCMD is a comprehensive directory of Earth and space science data sets collected worldwide (and interplanetary) over the past century. However, you must remember that you cannot get the actual data from the GCMD. It is a directory of information, not a data archive. The listings provide detailed descriptions of the data including, resolution, spatial and temporal coverage, location, and information about the data providers. In some cases, there are hyperlinks that will take you directly to a data ordering system for the data set or collection.

The GCMD contains more than 13,500 Earth science data set descriptions. Approximately 2,500 new descriptions are added annually in one of thirteen classifications. More than 25 percent of the data listings refer to data from NASA's Earth science missions and from the Federation of Earth Science Information Partners (Federation of Earth Science Information Partners). More than 1,200 data providers contribute to the GCMD. If you have a data set that you want listed then you can fill out a data set description form online at the GCMD website – gcmd.nasa.gov.

The directory can be searched in many different ways. Users may choose from one of the topics listed in Table 1 and "drill down" by successively clicking on additional parameters that appear at each level. For example, clicking on "Solid Earth" will take you to a new page with parameters such as geochemistry, geomagnetism, seismology, volcanoes, etc. Then clicking on volcanoes will take you one level lower with parameters like lava, magma, volcanic ashes, etc. At each level, the number of listings for each parameter is given. Table 4.1 shows the total number of listings as of July 2005 for the top level parameters (State of the GCMD 2004).

Listings in the directory may also be displayed and searched by platform (or spacecraft), instrument, data center, geographic location, or project. For example, a user can search for all data sets from the Terra satellite or for data sets collected by the MODIS instrument aboard Terra. Location listings are not strictly for geographic regions either. There are choices such as oceans, space, and vertical location (like troposphere). To aid in narrowing a search, users may enter a keyword at any point. For example, if a user has selected "Solid Earth" followed by "Volcanoes," all listings for Mount St. Helens can be located by typing "Mount St. Helens" into the

Table 4.1. GCMD Listings by Earth Science Topic

TOPIC	2003	2004
AGRICULTURE	139	130
ATMOSPHERE	617	728
BIOSPHERE	798	545
CLIMATE INDICATORS	21	80
CRYOSPHERE	218	145
HUMAN DIMENSIONS	398	302
HYDROSPHERE	332	382
LAND SURFACE	553	452
OCEANS	702	468
PALEOCLIMATE	93	203
SPECTRAL/ENGINEERING	160	156
SUN-EARTH NTERACTIONS	13	53
SOLID EARTH	322	140

free text box. Fig. 4.1 shows the results of this search.

Users may also choose the free text search option that operates similar to many Internet search engines. Users may enter keywords with Boolean operators and wildcard symbols. Spatial extent may be expressed by typing in latitude/longitude values or by drawing a rectangle on a world map. A time range may also be specified. This search technique is very powerful, producing listings of interest with only a few mouse clicks.

Data set listings are only part of the capabilities of the GCMD. In addition, the GCMD also maintains listings for more than 750 applications that operate on data sets (software, analytical tools, educational resources, etc.). These applications are known as data set services and can be invaluable when working with data sets. Examples range from specialized tools for browsing, manipulating, and visualizing data products to Earth science educational products and environmental hazard advisory services. Users can search for services using controlled keywords or free-text to discover data-set-specific tools. Table 4.2 shows the service classifications and number of listings for each (State of the GCMD 2004).

4.2.2 Federation Interactive Network for Discovery (FIND)

Another useful directory is the Federation Interactive Network for Discovery or FIND (Federation Interactive Network for Discovery). This directory maintains listings of data sets produced and held by the Federation of

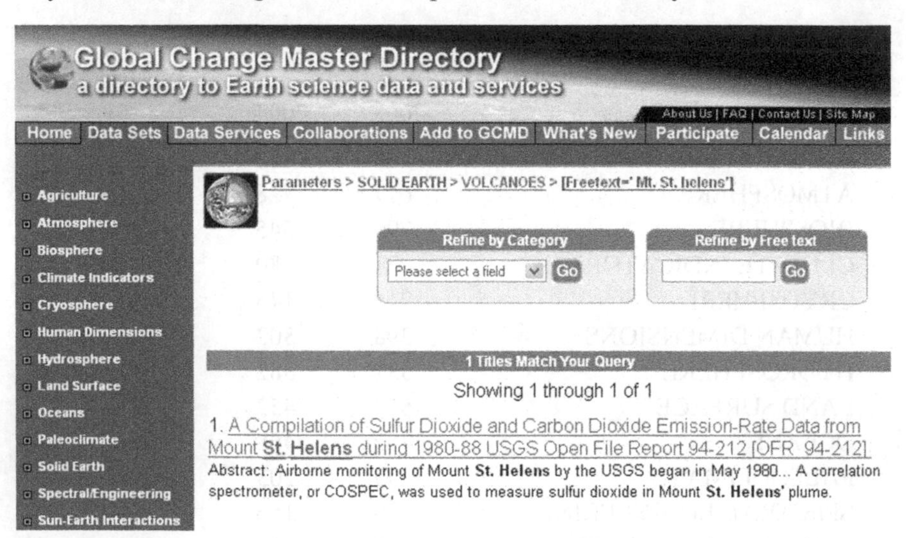

Fig. 4.1. GCMD search results page.

Earth Science Information Partners (ESIP Federation). There are fewer listings than that of the GCMD, but the focus is strictly on Earth science data sets. The Federation partners (over 75 at this writing) bring together government agencies, universities, nonprofit organizations, and businesses, in an effort to make Earth science information available to a broader community. One objective of the Federation is to evolve methods that make Earth science data easy to preserve, locate, access, and use for all beneficial applications. The Federation maintains a comprehensive inventory of information about its data holdings — over 3,500 data sets and services (Federation of Earth Science Data Partners – Data Center). Data set and service listings of FIND are closely matched with the GCMD. In fact FIND utilizes metadata from the GCMD in its search domain. FIND is able to search in many databases (GCMD, ESIP Partners and ESIP Partner websites) simultaneously via a single query, using the underlying Mercury (Mercury Search Engine) search system. Users may search for data sets by specifying a keyword (default search) or define a more sophisticated query using the spatial, temporal, and data center search screens.

The keyword search is very simple to use. Just enter a parameter or geophysical term into the text box and hit enter. The keyword search allows you to search for words or phrases within a specified field or within the entire metadata record. As with the GCMD, Boolean operators and wildcards are supported. There are optional temporal and spatial search capabilities also. Pull-down menus can be used to specify a time interval. Geographic regions can be specified by entering latitude/longitude pairs or by drawing a rectangle on a world map. FIND also has a list of over 175 data centers. By selecting a specific data center, this option produces a listing of all ESIP Federation data sets held there. This is useful when you know that

Table 4.2. GCMD Service Listings by Earth Science Topic

Topic	2003	2004
Data Analysis & Visualization	374	499
Data Management/Data Handling	206	262
Education/Outreach	199	251
Environmental Advisory	78	115
Hazards Management	28	47
Reference and Information Services	96	137
Metadata Handling	25	41
Models	126	192

data of interest reside at a particular archive. More information on data centers appears later in section 4.3 below.

Once the search parameters have been specified the search begins. Results are returned (Fig. 4.2) as a page with lists of: successful searches, databases searched but with no hits returned, and databases with failed connections. By clicking on the search results a short summary of each record is displayed. More in-depth information can be retrieved by clicking on the title.

4.2.3 The Geospatial One-Stop

The data resources listed by the GCMD and FIND are in a wide variety of data formats. If your data interests fall strictly within the area of geospatial information systems, then the Geospatial One-Stop (GOS) (Geospatial One-Stop) may be the best choice. GOS serves as a public gateway for improving access to geospatial information and data under the United States' e-government initiative (United States' Electronic Government Initiative). Located on the Web at geodata.gov, GOS is designed to facilitate communication and sharing of geospatial data.

The second version of GOS (GOS2) was released in July 2005. It employs new portal technologies, the Google search engine, and the latest

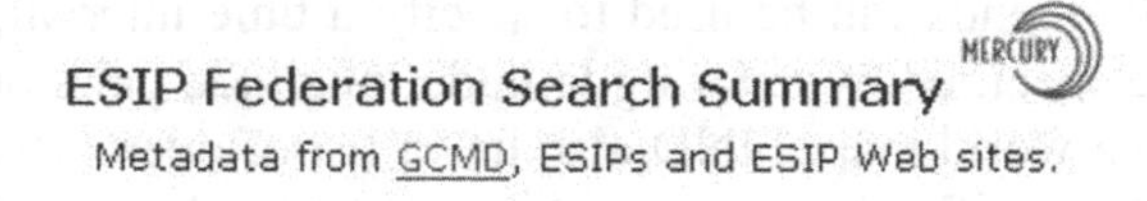

Help

QUERY:
Text = Temperature

Elapsed Search Time (MM:SS): 0:3		
Database Name	**Status**	**Count**
ESIP Federation Web Site	Search Complete	35
ESIP Web Pages	Search Complete	2738
ESIP Collection Data	Search Complete	2013

For successful searches click on the database name to get the corresponding results

Fig. 4.2. Search results from the FIND search engine using parameter "temperature."

GIS map creation and viewing software. This makes GOS2 very powerful. Users can register and create their own custom portal. They can use the powerful search interface to access a wide variety of geospatial information. Maps can be created and saved from one session to the next. Search parameters can also be saved. GOS2 has a catalog of thousands of metadata records and links to web map servers, web feature servers, web coverage servers, and catalog services. Data sources are classified by the communities shown in Table 4.3. Additional classifications can be suggested such as the special community set up following the December 2004 tsunami in Indonesia. Users can download data sets, images, clearinghouses, map files, and more. Registered users can easily contribute their own resources making the Geospatial One-Stop a rapidly evolving system.

As of this writing the geographic coverage of almost all resources within the GOS are within the United States. As more resources are contributed this limitation is expected to ease.

4.3 Accessing Data

Once the data has been identified, the next step is to access a data archive and order or download the data. Fortunately, the vast majority of data from NASA's Earth orbiting satellites, aircraft, ocean and land based instruments is located in one of nine cooperating data centers known as the

Table 4.3. GOS Communities

1	Administrative Boundaries
2	Agriculture
3	Atmosphere
4	Biology
5	Business
6	Cadastral
7	Demographic
8	Elevation
9	Environment
10	Geology
11	Health
12	Imagery and Basemaps
13	Inland Water
14	Locations
15	Oceans
16	Transportation
17	Utilities

DAAC Alliance (DAAC Alliance). The DAACs (Distributed Active Archive Centers) were formed in 1991 as part of NASA's Mission to Planet Earth Program, now known as the Earth Science Enterprise (NASA ESE). Their primary objective was to make the vast amount of NASA's data holdings readily available to researchers, educators, students, and the general public. Today the DAAC Alliance is home to over 3000 data sets. They continue to ingest data at over 1000 GBytes per day, mainly from the instruments aboard the suite of Earth Observation Satellites like Terra and Aqua. In addition to holding the actual data files, the DAAC Alliance provides a vast selection of software applications, documentation, and user services assistance for interpreting and using the data. And, except for a very few special cases, the data are free.

There are several ways to access data from the DAAC Alliance. Each data center operates a data search and order system tailored to the respective data holdings. All nine data centers can be searched at one time by using the EOS Data Gateway (EDG) (Earth Observing System Data Gateway). The result of using these search and order systems compared to the GCMD or FIND is that the user is given a list of actual data files instead of descriptions of the data. But, it should be pointed out that the DAAC Alliance data search and order systems are not comprehensive as compared to the GCMD and FIND.

In the following sections, I will describe the EOS Data Gateway and give a summary of each data center in the DAAC Alliance. Before venturing into those topics, a quick tutorial on data product processing levels and formats will help the first time user of those systems.

4.3.1 Earth Science Data Terminology and Formats

Data Processing Levels

Data products from the DAAC Alliance exist at five processing levels ranging from level zero to level four. These processing levels were defined by the Data Committee of the Earth Observing System Data and Information System (EOSDIS) and are consistent with the Committee on Data Management, Archiving, and Computing (CODMAC) definitions (NASA 1993).

Level zero products are raw data directly output by instruments at full resolution with engineering and communications information removed. Level zero data is the base from which all other data products are produced. The unprocessed level zero data are always archived because it represents the purest form of the measurements. This level of data is rarely used for research.

Level 1A data products are constructed by adding a time reference and annotations to the level zero data. The annotations include information such as radiometric and geometric calibration coefficients and georeferencing parameters.

Data from some instruments are processed to level 1B by adding a step that makes the data conform to sensor units. Data products at levels zero, 1A, and 1B generally are very large in volume since individual wavelengths or bands have not been separated out.

Level two data have derived geophysical variables at the same resolution as that of level one. These data sets can directly yield information in terms of geophysical parameters such as temperature. Level two data sets are commonly used in research where the maximum resolution and completeness of geographical coverage is desired.

Perhaps the most recognizable form of data is level three. Level three data products are normally mapped onto a uniform space and time grid. This means that they can readily be viewed as imagery or other visual product. Level three data best matches data products normally affiliated with GIS systems.

There is a special data level (level four) for data products generated from models and simulations based on analyses of lower level products or fused from multiple instruments.

Data Format Descriptions

Data from EOS instruments are primarily formatted using either HDF (Hierarchical Data Format) or HDF-EOS (Hierarchical Data Format for the Earth Observing System). The Hierarchical Data Format (HDF) is designed to facilitate sharing of scientific data. Its features include platform independence, user extendibility, and embedded metadata for units, labels, and other descriptors. Standard data types include multidimensional array, text, table, raster image, and palette. HDF files are portable and can be shared across most common platforms, including many workstations and high-performance computers. An HDF file created on one computer can be read on a different system without modification. HDF was developed by the National Center for Supercomputing Applications (NCSA). This format is extensible and can easily accommodate new data models, regardless of whether they are added by the HDF development team or by HDF users.

The HDF for the Earth Observing System (HDF-EOS) data format is standard HDF with EOS conventions, data types, and metadata. HDF-EOS adds three data types (point, grid, and swath) to the HDF structure that allow file contents to be queried by Earth coordinates and time. An HDF-EOS file also contains metadata essential for search services. An HDF-EOS file can be read by any tool that can read standard HDF files. A data product need not fit any of the grid, point, or swath models to be considered HDF-EOS. HDF-EOS is implemented as a C library extension of the standard HDF library. Use of HDF-EOS can eliminate duplication of software development efforts, especially for analysis and visualization software.

Another widely used format is the network Common Data Form (netCDF) (Network Common Data Format). The netCDF software was developed at the Unidata Program Center in Boulder, Colorado, and augmented by contributions from other netCDF users. The netCDF libraries define a machine-independent format for representing scientific data. Together, the interface, libraries, and format support the creation, access, and sharing of scientific data. The HDF and HDF-EOS data formats include features of netCDF.

4.3.2 The EOS Data Gateway

The EOS Data Gateway system provides a single interface through which users can find and order data from multiple participating archives. You may use the EDG as a guest or you can create an account. Signing in as a guest will create a transient session. All information will be deleted after you sign out. It is better to create a user account (there is no cost), so that you can set up preferences and preserve your information. You can log out, and sign back in at anytime and pick up where you left off. Click on the "Create Account" button to create a new account. You will be asked to select a password and enter user contact information. Once you have submitted the information, you can then sign into your user account. The EDG is an old system and as such does not have a good intuitive user interface. It is best to take an hour and read the tutorial before beginning. The time spent in doing so will pay off later.

The EDG offers the following capabilities: a) search; b) browse; c) order; and d) guide. The search feature allows the user to search for data based upon specific criteria such as location, time, and geophysical parameter (e.g. temperature). Searching should be the first step of your EDG experience. After the search completes, returning a set of data files (known as granules), the user may browse the results, by viewing a reduced resolution image depicting the data, or inspect metadata about the granule. Following this inspection, the user may choose to do another search – likely in most cases since a typical search will return hundreds of granules – or proceed to place an order for selected granules. The order feature allows the user to select data for ordering, choose packaging information, enter ordering information (such as shipping address), and place the order. Since the EDG contains complete information from all nine data centers a data order may be sent to multiple data centers for filling. In this case, the user will receive packages (or e-mail notifications) from several data centers. In addition to the raw data, users will require documentation to help them decode the granules and use the data. Every data set is required to have a set of guides for this purpose, containing information on data sets, platforms, instruments, etc. Depending on the sophistication of the user, different guides will be needed. The EDG allows users to locate this information and have it delivered with the data package. The EDG may be accessed at the following URL: http://redhook.gsfc.nasa.gov/~imswww/pub/ims-welcome/

4.3.3 The DAAC Alliance

Each of the nine data centers of the DAAC Alliance serves a specific Earth science discipline. However, this should not be taken as a strict classification and separation of data sets. In every case, the data centers hold a rich variety of data products. In fact the data inventories are of such magnitude, it is not possible to list all of them in this chapter. In the sections below, a short summary is given that highlights the main data products and data access services. The URL for each data center is included for more information.

The Alaska Satellite Facility (ASF)

The ASF (Alaska Satellite Facility) is located in the Geophysical Institute at the University of Alaska in Fairbanks. It maintains a variety of data sets applicable to the geophysical phenomena of sea ice, polar processes, and geophysics. Its primary data sets are those from the Synthetic Aperture Radar (SAR) instruments aboard the RADARSAT-1, the European Sensing Satellite-2 (ERS-2) and legacy SAR data from the ERS-1, and the Japanese Earth Resources Satellite -1 (JERS-1). All SAR data sets are restricted and available only to NASA-approved researchers.

The ASF does have several mosaics derived from SAR data that are generally available. The mosaics show complete detailed views of the Antarctic, boreal forests of Alaska, Canada, and portions of the Northeastern

United States. Other SAR imagery reveals rainforests in the Amazon, Central America, Africa, and the Pantanal region.

Sea ice imagery provides a detailed view of sea ice movements. The Glacier Power CD, designed for the classroom, uses SAR data to lay a foundation for the study of glacier dynamics through the use of imagery and cartoon characters.

Data products can be obtained through the ASF website, via ftp or through the EOS Data Gateway. For more information: www.asf.alaska.edu

The Goddard Earth Sciences Data and Information Services Center (GES DISC)

The GES DISC (Goddard Earth Sciences Data and Information Services Center) provides an immense volume of data. It offers data sets that pertain to the study of the upper atmosphere, atmospheric dynamics, global precipitation, global biosphere, ocean biology, ocean dynamics, and solar irradiance. It also provides services that enable users to fully realize the scientific, educational, and application potential of the data sets. The list of data products is far too long to fit into this chapter, see their website at daac.gsfc.nasa.gov for further information. Here are a few highlights.

The GES DISC archives and distributes data products from many very popular instruments including the Moderate Resolution Imaging Spectroradiometer (MODIS) on both the Terra and Aqua satellites. MODIS acquires data in 36 discrete spectral bands, and the data almost cover the Earth every day. MODIS data has a high radiometric resolution (1KM, 500M, and 250M), global coverage, and accurate calibration. MODIS data are useful for long-term climate and global change studies, as well as for short-term monitoring of natural disasters. The GES DISC distributes level one radiometric and geolocation data products and other ocean and atmosphere products at higher levels

Data products are also available from the Atmospheric Infrared Sounder (AIRS), Sea-viewing Wide Field-of-view Sensor (SeaWiFS), Total Ozone Mapping Spectrometer (TOMS), and Tropical Rainfall Measuring Mission (TRMM) to name a few.

The GES DISC offers five methods for locating and ordering data products. There is a simple search interface built into the index page of its website. It offers a massive amount of data for direct download through the data pool, which includes its own search capability. Data sets can also be located through parameter based searches. All data are also orderable through the EOS Data gateway. For more information: daac.gsfc.nasa.gov

The Global Hydrology Resource Center (GHRC)

The GHRC (Global Hydrology Resource Center) offers data products that focus on the study of the global hydrologic cycle, severe weather interactions, lightning, and convection. The GHRC is the national repository for lightning data, holding data sets from the Lightning Imaging Sensor (LIS) and its predecessor the Optical Transient Detector (OTD). Available data sets include sea surface temperature, atmospheric water vapor, wind direction, and atmospheric temperature derived from passive microwave instruments like the Special Sensor Microwave Imager (SSMI) and the Advanced Microwave Sounding Unit (AMSU). The GHRC also has many data products relating to hurricane structure, dynamics, and motion from the series of Convection and Moisture Experiments (CAMEX) field campaigns.

You can locate data products from the GHRC through a local search and order system known as HyDRO or through the EOS Data gateway. Passive microwave data sets may be downloaded directly from the GHRC data pool. For more information: ghrc.msfc.nasa.gov

The Atmospheric Sciences Data Center (ASDC)

The ASDC (Atmospheric Sciences Data Center) has more than 800 data sets relating to radiation budget, clouds, aerosols, and tropospheric chemistry. Radiation budget data sets pertain to the variability of total solar irradiance, radiation properties of the atmosphere and of the Earth's surface. Cloud data sets contain information on the radiative properties of clouds; cirrus, marine stratus, and arctic cloud field studies and subsonic aircraft effects on contrails and other cloud systems. Aerosol data sets contain information on the spatial and vertical distribution of aerosols, as well as their chemical, physical, and optical properties. Tropospheric chemistry includes biomass burning, concentrations of atmospheric chemicals, and the distribution and behavior of carbon monoxide, ozone, and water vapor.

The ASDC, like the GES DISC, maintains massive amounts of data mainly due to data products derived from instruments aboard the EOS satellites such as Aqua and Terra. Radiation budget data sets are derived from several instruments. There are five Clouds and the Earth's Radiant Energy System (CERES) instruments active at the time of this writing. There are two on Aqua, two on Terra, and another aboard the Tropical Rainfall measuring Mission (TRMM) satellite. CERES data is also used to develop data products for cloud studies.

Massive amounts of data are collected each day from the Multi-angle Imaging SpectroRadiometer (MISR) aboard Terra in support of cloud and aerosols studies. Aerosol data sets are also derived from data measured by the Stratospheric Aerosol and Gas Experiment (SAGE), MISR, and others. Data sets for tropospheric chemistry investigations are currently produced from data from the Measurements Of Pollution In The Troposphere (MOPITT) instrument aboard Terra and will soon be supported by results from the planned Cloud-Aerosol Lidar and Infrared Pathfinder Satellite Observations (CALIPSO) mission.

Users may use the local ASDC search interface to locate and order data, or they may access the massive on-line data pool to download data directly. All data sets held by the ASDC may be ordered through the EOS Data gateway. For more information: http://eosweb.larc.nasa.gov/

The Land Processes Distributed Active Archive Center (LP DAAC)

The LP DAAC (Land Processes Distributed Active Archive Center) maintains data sets that are focused on investigation, characterization, and monitoring of biologic, geologic, hydrologic, ecologic, and related conditions at the surface of the Earth. It is this focus that matches the LP DAAC with users of geospatial information systems. GIS users are certainly frequent users of imagery from Landsat and MODIS. The LP DAAC also offers imagery from the Advanced Spaceborne Thermal Emission and Reflection Radiometer (ASTER) instruments aboard Terra and Aqua. ASTER offers the highest resolution image data in visible and near-infrared (15M), shortwave infrared (30M), and thermal infrared (90M) wavelengths. AVHRR as well as aerial photography, are also available. Related data sets include the Global 30-Arc-Second Elevation Data and the NASA SIR-C Precision Data.

Two of the most desirable data sets – Landsat and ASTER – are not free. Pricing changes so you will need to check with the LP DAAC for actual costs. At this writing, the cost for ASTER was $80 per scene via FTP or $91 on CD or DVD media.

Data products are available from the LP DAAC through an online data pool or through the EOS Data Gateway. For more information: edcdaac.usgs.gov

The National Snow and Ice Data Center (NSIDC)

The NSIDC (National Snow and Ice Data Center) provides data sets that focus on the study of snow and ice processes, particularly interactions among snow, ice, atmosphere, and ocean that influence global change detection and model validation. Data subjects include permafrost, frozen ground, glaciers, ice shelves, icebergs, ice sheets, snow cover, ice velocity, and ocean chemistry and temperature. It offers a CD-ROM collection containing millions of vertical soundings of temperature, pressure, humidity, and wind from Arctic land stations at 100KM resolution from 1950 to 1996. The NSIDC distributes a host of ancillary sea ice products, including ice extent, melt onset data, ice persistence, total ice-covered area, and ocean masks.

The extent of snow and ice cover is given by MODIS data products from both Terra and Aqua. The Advanced Microwave Scanning Radiometer-EOS (AMSR-E) on Aqua yields data products that include soil moisture, ocean products (water vapor, cloud liquid water, sea surface temperature), rain, snow, and sea ice.

The NSIDC also maintains an extensive suite of passive microwave data products. Near-real-time maps, a best estimate of current ice and snow conditions, are available at 25KM resolution. The Advanced Very High Resolution Radiometer (AVHRR) data set provides nearly complete coverage of sea ice, land ice, and land in polar regions at 1.1-kilometer resolution for all 5 bands of the AVHRR sensor. Data from the Ice, Cloud, and Land Elevation Satellite (ICESat) Geoscience Laser Altimeter System (GLAS) provide ice sheet elevations and changes in elevation through time. GLAS data sets also include measurements of cloud and aerosol height profiles, land elevation, vegetation cover, and sea ice thickness.

Data orders may be placed at the NSIDC through the EDG data search-and-order system or directly through the online data catalog. For more information: nsidc.org

Oak Ridge National Laboratory (ORNL)

The ORNL (Oak Ridge National Laboratory) data center specializes in data about the dynamics between the biological, geological, and chemical components of the Earth's environment. These dynamics are influenced by interactions between organisms and their physical surroundings, including soils, sediments, water, and air.

ORNL also archives and distributes data from a number of field campaigns. The list is too extensive to fully provide here but a few examples are: the Boreal Ecosystem-Atmosphere Study (BOREAS), Large-Scale Biosphere-Atmosphere Experiment in Amazonia (LBA), and the Southern African Regional Science Initiative (SAFARI 2000). The BOREAS project investigated exchanges of energy, water, heat, carbon dioxide, and trace gases between a boreal forest and the atmosphere. LBA data include measurements of precipitation in Bolivia, Brazil, and Peru, plus Synthetic Aperture Radar (SAR) imagery from the rain forest region during 1995 and 1996. SAFARI 2000 studied the linkages between land and atmosphere processes in southern Africa, especially the relationship of biogenic, pyrogenic, and anthropogenic emissions and the functioning of the biogeophysical and biogeochemical systems.

There is a significant number of historical data sets available from ORNL. Historical climatology, mean climatology, and precipitation data date back to 1753. Hydroclimatic data collections such as streamflow, wetlands, precipitation, and temperatures exist from 1874 to 1988. River Discharge data from 1807 to 1996 containing long-term monthly averaged values for river discharge measured at various stations is available. Soil characteristic data measured at sampling sites or estimated for grids of various sizes are available from 1940 to 1996. Holdings pertaining to vegetation characteristics, including the distribution of vegetation types, as well as leaf area index calculated from field measurements, can be obtained from 1932 through 2000.

ORNL DAAC data are available through an online search-and-order system at www.daac.ornl.gov and through the EDG data search-and-order system. For more information: daac.ornl.gov

The Physical Oceanography Distributed Active Archive Center (PO.DAAC)

The PO.DAAC (Physical Oceanography Distributed Active Archive Center) provides global oceanographic data derived from NASA satellites. Its primary holdings include data sets about ocean surface topography, ocean winds, and sea surface temperatures. Other data sets include ocean wave height, electron content of the ionosphere, atmospheric moisture, and heat flux.

Ocean surface topography data holdings are derived from the TOPEX/POSEIDON and Jason-1 missions. These data products include sea surface height, wind speed, significant wave height, tropospheric water vapor, electron content of the ionosphere, and ancillary information along the track of the TOPEX/POSEIDON satellite. Jason-1 is a follow-on mission to TOPEX/POSEIDON primarily yielding surface topography along a 10-day repeated ground track.

Ocean vector winds data are available from the SeaWinds instruments on board the QuikSCAT satellite and the Advanced Earth Observing Satellite II (ADEOS-II). Twenty-five kilometer wind vector data is also available from the NSCAT scatterometer on a daily 0.5- by 0.5-degree map.

Sea Surface Temperature data from AVHRR are available as daily, 8-day, and monthly averages. Daily, weekly, and monthly sea surface temperature data, from the very popular MODIS instrument on board Terra and Aqua, are available in thermal infrared or mid-infrared mapped products.

The PO.DAAC also provides many other surface and multi-parameter products, including significant wave height, chlorophyll concentration, near-surface currents, atmospheric moisture, brightness temperatures, and heat flux.

Data may be accessed through the PO.DAAC online search-and-order service. Data is also available through the EOS Data gateway. For more information check the PO.DAAC catalog of products at the web site: podaac-www.jpl.nasa.gov

The Socioeconomic Data and Applications Center (SEDAC)

The SEDAC (Socioeconomic Data and Applications Center) data focus is much different than the other data centers described above. It specializes in data sets that focus on human interactions in the environment. They combine satellite data with socioeconomic data to create an "Information Gateway" between the socioeconomic and Earth science data and information domains.

Much of the SEDAC data relates to population statistics. In the Gridded Population of the World (GPW) data set, land area, population counts, and densities are available for the entire globe and six continental regions. Land data and population counts are also available for each country. The Population, Landscape, and Climate Estimates (PLACE) data set gives population and territorial extent overlaid with biophysical parameters such as biome, climate, coastal proximity, elevation, population density, and slope, resulting in a data set of population estimates and area. This is valuable for researchers who require tabular data aggregated to the national level.

Another category of information from SEDAC is hazard and impact reports. They include “Potential Impacts of Climate Change on World Food Supply: Data Sets from a Major Crop Modeling Study,” which provides data on projected crop yield changes for major world regions based on climate model estimates, increased atmospheric carbon dioxide concentrations, and alternative adaptation scenarios. The “Central American Vegetation/Land Cover Classification and Conservation Status” report assesses the degree to which both existing and proposed terrestrial protected area networks protect or would protect landscape-level biodiversity, which is represented as vegetation types delineated from remotely sensed imagery.

Also of interest is the Environmental Sustainability Index (ESI). The ESI is a measure of overall progress toward environmental sustainability developed for 142 countries. The ESI permits cross-national comparisons of environmental progress in a systematic and quantitative fashion. This index represents a first step toward a more analytically driven approach to environmental decision making.

SEDAC has developed an electronic gateway to provide access to the catalogs of a diverse international group of data archives and other institutions. For more information: sedac.ciesin.org

4.4 Other Data Centers

The data centers described above offer a wealth of data sets and services. However there are many, many others. In this closing section, I will provide a list of selected data centers that users may wish to peruse.

4.4.1 Data Centers of the National Oceanic and Atmospheric Administration (NOAA)

The NOAA Satellite and Information Service (NESDIS) (NOAA Satellite and Information Service) operates four national data centers for climate, geophysics, oceans, and coasts. The National Climatic Data Center (NCDC) is the world's largest active archive of weather data. It produces numerous climate publications and responds to data requests from all over the world. The National Geophysical Data Center (NGDC) provides over 300 data sets describing the solid Earth, marine, and solar-terrestrial environment, as well as Earth observations from space. The National Oceanographic Data Center (NODC) holds global physical, chemical, and biological oceanographic data sets used by researchers worldwide, and the National Coastal Data Center (NCDC) has a diverse inventory of coastal data.

4.4.2 The United Nations Environmental Programme (UNEP)

Another massive source for data is the United Nations Environmental Programme (United Nations Environmental Programme). As expected these data sets are generated for decision makers worldwide and tend to be higher level products that may typically be displayed with GIS applications. The Global Environmental Outlook data portal is the authoritative source for data sets used by UNEP and its partners in the GEO. You may search the online database, holding more than 450 different variables, such as national, subregional, regional, and global statistics, or as geospatial data sets (maps), covering themes like freshwater, population, forests, emissions, climate, disasters, health, and GDP. The data may be displayed online as maps, graphs, or data tables, or you may download the data for desktop display.

4.4.3 International Data Centers

My limitation of speaking only the English language does not detract from pointing out the existence of data centers located in countries worldwide. From Argentina (National Antarctic Data Centre) to Uruguay (Centro Uurguayo de Datos Antarticos) data sets are available. The GCMD, introduced in section 4.2 at the beginning of this chapter, contains data set listings from 45 countries. There is indeed a world of data available.

References

Alaska Satellite Facility. http://www.asf.alaska.edu/

Atmospheric Sciences Data Center. http://eosweb.larc.nasa.gov/

Centro Uurguayo de Datos Antarticos. http://www.iau.gub.uy/cuda

DAAC Alliance. http://nasadaacs.eos.nasa.gov/

Earth Observing System Data Gateway. http://redhook.gsfc.nasa.gov/~ims-www/pub/imswelcome/

Federation of Earth Science Data Partners – Data Center. http://www.esip-fed.org/data_center/index.html

Federation of Earth Science Information Partners. http://www.esipfed.org/

Federation Interactive Network for Discovery, http://www.esipfed.org/inter-com/index.html

Geospatial One-Stop. http://www.geodata.gov/

Global Hydrology Resource Center. http://ghrc.msfc.nasa.gov/

Global Change Master Directory. http://gcmd.nasa.gov/

Goddard Earth Sciences Data and Information Services Center. http://daac.gs-fc.nasa.gov/

Hierarchical Data Format. http://hdf.ncsa.uiuc.edu/

Hierarchical Data Format for the Earth Observing System. http://hdf.nc-sa.uiuc.edu/hdfeos.html

Land Processes Distributed Active Archive Center. http://edcdaac.us-gs.gov/main.asp

Mercury Search Engine. http://mercury.ornl.gov/esip/freetext.html

NASA Current Missions. http://www.nasa.gov/missions/timeline/current/cur-rent_missions.html

NASA (1993) Earth Observing System (EOS) Reference Handbook. Asrar G, Dokken DJ. (eds) Washington, DC, National Aeronautics and Space Administration, Earth Science Support Office, Document Resource Facility.

NASA Earth Science Enterprise, http://www.earth.nasa.gov/

National Antarctic Data Centre. http://www.dna.gov.ar/

National Snow and Ice Data Center. http://nsidc.org/

Network Common Data Format. http://my.unidata.ucar.edu/content/software/netcdf/index.html

NOAA Satellite and Information Service. http://www.nesdis.noaa.gov/datainfo.html

Oak Ridge National Laboratory. http://daac.ornl.gov/

Physical Oceanography Distributed Active Archive Center. http://podaac-www.jpl.nasa.gov/

Socioeconomic Data and Applications Center. http://sedac.ciesin.org/

State of the GCMD – CY (2004) http://gcmd.nasa.gov/Aboutus/state/State_GCMD_2004_FINAL.pdf

United Nations Environmental Programme. http://www.unep.org/

United States' Electronic Government Initiative. http://www.whitehouse.gov/omb/egov/

5 Exploring the Use of a Virtual Map Shop as an Interface for Accessing Geographical Information

William E. Cartwright

Abstract. The need for designers to consider the wide range of user preferences and how they (the users) interact with contemporary map information packages is paramount when developing effective interfaces for information access, retrieval and use. It is argued that a suite of metaphors, allowing users to choose the most effective access method for their application makes for a more effective package. This chapter describes the theory of the Map Shop, which could be provisioned with maps, videos, books, guides, games and databases of facts and could provide expert tips. The Map Shop can be linked locally or internationally through the Internet and, more specifically, the World Wide Web. From the users' perspective the boundary between discrete and distributed multimedia would be transparent, presenting them with the most current and customised information possible. It also describes the use of the metaphor suite developed as part of the GeoExploratorium, a tool for the provision of geographic information in a manner that is complementary to the map metaphor. Finally, it describes the building of a prototype Virtual Map Shop, a discrete / World Wide Web tool for exploring its use as an innovative geographical information access (virtual) resource.

5.1 Introduction

With the sheer number of publications readily available through contemporary communications and multimedia publishing systems the way in which we access information has changed forever. The geospatial sciences are no different. It is argued that a new genre of spatial artefact has now stabilised and become an accepted tool for exploring geography and for mining geographical information. This has resulted in adapting new ways to use these products and new ways of assembling data into a personalised cartographic product.

Traditionally, data and user were 'merged' by the provision of a particular mapping product that was generated to meet a certain usage requirement for viewing geographical information within a designated area. Contemporary products have changed the process. Users can become the map drawer, data can be assembled from many discrete and geographically dispersed sites and visualization products can be generated using a plethora of depiction techniques that interpret data into usable maps using software that is both available and inexpensive.

According to Fisher (2003), changing trends in Media Technology, with the use of multi-sensory display systems, allow the viewer's movements to be non-programmed - they are free to choose their own path through available information rather than being restricted to passively watching a 'guided-tour'. The advantages are that the viewer's access to greater than one viewpoint of a given scene allows them to synthesise a strong visual perception of geography from many points of view. This enlarges and enhances the traditional view of geography, through only one visualization window. The multiple points of view places an object in context, animating its meaning. Providing different views of reality, through, say maps, books, (virtual) field trips and videos, can enlarge a somewhat restricted view and enhance the user's perception of reality. There are a number of possible views with existing approaches, but these can be enhanced with more intuitive tools and strategies that guide designs of products that control how users use a product, without removing any perceive freedom of use. This is illustrated in Fig. 5.1.

So how can we best use these tools? How much control over the use of the product can we implement? And, how can we ensure that, when New Media installations are employed, that users see essential views of geography?

The following sections develop the idea of the related *Map Shop*, and how it might be further explored as a means of allowing users perceived freedom when using New Media cartographic products, but ensuring that they do not overlook essential information. The current stage of this ongoing research has developed the concept of the Virtual *Map Shop* and then built a prototype that allows for a hybrid discrete / Web multimedia product to be delivered. The following section begins with discussing the theory behind how such tools might be used for exploring and accessing geographical information.

5.2 Underlying theory

The underlying theories behind this research are:

- The concept of the *Map Shop* and the *Literate Traveller*, was developed by almost a decade ago (Cartwright 1997, 1998) as part of a product

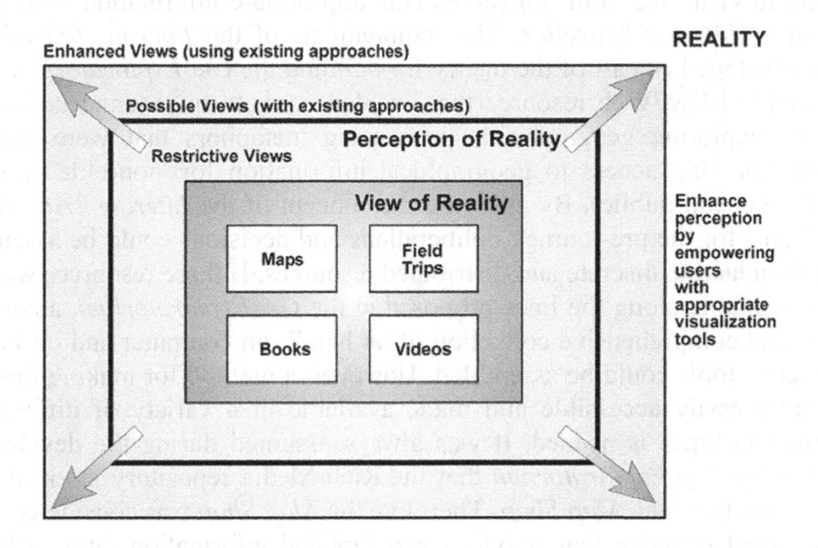

Fig. 5.1. Building more intuitive tools.

called the *GeoExploratorium,* a system for providing access to Rich Media that could aid the building of Geographical Knowledge.

- The use of a metaphor suite (also developed under the general 'umbrella' of the *GeoExploratorium)* for providing information in a user-preferred way.
- Engineered Serendipity – ideas that related to design of interactive multimedia products that allow users to explore information in a serendipitous way, but having some designer control over what they see and how they collect information.

The following sections elaborate on these foundation concepts.

5.3 Map Shop concept

The idea behind designing the method of information enhancement through the provision of Rich Media via the *Map Shop* is that of the *Literate Traveller.* As consumers we use real-world images as artifacts for constructing mental images of places that we intend to visit. We are used to using many artifacts to enable us to build a better mental map of places we are yet to visit. We ‘arm’ ourselves with appropriate information so as to become a *Literate Traveller*. The requirements of the *Literate Traveller* were developed as part of the theory for building the *GeoExploratorium (*a hybrid CD-ROM/Web resource that enabled users to understand geography by exploring geographical space using metaphors that were user-driven, enabling access to geographical information for 'non-elite' users and the general public). By applying the concept of the *Literate Traveller* provisions for the pre-journey deliberations and decisions could be assembled from tactile, discrete and distributed resources. If these resources were made available along the lines proposed in the *GeoExploratorium*, a composite and comprehensive collection of ‘at hand’, on computer and on-line interactive tools could be assembled. However, a method for making these resources easily accessible and made available in a variety of different multimedia types is needed. It was always assumed during the development of the *GeoExploratorium* that the Rich Media repository associated with it would be the *Map Shop*. Therefore the *Map Shop* was considered to be a virtual resource that provides geographical information through the conventional tools of maps, air photographs etc., but also via books, games, videos and expert advice. This concept is illustrated in Fig. 5.2.

The *Literate Traveller* is seen to be the ultimate user of the *GeoExploratorium* and the consumer who uses the *Map Shop* as a source of Geographical Knowledge.

To enable efficient access to external information, the *Map Shop* would

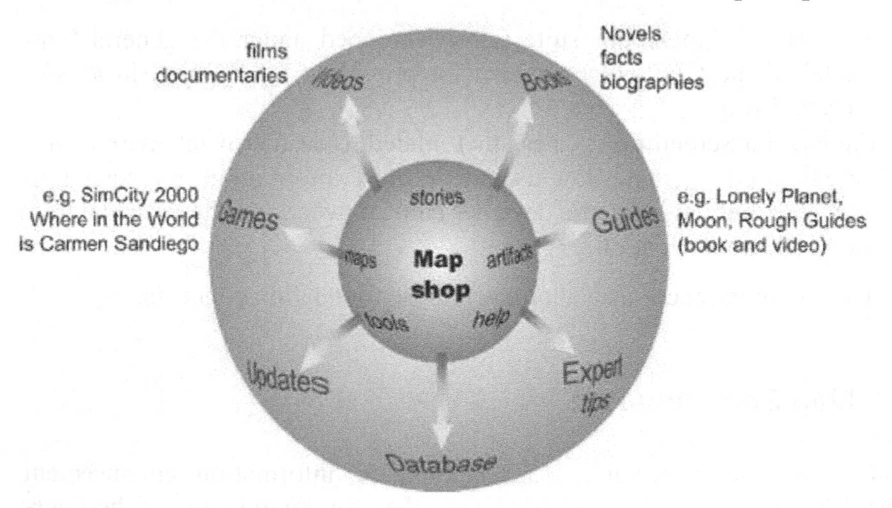

Fig. 5.2. The Map Shop concept.

need to go online. Then the product could connect to on-line interactive books, current travel information via travel guide publisher Web sites, get expert tips and information via email, update databases and offer distributed multi-player games. Fig. 5.3 illustrates this enhanced version of the *Map Shop*.

5.4 Metaphors

Using different metaphors to access geographical information does this. The metaphor set includes the Storyteller, the Navigator, the Guide, the Sage, the Data Store, the Fact Book, the Gameplayer, the Theatre and the Toolbox (Cartwright 1999). This concept is shown in Fig. 5.4, which illustrates how the metaphor set contributed to the information delivered via the *GeoExploratorium*.

These metaphors were acknowledged by Laurini (2001) as suitable access genres for visualising and access urban information. The combination of these metaphors, when used with the map metaphor would provide the means to deliver the contents of the *GeoExploratorium* using multimedia components.

To enable the *GeoExploratorium* metaphors to be applied to developing

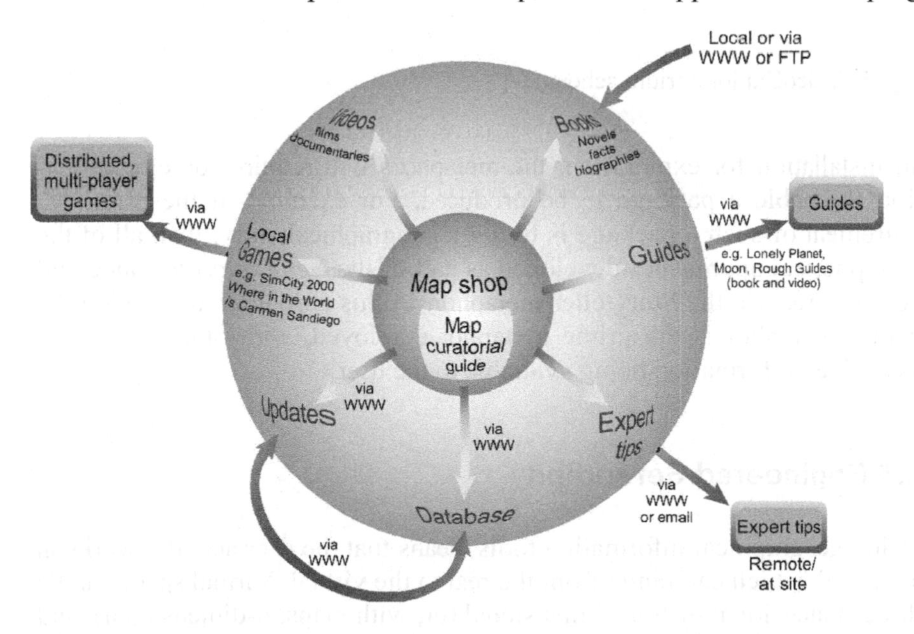

Fig. 5.3. Layout and connections proposed in the on-line *Map Shop*.

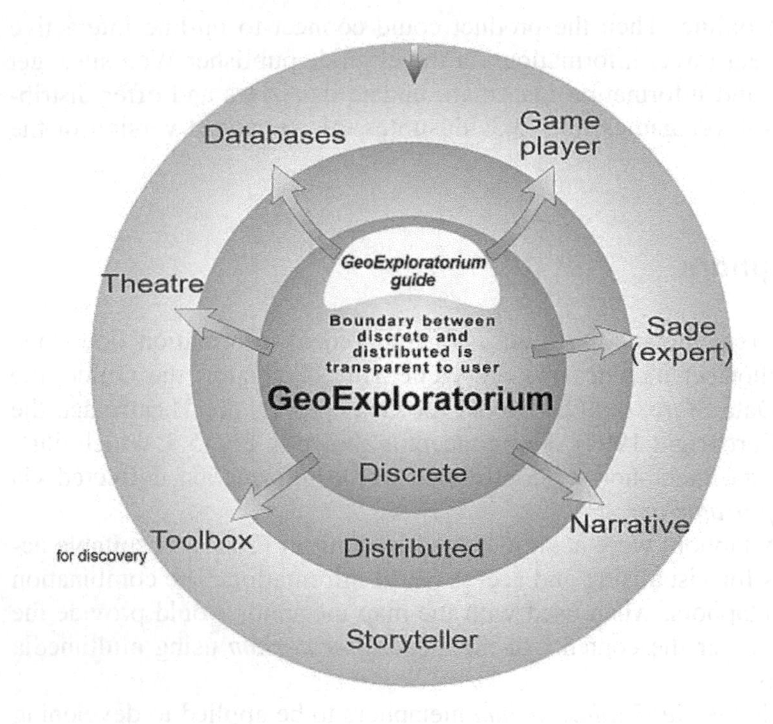

Fig. 5.4. GeoExploratorium schema.

an installation for exploration, the metaphors will require some modification to enable a package to be produced. For example, if the prime requirement of such a package is to tell a geographical story, than all of the metaphorical approaches developed as part of the *GeoExploratorium* could be delivered via the Storyteller metaphor. In this particular application the games metaphor is the prime metaphor employed, with other metaphor-assembled information being available to the user.

5.5 Engineered Serendipity

Using geographical information tools means that the user actually works in a 'space', which can range from the real to the virtual. Virtual space can be three-dimensional or four-dimensional (or, with maps, n-dimensional), and it can be modified by new or generated data plus 'things' can 'exist' which do not in the real world. A user can become confused and, at worst, lost in an interactive package. So, there needs to be a method provided for controlling exploration. According to Laurel (1993), computer and machine are two distinct parties, and dialogue between them is complex and something that should not be thought of in terms of linearised turn taking. Entrances into and exits from interactive processes should be synchronised and a ‘common ground’ for exploration should be found.

Designers of interactive multimedia products for geographical information provision face a major problem: the balance between ‘Free-range’ vs specific whey-points to guide exploration of geographical information. On one hand users need to be guided to specific information to ensure that they are provided with a complete picture of the area of geography being studied. But, on the other hand, information ‘discovered’ as users wander in a serendipitous manner, ‘falling upon’ information that they were not looking for. And, in this way a more interesting, and potentially more effective way of exploring geographical information can be provided. But, how to provide some kind of author control, whilst enabling the user to make their own explorations in a serendipitous way? Hence the concept of 'Engineered Serendipity' (Cartwright 2004a).

‘Engineered Serendipity’ could be used to extend and control geographical exploration and access in a multimedia cartography environment. Engineered Serendipity might be used to ensure that products that include both Geographical Visualization (GeoViz) tools and New Media artefacts are presented to users in such a manner that different information prospecting methods can be offered. Certain, different, design elements should be considered if such a hybrid as geoinformation product is to provide this information in such a way that serendipitous discovery is supported.

Engineered Serendipity was therefore a core idea that was incorporated into the overarching design of the prototype. It was envisaged that the application of this concept would be done through the application of games-like tools. The research is designing and producing interactive multimedia cartographic artefacts to complement the concept of *Engineered Serendipity* so as to be better informed about how to best design and deliver such geographical information exploration tools via computer game devices, with use strategies built around the methods and theory of gameplay using computers, and particularly connected computers.

5.6 Concepts Behind the Interface and Exploration Tools

In previous research the concepts of the *GeoExploratorium*, the *Map Shop* and the *Literate Traveller* were developed theoretically. Both the *GeoEx-*

ploratorium and the *Literate Traveller* concepts were further developed, prototypes built and subsequently evaluated. This research focusses on information provision via the use of the *Map Shop*, a virtual repository of (geo)information that can be delivered via interactive multimedia. The overarching framework for this research is to provide an exploration tool by using New Media. As stated earlier, this was first realised through the development of the *GeoExploratorium*a and then the *Literate Traveller*. Now the elements that form the theoretical information provision model of the *Map Shop* have been developed, a prototype built and this information provision approach will be evaluated.

The theory underwriting the design of a suitable prototype for evaluation is the Virtual *Map Shop*. As explained earlier, it was conceived as a Web resource that would provide enhanced interactive multimedia products so that users could appreciate and explore geography in the manner that is most appropriate for their particular usage patterns and preferences. The *Map Shop* was conceived as a virtual resource that provides geographical information through the conventional tools of maps, air photographs etc., but also via books, games, videos and expert advice.

To actually design and develop this product, research conducted with earlier, but related, interface tools was used as a general guide for what was appropriate, and could be built. Previous efforts related to the development of a suitable interface tool using the *Doom* games engine, 3D Web-delivered information using VRML and a hybrid Web product that contained a number of evaluation elements, the *Townsville GeoKnowledge* project (Cartwright *et al.* 2003a). Experience from these previous products, and associated evaluations, allowed a number of decisions to be made relatively quickly. These were related to the actual development of the prototype in terms of using the metaphor set as a general concept, designing the product to have the 'look-and-feel' of a computer game and using VRML as a development tool.

Previous evaluations about the use of metaphors found that they help to better understand the information depicted (Cartwright 2004b). During this research a prototype was developed and a questionnaire was used to obtain feedback from a so-called Nintendo Generation user group (Ormeling 1993), that group of users that have been brought-up in a world where computer games pre-existed. This generation are comfortable using computer games and adapt quickly to packages that 'work' like computer games do. Candidates were first asked to complete a simple profile information section and then to answer questions specific to their use of the test product. Candidates thought that metaphors helped to better understand the information depicted. They did not think that maps alone are best to gain geographical information, and that the metaphor set illustrated is a useful adjunct to simply using maps.

As noted earlier, the *MapShop* interface was deliberately designed to have the look-and-feel of a computer game. Again, this decision was made from experience gained from previous research. Test candidates who evaluated the games interface (Cartwright 2004b) found that the games-like product could be used with little prior experience and that this type of interface was easier to use than 'conventional' geographical information product user interfaces. Satisfying results were the facts that the interface could be used with little instruction and it was immediately obvious how to use the product (for this user generation). Candidates generally felt confident using the product. They noted that they would choose this type of interface over a conventional one if they were offered such a choice. Finally, they saw this type of interface as being appropriate for a first-time user.

Summarising, the results from previous evaluations indicated that this particular user group deemed that an interface that appeared more like a computer game, and one that required navigation through a 3D information display, was in fact the preferred interface. In general terms, the focus group of used, a Nintendo generation expert user/producer group supported the concept of using a different metaphor, and having a metaphor 'suite' delivered using a games metaphor.

From these explorations of the use of New Media for the provision of geographical information the idea was developed that using a 3D product that is designed to appear to be more game-like might provide an appropriate tool schema, and one that the further development of the Virtual *Map Shop* (and associated research) should take. The interest of the author here is related to developing and testing strategies that would provide innovative access to geographical information, but also ensure that users acknowledge that they are in fact using a scientific product. The following sections describe how this next phase of the research is being conducted – designing, building and evaluating the effectiveness of a Virtual *Map Shop* built using a 3D games-like interface.

5.7 VRML Map Shop Prototype Design Ideas

The 'proof-of-concept' product was designed to emulate what the user might like to find by exploring the virtual space in a serendipitous way, with the actual information access and subsequent provision engineered so as to ensure that all needed information was assured. It was envisaged that the user would access the information access 'cues' in a serendipitous way and in so doing, 'discover' information about the town. The initial ideas about how the Web-delivered 3D *Map Shop* might look are shown in Fig. 5.5.

The ideas about how the product would work guided the actual development of the prototype. The concepts behind the design were that a prod-

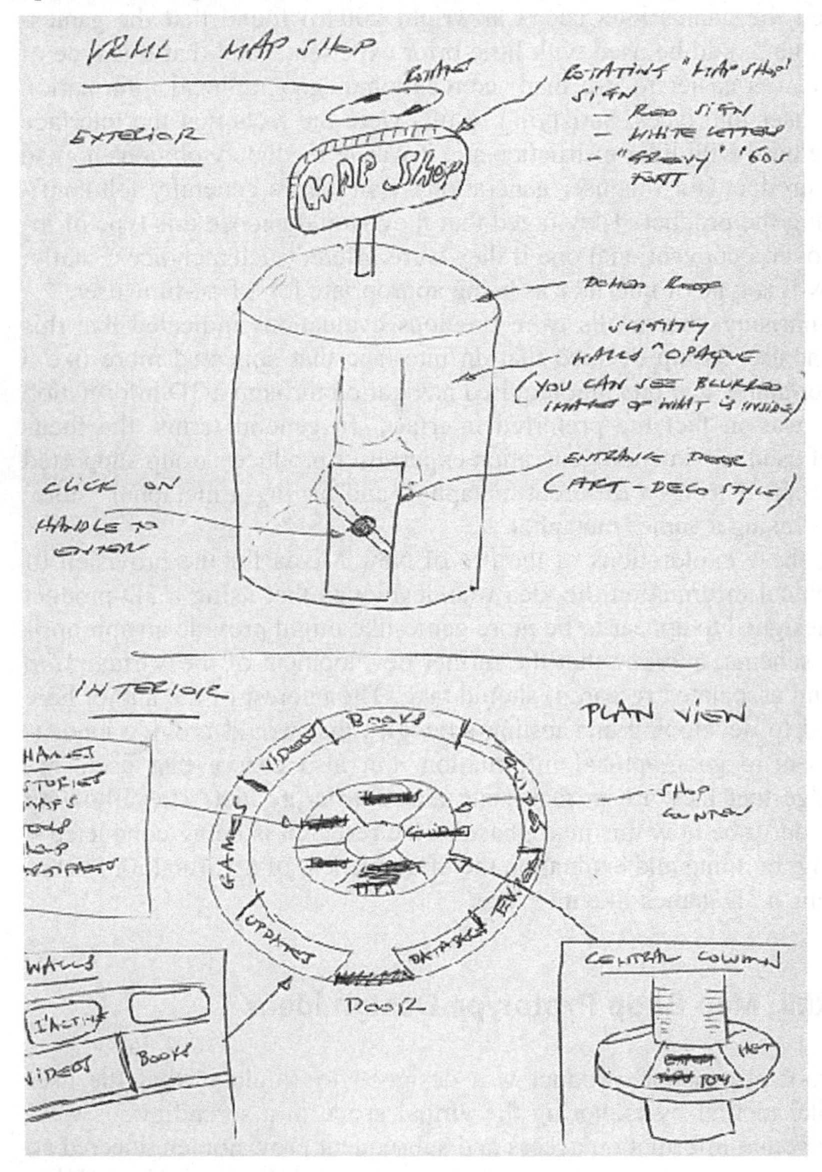

Fig. 5.5. Initial concept for the 3D On-line Map Shop.

uct should be built to have the look-and-feel of entering and exploring what information is available in a contemporary map shop. Using controls provided with standard VRML browser plug-ins the user moves towards the Map Shop and then opens the door to the shop by clicking the mouse button on the door. The user can then move into the 'shop', but only via the front door. Just inside the front door is a map curator, willing to guide and assist. The shop comprises a central pillar that has lists of information under selected categories. The lists are interactive and, once clicked, lead to related information via appropriate Web sites. Users can also access packaged information, available on another Web site built by the author – the Townsville GeoKnowledge product) - via maps and graphics that are 'placed' (virtually) on tables directly below the information lists. It also contains a number of map drawers. These map drawers have interactive products available on the tops of the drawers and also inside the drawers. The user 'explores and discovers the information resources in a serendipitous way.

In the completed prototype the information provided to the user was designed to contain information about Townsville, Australia. The *Townsville GeoKnowledge* product contains similar, complimentary (but different) information. Web-linked information can be accessed via a hot links on the central 'information' pillar, by clicking on the items or 'link' buttons placed atop of the map drawers or in the drawers. These link to additional, on-line information accessible when a new Web browser is activated, providing that information. As well as the map drawers that have map information viewable on top of the drawers, additional and video information is provided on the top and within one of the drawers. The drawers can be 'opened' by the user and their contents inspected (this is explained later in the chapter). The map drawers were provisioned with information catalogued according to the metaphor set defined previously in this chapter, allowing the user to 'browse' this information according to their preferred metaphorical exploration choice. Around the walls are images depicting panoramas of the town. A hot-spot on the panorama wall links from the *Map Shop* browser to another browser containing the selected panorama.

5.8 Building the Map Shop prototype

A decision needed to be made about the development media and tools to be used. For the tool to be effective it needed to be:

- 3D;
- Allow users to explore information in a serendipitous manner;

– Games-like;
– Delivered as a discrete / Web hybrid product; and
– Developed using Open Source coding (to conform to the general developmental objectives of the author).

Web delivery was essential to ensure that the Web-facilitated components of the *Map Shop* could be provided (see Fig. 5.3). Delivering a 3D games-like tool via the Internet thus required the use of a 3D coding application.

VRML (Virtual Reality Modelling Language) was chosen to develop the proof-of-concept prototype. Some may argue that X3D should have been used, rather than VRML. However, many developers take the VRML 'track', rather than the X3D track. If the use of XML is not required, then the use of X3D for a prototype cannot justified. A useful product can be developed with VRML. Also, again as this was a prototype, the use of VRML provided greater confidence that a suitable browser/plug-in could be employed. If, later in the research project, XML is incorporated, the VRML code can easily be transported into X3D.

Finally, to implement games-like controls, a *Nostromo* controller was employed. This games controller provides the facility to code the controller keys, wheels and toggles to emulate the mouse/keyboard controls needed to navigate through a 3D VRML world. It also enables the speed of travel through the virtual world to be 'dialled-in' via a wheel that is part of the device.

For the actual delivery of the product the MicroSoft *Internet Explorer* browser was employed, and combined with the VRML browser plug-in from Bitmanagement Software (Germany). It has been found that this combination provides the most effective tool delivery, considering that this is such a mid-resolution image-laden product. The browser plug-in also provides the facility to move through the model in a games mode, allowing the programmed functions of the *Nostromo* controller to be employed. The browser can also include an avatar.

5.9 Exploring the Product

How this was realised can be seen in Fig. 5.6 (outside the *Map Shop*), Fig. 5.7 (the front door, with Map Curator) and Fig. 5.8 (the core display 'pillars). The outside of the *Map Shop* was developed to give the impression of a resource that invited further investigation and exploration. The walls are opaque, providing a glimpse of what is inside the product. Using the browser controller tools the user can approach the door, click on the door

Fig. 5.6. The Map Shop.

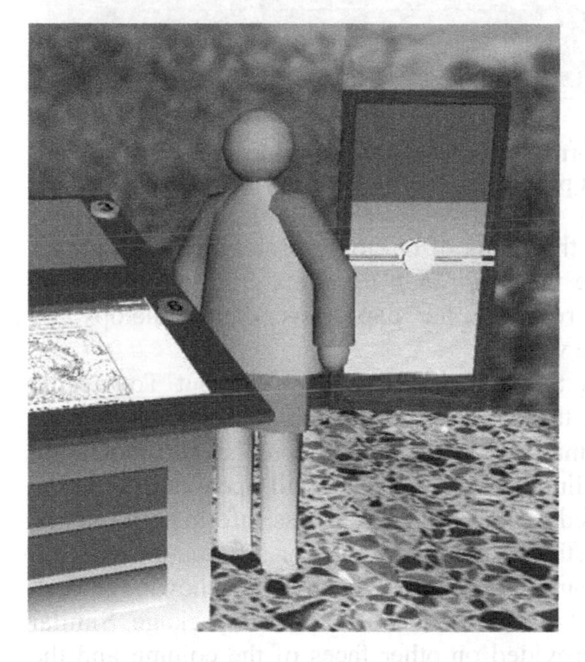

Fig. 5.7. Map Curator at the front door.

handle and then enter the shop by navigating the product with mouse or selected keyboard keys. Users can also move from one viewpoint to another, in a pre-programmed manner or by selecting particular viewpoints of interest. At the door has been placed a Map Curator avatar. This is not yet

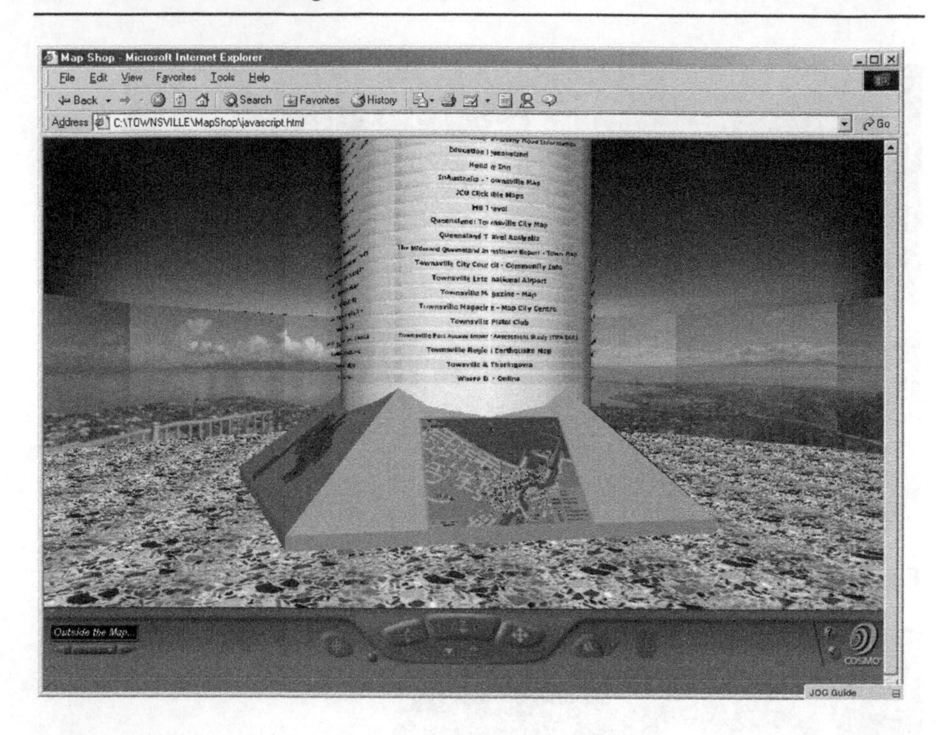

Fig. 5.8. Map Shop central information column and map table. The walls have images from the available product panoramas.

functional, but it is will be the focus of on-going prototype development and evaluation. Once inside the *Map Shop* users are free to explore the central column information resource, the map drawers and tabletops and the panorama that adorns the walls.

The central column lists the information available about Townsville (Fig. 5.8). By zooming into the column and moving the cursor over individual 'entries' on the information list, the cursor icon changes, indicating that by clicking the named link a new Web page will open and the information displayed. Users need to be on-line to access information outside the package. At the base of the column is a table, or information plateau. Here, the maps link to the complimentary Townsville GeoKnowledge Web site (Cartwright et al. 2003b), a 2D interface information package. Similar information resources are provided on other faces of the column and the plateaus – for databases, expert tips and images.

The *Map Shop,* which already provides information via the central column, was enhanced via a number of 'workstations' situated on the floor. Each of these workstations provides information access using these different metaphorical approaches. At the design stage it was envisaged applica-

Fig. 5.9. Map Shop drawers – contain information arranged by metaphor.

tions 'built' using these different metaphors would allow for 'everyday' tools and computers to empower users to more easily get access to comprehensive geographical databases using methods that they feel most comfortable with. The way in which the workstations would be placed in the *Map Shop* is illustrated in Fig. 5.9. 'Tools' and other resources placed on top of the workstation, and other artefacts made available in the drawers of the workstation provide information access tools.

Discrete interactive multimedia objects 'reside on the top of the workstation (Fig. 5.10). Items located in each of the drawers (that can be opened by 'pulling' the drawer handle with a mouse held down (Fig. 5.11)).

If the user wishes to view a larger version of the information that resides on top of the drawers, a mouse click enables an enlarged (and higher resolution) view to be viewed (Fig. 5.12). Users can also explore the contents of the drawers. The images of the maps and artifacts inside the drawers have been 'collected' and arranged in metaphorical sets. If a map or artifact in the drawer is of interest the user can click on the mage and, once clicked, this will link to and open a browser to outside information.

The manner in which it is envisaged that users will use these drawers, and the items atop the drawers, to explore the information in a serendipitous manner. The arrangement in metaphorical sets and providing links to pre-determined information sites allows the users serendipitous 'wanderings' within the *Map Shop* to be Engineered. Serendipitous movement

Fig. 5.10. Zooming in to the map drawers showing 'clickable' links to external Web resources on the drawer tops.

within the package + engineered information provision = Engineered Serendipity!

As well as static images, the VRML World contains links to videos. To access these, the user needs to move to the map drawers with the video images on the top (Fig. 5.13). They are activated by clicking on the button to the right of the drawer top image. Once the button is clicked a movie player is activated and the user can view the video using the standard controls. Note - clicking on the image itself provides an enlarged vertical view of a still from that particular video (Fig. 5.14).

Surrounding the *Map Shop* floor is an image collection from one of the panoramas available. The panorama image has been 'sliced' to manufacture wall panels. Hot spots (with the camera icon) have been placed to indicate the further links to panoramas in the package. A click on the selected icon leads to a panorama (Fig. 5.16), created with the package *PixAround*. It is envisaged that eventually these panels could be changed via a user-selected graphic device.

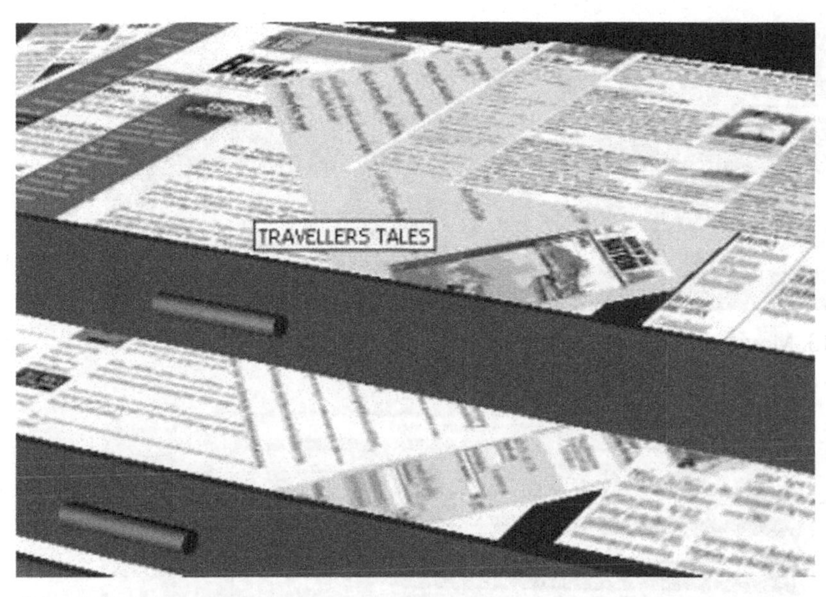

Fig. 5.11. Information links inside drawers.

Fig. 5.12. Zooming in to the map drawers showing 'clickable' links to external Web resources on the drawer tops.

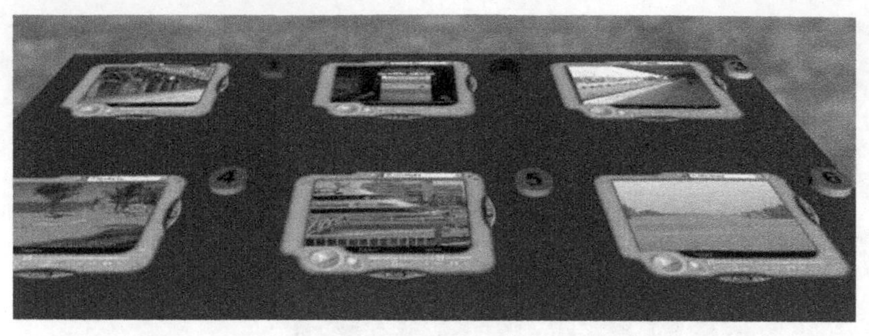

Fig. 5.13. Movie top.

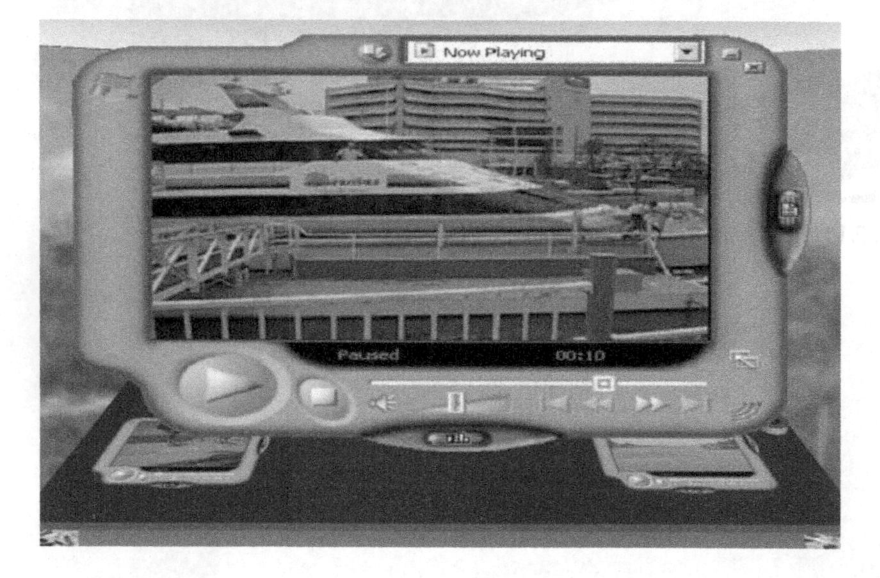

Fig. 5.14. Clicking on image enlarges image. Clicking on button to right of image on tabletop opens and plays movie in new browser window.

At the side of each drawer cabinet is an icon illustrating the metaphor that was used to define the 'collection' of online artifacts available (see Fig. 5.16). Clicking on this item leads to another Web page that describes the icon, for example the Navigator metaphor in Fig. 5.17.

5.10 Further Development of the Package – Adding Map Curator

The links to all of these resources need not require the user to be physically in the *Map Shop*, nor should it be imperative that all of the participants in the total exploration of all of the resources be there as well (or at all). Communication with experts could be made through electronic means, and access to other collections and artifacts could be made possible via electronic links. However, what happens to the novice user of this enhanced

Fig. 5.15. Panorama and metaphor hot-spots

Fig. 5.16. Panorama.

Map Shop? Are they allowed to wander aimlessly around, looking for information that is irrelevant to their particular application, or not available or completely misleading? Or should a map curator (either real or virtual) be provided to assist them with their quest? There are many curators of map collections who are expert in both their own collection and have knowledge about other collections, and the availability of maps from international providers and private repositories. The expanded Map Shop should be manned with a curator to provide advice and guidance on finding the most appropriate resources and assistance in the best way to use these products and resources. The curator can 'reside' (as an avatar) in an on-line shop or be a real curator, available at the end of an Internet connection. The curator would be an invaluable component of the Map Shop.

A curator could assist users in selecting the appropriate metaphors to use and find the required information through links to distributed resources. When using the World Wide Web for resource access, this curator could be developed as an Agent. Agents, according to Maes (1996, p. 1) are "active entities that can sense the environment - the digital world - and perform actions in the digital world and interact with us". These active entities can be called software agents. They can be semi-intelligent, and they can be born, reproduce or die. Shapiro (1992) saw the need to pursue the 'Intelligent Agent' approach to user interfaces for GIS. In an ideal package, the human would be able to interact with such a system through a mix of natural language and gestures. Agents could be used as a surrogate curator, constantly seeking out information and 'directing' users to its location. Rodrigues *et al.* (1995) noted that intelligent agents have been used for information retrieval, assisting in the operation of intelligent networks, for task execution and for collaborative processing. They further noted that spatial intelligent agents could be used for locating and retrieving spatial information, to automatically generate templates for spatial modelling, to automatically monitor spatial model execution and to allow for collaboration between GIS and other specific use packages. The *Map Shop* Curator could ensure that the user obtained the required or ideal mixtures of resources from the *Map Shop* (discrete and on-line resources).

As explained earlier, the Map Curator can be developed for the package by developing the 3D character located just inside the front door of the Map Shop. Alternatively, the browser plug-in provides a built-in avatar. This would provide another route for Map Curator development. Both the Map Curator and the browser-facilitated avatar can be seen together in Fig. 5.18.

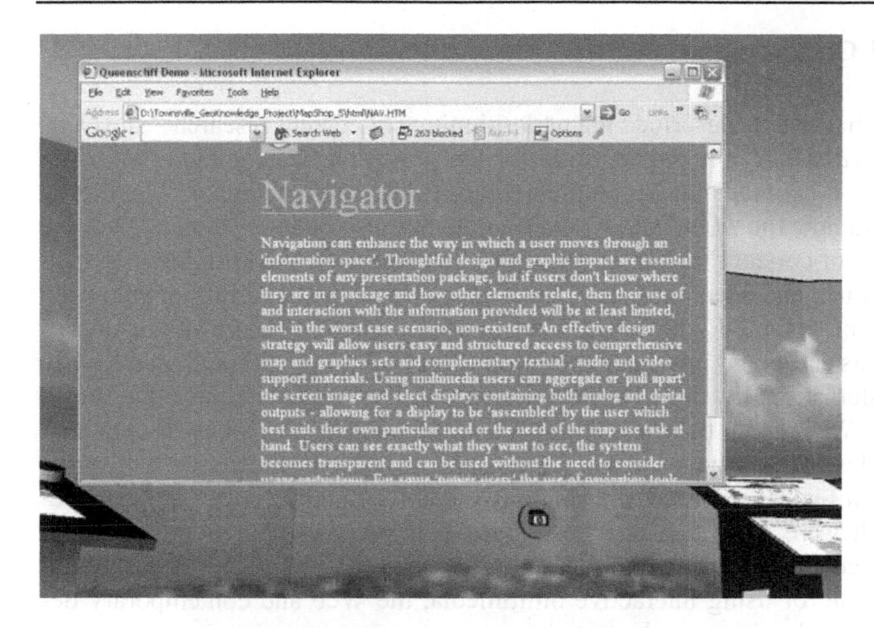

Fig. 5.17. Metaphor information.

Fig. 5.18. Map Curator and browser-facilitated Avatar.

5.11 Conclusion

The chapter has described the underpinnings of current research – the *Geo-Exploratorium*, a metaphor suite for the provision of geographic information in a manner that is complementary to the map metaphor. The theory behind how the *Map Shop* would operate has also been explained. And, the 'proof-of-concept' prototype, being built to examine the usefulness of such a product, has been outlined. This is being developed and further enhanced to be provisioned with maps, videos, books, guides, games and databases of facts and could provide expert tips. The *Map Shop* will be linked locally or internationally through the Internet and, more specifically, the World Wide Web. From the users' perspective the boundary between discrete and distributed multimedia would be transparent, presenting than with the most current and customised information possible. Finally, areas of future research were identified and described. It is believed that if this research were continued to include the addition of the Map Surator, then the real potential for using interactive multimedia, the Web and contemporary delivery mechanisms can be tested.

Acknowledgements

Initial development of the Townsville GeoKnowledge project was supported through a research grant from the Command and Control Division, Defence Science and Technology Organisation, Department of Defence, Australia, Edinburgh, South Australia.

Initial VRML programming for the *Virtual Map Shop* by Tim Germanchis. Additional VRML programming by Dominik Ertl, Adam Anderko and Dane McGreevy.

References

Cartwright WE (1997) The application of a new metaphor set to depict geographic information and associations. In: Proceedings of the 18th International Cartographic Conference, International Cartographic Association, Stockholm, Sweden, pp 654-662

Cartwright WE (1998) The development and evaluation of a web-based 'GeoExploratorium' for the exploration and discovery of geographical information. In: Proceedings of Mapping Sciences '98, Fremantle: Mapping Sciences Institute, Australia, pp 219-228

Cartwright WE (1999) Extending the map metaphor using web delivered multimedia. International Journal of Geographical Information Science, vol 13(4), pp 335-353

Cartwright WE (2004a) Engineered serendipity: thoughts on the design of conglomerate containing geoviz tools and geographical new media artifacts. Transactions in GIS. Blackwell Publishing Ltd, vol 8(1), pp 1-12

Cartwright WE (2004b) Using the web for focussed geographical storytelling via gameplay. In: Proceedings of UPIMap 2004, the 1st International Joint Workshop on Ubiquitous, Pervasive and Internet Mapping, International Cartographic Association Commission on Ubiquitous Cartography, Tokyo, Japan, pp 89-109

Cartwright WE, Williams B (2003a) Exploration of the potential of using the concept of the GeoExploratorium for facilitating acquisition of geographical knowledge: first cut. In: Proceedings of the 20th International Cartographic Conference, International Cartographic Association, Durban, South Africa

Cartwright W, Williams B, Pettit C (2003b) Realizing the 'Literate Traveller'. In: Spatial Sciences Institute Conference Proceedings 2003, Spatial Sciences Institute, Canberra

Fisher S (2003) Media technology and simulation of first person experiences. In Depth, ArtM useum.net, http://www.artmuseum.net/w2vr/archives/Fisher/Environment.html#Sensorama

Laurel B (1993) Computers as Theatre, Reading. Addison Wesley

Laurini R (2001) Information systems for urban planning - a hypermedia cooperative approach. Taylor & Francis

Mayes JT (1992) The 'M' word: multimedia and their role in interactive learning systems. In: Edwards AND, Holland S (eds) Multimedia interface design in education, Springer-Verlag, pp 1-22

Ormeling F (1993) Ariadne's thread - structure in multimedia atlases. In: Proceedings of the 16th International Cartographic Association Conference, International Cartographic Association, Köln, Germany, pp 1093-1100

Rodrigues A, Raper J, Capitao M (1995) Implementing intelligent agents for spatial information. In: Proceedings of the 1st Joint European Conference on Geographical Information, The Hague, The Netherlands, pp 169-174

Shapiro SC (1992) User interfaces for geographic information systems. In: Mark DM, Frank AU (eds) Research Initiative 13, Technical Report 92-3, NCGIA, pp 153-154

6 Atlases from Paper to Digital Medium

Cristhiane da Silva Ramos, and William Cartwright

Abstract. This chapter analyses the concept of digital atlas, analysing different definitions found in the literature. Additionally, the characteristics and different classifications of digital atlases are analysed. The lingering question about the difference between GIS (Geographical Information Systems) and digital atlases was also addressed in the chapter. It is believed that a better understanding of what a digital atlas can be will facilitate the development of future digital atlas projects.

6.1 Introduction

The main purpose of this chapter is to define atlases in general, and particularly digital atlases. Additionally, the chapter focuses on gaining an understanding about what constitutes digital atlases: defining their characteristics, classification and delivery medium. It is believed that this is important to better understand atlases in this contemporary context, as atlases are no longer restricted to just paper and, as well, they are not necessarily comprised of just maps.

Many researchers have reviewed the definition of what an atlas is. However, although many definitions of digital atlases have been made, digital atlases encompass such a wide range of features and technologies, it is necessary to extend these by defining their characteristics and different atlas types. Also, the recent advances in atlases production have raised the question of: are digital atlases a kind of GIS? This question is inevitable as, in some ways, atlases cannot just be comprised of functions for spatial analysis, but also GIS objects as well. If there was a dividing line separating the definitions of GIS and digital atlases, one could say that it is becoming fuzzier.

Atlases are, probably, the best known and the most flexible of popular cartographic products. Atlases can be used to address different issues and to target different audiences. Historically, atlases have played different roles - from instruments of power, in the renaissance to current decision and planning support tools. Atlases can be used for general reference, education and business. As they evolved, atlases were produced in different ways, from the initial manual production to current computer-generated applications. Atlases have experienced many changes in the way they are conceived, produced, disseminated and used.

6.2 Technology and Cartography

Technology can also be considered to be an underlying concept for understanding the development of cartography. Traditionally, cartography has been quick to adopt technological advances from other fields. However, as stated by Robinson *et al.* (1995), it seems that every new technology that has been introduced has caused greater changes in cartography than the previous one, and, as well, it can be argued, the time between technological changes seems to be diminishing.

These advances changed the way cartographic products were produced and distributed. As humanity evolved technologically, these advances were adopted by cartography. Initially, mediums such as clay tablets, leather and papyrus were used to convey maps. Each of these maps was unique, just like the first map drawn in the sand. Maps had to be reproduced one by one using a painstaking process mostly done by high-skilled workers. Therefore, even after the use of durable media, maps were still not widely available.

During the 20th century many other advances were introduced to cartography from field equipment to the use of satellite images. The introduction of computers represented a significant revolution to the cartographic process, because they not only changed the way that cartography is produced, but they became a medium for cartography.

Considering changing technologies as a guiding concept to understand cartography, Robsinson *et al.* (1995) emphasised the technological periods in cartography. This is summarised in Table 6.1.

Due to the advances summarised above, maps became increasingly present in everyday life, not only for decision makers, but also for ordinary people. In the news, in weather forecasts, in multimedia kiosks in shopping centres, in street directories, in tourist brochures, schoolbooks, maps are omnipresent.

6.3 Atlases: A Definition

Atlases are probably one of the first cartographic products that people use, because they are introduced to students in early at school. Atlases can be considered the most widely known cartographic products (Kraak, 2001b)

Table 6.1. The six major technological revolutions that affected cartography, after Robinson *et al.* (1995).

Technology	Beginning	Effects on Cartography
Manual	Pre-history	The beginning of cartography. Maps were result of field observations and were made by skilled hand-workers, one by one. The authors highlight that the techniques developed during this period remained in use in the following periods.
Magnetic	12th Century	The introduction of the magnetic compass, brought from China, represented a major change to cartography. After this technology, angles could be measured.
Mechanical	12th Century	The introduction of the printing process, first in China during the 12th Century and later in the Western culture, by Johannes Gutenberg in Germany during the 15th Century, made possible to reproduce maps more rapidly and efficiently
Optical	17th Century	The introduction of the telescopic and magnifying lenses enhanced the human sight. Other important optical technologies such as the stereoscope and CD-ROMs for data storage were also introduced in the following centuries.
Photo-chemical	18th Century	Lithography, which is a chemical printing process created in the late 17th Century, make it easier and cheaper to reproduce maps. Another remarkable advance in this period was the introduction of photography, first for field observation and later the use of aerial photographs made possible perpendicular view of geographical space.
Electronic/ Digital	20th Century	Electronic technology changed cartography as a whole. Electronic devices, mainly computers, are used to collect data, to produce and print maps and even as the medium in which the map is to be used. Digital technology is used to store data, and millions of maps as digital files are exchanged daily over the internet.

and a high point for cartography (Kraak and Ormeling, 1996, Kraak, 2001b).

The original meaning of the word atlas refers to ancient Greek mythology. Atlas was one of the twelve Titans, sun of Uranus (God of the sky) and Gaea (Goddess of the earth), brother of Zeus, the supreme governor of the universe, and father of Pleiades and Hesperides. According to the ancient Greek beliefs, Atlas was the God who was condemned to hold the vault of the sky, represented by a globe, on his shoulders. (Aghion *et al*, 1996).

According to Thrower (1972), the use of the word atlas to describe a collection of maps was introduced by the famous cartographer Gerhardus Mercator, who, after publishing his world chart in 1569, dedicated his final years to produce a large atlas, which was released in 1595, one year after his death. Thrower (1972, p. 56) claimed that "it is undoubtedly because Mercator used the term 'atlas' for a book of maps that it is in use today".

In general, atlases can be understood as a collection of maps with a specific purpose and organized in the form of a book, which usually includes tables, graphs and text. The word "atlas" can also be used to describe a collection of information that covers a field of knowledge, for example, an Atlas of Anatomy or History. However, in this chapter the word refers to geographic atlases.

'Traditional' atlases, such as books, are bound publications and therefore have a fixed linear structure. Topics are developed linearly throughout the publication. The maps are developed according to a fixed format, limited to the size of the page. However they cannot be considered to be books, as noted by Alonso (1968):

> "To the layman, any book consisting mainly of maps is an atlas, but technically to the geographer, no cased collection of maps deserves the name unless it be comprehensive in its field, systematically arranged, authoritatively edited and presented in a unified format." (Alonso, 1968, p. 108)

According to Keates (1989) the use of maps in atlases is very flexible, he argued that "Although the term 'atlas' is frequently associated with the concepts of 'world' and 'small scale', there are atlases with large-scale plans (city street guides), special-purpose atlases (such as road atlases for motorists), and special-subject atlases (such as an atlas of agriculture)" (Keates, 1989, p.235). Generally, Keates (1989) highlighted that one could discern atlases by its scale; topic; and target audience.

Maps have been present throughout the human history and their evolution can be looked at from different viewpoints. Atlases can be classified by several aspects. Ormeling (1995, p. 2128) classified traditional atlases in regarding to their contents with respect to: geographical atlases; historical; national/regional; topographic; and thematic atlases. Considering communication purposes, the following types of atlases that could be identified are: educational, navigational, physical planning, reference, and management/monitoring. Furthermore, Borchert (1999) stated that different categories of atlases can be distinguished according to format, geographical coverage, thematic content, information level, purpose, publisher, quality and price.

Focussing on how atlases were used interactively, Peterson (1998) has stated that, despite new advances in interactive cartography, the concept of interactivity itself is not a new one for cartography. He divided the history of cartography into three periods. The first is characterised by interactivity, but consisted of ephemeral, maps. According to Peterson it is very probable that the first maps were results of conversations. Maybe, these maps could have been the consequence of discussions about the location of enemy tribes or better sites for hunting. Perhaps, they were drawn on sand. What is clear is that they were not meant to last, but to enrich communication, at the same time providing additional information – the result of the conversation. About interactivity in early cartography Peterson (1995, p. 10) had this to say:

> "That map in the sand was not the kind of static map that we find today on paper. The meaning of the symbols would have been explained as the map was created. The second person would have asked questions that influenced how the map was drawn. As new features were added, they may have obscured part of the existing map. Indeed this first map would have been very much like the kind of interactive and dynamic map that we are attempting to create today with the computer."

About the second atlas period, Peterson (1998, p. 3) argued that "A major shift occurred long ago as a stable medium was used and maps were transformed into static objects, first on clay and later on paper", in this way, it was possible to create durable maps that could also be distributed to map users in different locations. Therefore, the cartographer was necessary just to create the map. Once the map was ready it could be distributed and used by anyone, anywhere. This was an important transition for cartography and was crucial for the history of the mankind. However, this evolution separated cartographer from map user and, hence, the interactive factor was removed. Cartography, as a result, dived into a static period".

The third period has seen the introduction of computers. They were initially used as a tool and later as a medium, bringing back interactivity to cartography. However, as Peterson has highlighted, it is a different kind of interactivity - human vs. computer. Nevertheless, after centuries of static maps (mainly on paper) people got used to think maps as static representations. This thought, which Peterson (1998) called 'paper thinking', now provides one major drawback for interactive cartography - it is hard for the present atlas-user generation to overcome the way they were initially taught to conceive maps. But, as technology and interactive maps become ubiquitous, this obstacle should tend to vanish in the near future.

6.4 Digital Atlases: the Search for a Definition

The technological advances represented by the use of computers for the production and distribution of maps have had a major impact since the early eighties and, therefore, created a new category of atlases: digital (or electronic) atlases. Atlases, as described previously, were studied and produced for centuries, however, digital atlases have a shorter history and, consequently, concepts have been developed over the last few years. The first digital atlases were developed during the eighties and an increasing research effort in the field has been carried out since then (Rystedt, 1996).

Ormeling (1995) and Kraak (2001b) also noted that the development of digital atlases started in the late eighties. The authors consider the *Atlas of Arkansas*, presented in 1987 during the 13th International Cartographic Conference of the International Cartographic Association (ICA), the first digital atlas to be developed.

According to Siekierska and Williams (1996), the first digital atlas developed was the *Electronic Atlas of Canada,* in 1981. This pioneer atlas was the result of a long Canadian tradition in producing national atlases. Afterwards, the Canadian government created the *National Atlas Information System* (NAIS), aiming to create digital databanks for mapping as well as facilitating the digital production of paper maps. Thenceforth the following years were distinguished by the development of many other digital atlases either by governments, universities or private companies.

In 1986 the *Digital Atlas of the World* was created by Delorme Mapping Systems. Analysing the use of that atlas, Siekierska and Taylor (1991, p. 12) stated that "apart from scale change and the additions of overlays, the analytical capabilities are limited".

Kraak (2001b) considered this first digital atlas as an extension of a paper atlas, in digital media, as it was comprised by a set of static maps accessed via a menu. As stated by the author the development of digital atlases presents similarities with the history of computers in cartography. The beginning was characterised by hardware limitations, such as storage capacity, and software, represented by the lack of authoring tools for developing more interactive applications.

According to Kraak and Ormeling (1996, p. 183) atlases are "intentional combinations of maps, structured in such a way that given objectives are reached. In a way, atlases are similar to rhetoric: if a number of arguments are presented in a speech in a given sequence, a specific conclusion is reached.". This definition can be applied either to paper or digital atlases, however, the idea of defining an atlas as a bound collection of maps should be reviewed.

Siekierska and Taylor (1991, p. 11) tried to fill the existing gap between the traditional definition of paper atlases and digital atlases by creating a new definition of digital atlases, which stated that "The electronic atlas is a new form of cartographic presentation and can be defined as an atlas developed for use primarily on electronic media."

Koop (1993, p. 129), created a more flexible definition of atlas, as a "Systematic and coherent collection of geographical data in analogue or digital form, representing a particular area and/or one or more geographical themes, based on a narrative together with tools for navigation, information retrieval, analysis and presentation."

Elzakker (1993) claimed that a digital atlas, and particularly an analytical atlas, is a special kind of GIS. In his opinion the main difference between digital atlases and GIS is that digital atlases have a narrative faculty, once they are designed to attend a specific purpose. He defines electronic atlases as "a computerised geographic information system – related to a certain area or theme in connection with a given purpose – with an additional narrative faculty in which maps play a dominant role" (Elzakker, 1993, p. 147).

The same definition was adopted by Ormeling (1995, p. 2127). Other authors such as Richard (1999) and Borchert (1999), also stated that nowadays atlases could be considered a collection of maps distributed either on paper or by digital means.

Kraak and Ormeling (2003, p. 154) considered that "If paper atlases are considered intentional combination of maps, then not all electronic atlases might fit in this definition. Some could better be defined as intentional combinations of specially processed spatial data sets, together with the software to produce maps from them". In other words, maps do not comprise digital atlases necessarily, but according to the authors the narrative is a cornerstone concept to define atlases.

As can be seen by these many definitions of digital atlases, the development of digital atlases has fostered an increased research focus since the early nineties. It is important, however, to highlight that as the development of digital atlases constituted a new field at that time, the first definitions tried to explain digital atlas by comparing them to paper atlases. In this way the first, and most obvious, difference emphasised was the medium. Further definitions focussed not only the medium but atlas use as well, by putting the emphasis on aspects such as the narrative or navigation tools.

6.5 Digital Atlases: Characteristics

With the introduction of this new field of research in cartography, these researchers tried to summarise differences and similarities, as well as advantages and disadvantages that could be identified between digital and paper atlases.

As stated previously, Canada played a pioneer role in the field of digital atlases. Siekierska (1984), explaining the procedures adopted for elaborating the early version of the *Electronic Atlas of Canada*, stated that the important aspect of digital atlases is that they made it feasible to generate maps that were user-demanded or user-created. In this way the user's needs would be more likely fulfilled. The author argued, however, that the main difference between digital and paper atlases is that with digital atlases the analysis is made directly on the data set, the input data, instead of the traditional analysis being made directly in the final product, the printed map.

In other words, in the early eighties Siekierska proposed that digital atlases could provide an exploratory environment in which the user interacts with the information directly stored in the data bank, using the map only as an interface. In this way, she saw that the proliferation of digital atlases would be an important field of research for cartography in years to come.

Summarising the characteristics of electronic atlases, Siekierska and Taylor (1991) indicated that important advantages were the fact that those products provide tools for innovative displays and analysis such as queries, overlay, animation, interactive zoom, scroll and pan. In addition, they emphasised the reduced cost of producing a digital atlas as a significant advantage, once they could be reproduced and distributed to a wider audience with lower costs, when compared to paper atlases.

Bakker *et al* (1987), analysing national atlases, foresaw that digital production of national atlases would be an important trend in the future, however, it was very unlikely that national atlases on paper would disappear completely. It is impossible to verify, by official statistics, if the authors were right, nevertheless sixteen years later papers presented and published on the proceedings of the 21st International Cartographic Conference, supported what the authors predicted, that digital atlases have become paramount atlas products. At that conference, digital and hybrid atlases represented 83% of the atlases presented (Table 6.2).

The *National Atlas of Spain*, for example(Aranaz *et al.* 2003), was first published on paper, afterwards a digital version in CD-ROM was implemented and at that time a future version for the Internet was being developed. However, the paper atlas was still being produced and new formats were being released to provide better user handling. Other examples of national atlases produced both on paper and digital formats presented in that conference are the *National Housing Spatial Investment Potential Atlas of South Africa* (Biermann and Smit 2003); the *Census Atlas of the United States*(Brewer *et al.* 2003), the *Atlas of Oregon* (Buckley *et al.* 2003) and the *National Atlas of Russia* (Zhukovsky and Sveshnikov 2003). The atlases presented just in digital version were the *Atlas of Switzerland* (Huber and Schmid 2003) and the *Statistical Atlas of the European Union* (Pucher *et al.* 2003).

In a similar way to Bakker *et al.* (1987), Ormeling (1995) predicted that by the year of 2000 paper atlases would be published with a CD-ROM as a digital complement. Even though after a decade his prediction was not achieved, the number of digital and hybrid atlas, atlases available both on paper and digital formats, has increased sharply. The number of paper atlases presented, on the contrary, decreased considerably (Table 6.2).

Ormeling proposed digital atlases should have three main functions:

- *To provide background information:* which could be tables with the statistics used to create the maps, photos, texts, graphs or drawings;
- *To expose other geographical views of the data:* the digital complement should be able to produce maps different from those published, based on the same data. This function would be possible by using different classification systems and changing the number of classes or even different boundaries; and
- *To provide additional information:* paper atlases have a series of issues to consider regarding their development. However, the cost of the publication could be isolated as the main issue. In this way the digital counterpart could provide more information than that published, with re-

Table 6.2. Number of atlases presented in the International Cartographic Conference, in percentage.

Type of Atlas	1995	2003
Paper*	74%	17%
Digital	14%	50%
Hybrid	12%	33%

* Atlases presented with no reference to the medium were considered paper atlases.

duced cost.

Rystedt (1996, p. 1) considered that "An electronic atlas can contain data and software giving the user possibilities to more thoroughly investigate the topics presented in the book version of the atlas". The author claimed that the use of different media and GIS functions would improve the potential of the new kind of atlas and furthermore the challenge would be to develop digital atlases for the Internet.

It seems, despite the advantages highlighted, that both paper and digital atlases have theirZ advantages and disadvantages. Analysing this issue, Koop (1993) claimed that the main asset of paper atlases is the "lazy armchair function". In contrast the major drawbacks of paper products would be the high cost of such products and the time spent to update them. The author also noted the advantages of digital atlases: they are easy to update and provide new forms of cartographic communication.

Kraak and Ormeling (1996) considered that the advantages of digital atlases are: the possibility of creating customized maps; the immediate provision of geometric information; the possibility of going beyond static map frames by using interactive tools (such as pan and zoom); the use of animations to depict data over time; links provided with databanks to provide additional information regarding features highlighted or clicked by the user (and here interactivity is, once more, an important point); the integration of multimedia; and the possibility of improving manipulation and display of information by different levels of aggregation of data, from small to large regions.

Additionally, Peterson (1999, pp. 35-36) analysed the advantages of paper for cartography over digital media and saw two major advantages: paper maps are easy to carry and paper supports higher spatial resolution. However, Peterson argued that, although paper is better for cartography, it is incompatible to represent dynamic phenomena. On the other hand, he considered the major advantages of digital maps to be that: they are more effective in representing dynamic phenomena; when distributed by networks they can be delivered much quicker than paper maps; they are more current; and they change the traditional map use, as digital maps can be interactive, and therefore the user's attitude towards the map is different as they engage the information more deeply.

Comparing digital and paper atlases, Schneider (2002, p. 24) stated the following about the advantages of digital atlases over its paper equivalents:

> *"Sie können neue Karten erstellen oder bereits bestehende ihren eigenen Bedürfnissen entsprechend flexibel anpassen. Die Informationsübermittlung erfolgt dabei nicht linear, sondern über eine thematische oder räumliche Orientierung.*

Durch die Verwendung von Multimedia kann ein bestimmter Sachverhalt dem Atlaspublikum über verschiedene Medien präsentiert werden. Zudem besteht die Möglichkeit, mittels Animationen zeitliche und räumliche Prozesse zu simulieren. Die grossen Speicherkapazitäten von elektronischen Medien erlauben ferner, nicht nur eine Auswahl von Daten, sondern komplette Datensätze in verschiedenen Massstabsstufen anzubieten. Schliesslich können AIS leicht aktualisiert und erweitert werden."[1]

Analysing the same issue, advantages of digital atlases, Borchert (1999) assembled a comprehensive list that is summarised in Table 6.3. Also focussing on the differences between paper and electronic atlases, Ormeling (1996) summarised what he saw to be the main differences between them. These are shown in Table 6.4.

Ormeling (1996) considered view-only atlases and paper atlases in the same category because, according to his arguments, view-only atlases do not use technology "adequately", which can be understood as, although view-only atlases are digital atlases they do not take advantage of the resources available in the digital medium. He made use of multimedia and the dual concepts static vs. dynamic, passive vs. interactive to analyse the differences between paper and digital maps.

6.6 Digital Atlases: classification

Digital atlases are a type of atlas that can be distinguished by its digital format, either discrete (produced for distribution in floppy disks, CD-Rom or DVD) or networked. Besides the traditional classification of atlases, more specific classifications considering only digital atlases have been developed. Siekierska (1991) cited by Elzakker (1993) classified digital atlases in three groups, considering basically the level of interactivity and

[1] "They (the user) can generate new maps or can adapt existing maps according to their needs. The transmission of information takes place in real time, not linearly, but rather via a thematic or spatial-oriented navigation. The use of digital multimedia, over different media, allows presenting a specific issue to the atlas audience. Moreover, it is possible to simulate temporal and spatial processes by using animations. The large storing capacity of electronic media permits furthermore, not only a selection of data, but rather to offer complete records in different scale levels. Finally AIS (Atlas Information Systems) can be updated and expanded easily."

Table 6.3. Summary of the advantages of digital compared to traditional atlases, after Borchert (1999, p. 76).

Attribute	Description
Exploration	Exploration can be understood as the amount of freedom given to the user in order to explore the contents of the atlas. The use of GIS functionalities can be included in this concept. Interactivity is a key component in an exploratory atlas.
Dynamics/Animation	The use of animation and its new visual variables brings new forms of communicating spatial data.
Customisability	Once more the issue of interactivity is present, at this point the concept is to allow the user to customise the map as the interface of the information, by changing layers or visual variables for example, in order to attend individual requirements.
Integration with diverse media	It is possible to integrate the digital atlas with textbook, paper atlas, working sheet, wall map, and so on. In this way new didactic perspectives could be reached.
Current contents	A digital atlas can be easily updated; if the product is networked its contents are current and immediately available to the user.
Portability	The digital atlas is easier to transport when available in discrete media, moreover if it is available on the Internet portability is not an issue. However, the computers are still heavy.

Table 6.4. Differences between paper and electronic atlases, after Ormeling (1996, p. 33).

Paper Atlases/View-only Atlases	Interactive Atlases/Analytical Atlases
Static	Dynamic
Passive	Interactive
Maps only	Maps and multimedia
Limited/selective	Complete
Fixed map frames	Panning and zooming possible
Compromise for all types of use	Customised
Maps as final product	Maps as interface

the analytic potential provided. There, digital atlases were subdivided as being:

- View-only atlases;
- Atlases that generate maps on demand; and

- Analytical atlases based on GIS capabilities.

Also, analysing the 'state-of-the-art' in atlas production and describing the procedures adopted to develop the *Electronic Atlas of Canada*, Siekierska and Taylor (1991) implied that one could discern basically between two types of digital atlases:

- View-only atlases, which could be considered an extension of paper maps in digital form because this kind of atlas allows the user to access stored static maps; and
- Atlases with dynamic interaction and analysis, which would provide more flexible functions such as selection of particular features, addition of data, changes of scale and customisable interface, for instance.

Moreover, Kraak and Ormeling (1996) stated that one could discern the following types of digital atlases:

- View-only atlases, in the same way as Siekierska and Taylor (1991), the authors considered this kind of atlas to be an extension of paper atlases, however they noted three advantages as being: the reduced cost of reproducing digital atlases, the possibility of consulting more than one map at the same time (by dividing the screen) and random access to the maps (instead of the linear structure of paper atlases). It is believed, however, that bound publications can be considered hypertextual, because the reader can read them at random and the author can use footnotes.
- Interactive atlases that provide the user the opportunity of interacting with the data set, either by changing colour schemas, classification methods or the number of classes displayed.
- Analytical atlases, where queries to the databank can be made directly from the map, as the map is just a graphical interface to the data. In addition, data sets could be combined in order to create new data sets. Therefore the user is not restricted to the data provided within the atlas itself. Although this kind of atlas incorporates most GIS functions the authors re-state that the emphasis is on analysing the data and visualizing the results.

Which classification to choose? When considering the various classifications outlined previously, the work of Kraak and Ormeling (1996) is considered to be the most appropriate, because it not only considers and expands previous upon discussions and classifications; but it also takes into account the structure of the atlas, the level of interactivity and the technology employed. The inclusion of all of these elements provides what is considered to be the best classification.

6.7 Digital Atlas: Merely a Variant of a GIS?

It seems that some categories of atlases are somewhat similar to GIS, however the authors argued that the main difference is that atlases "give major importance to the presentation and display of spatial information while in most GISs is on information retrieval and analysis of data" (Siekierska and Taylor 1991, p. 14).

It is undeniable that there are similarities between Geographic Information Systems and some kinds of digital atlases. Conversely, as stated by Elzakker (1993) probably the main difference is that digital atlases emphasise the use of cartographic methods to improve the analysis and presentation of the maps and on the narrative, which intends to improve the user's comprehension of the atlas information.

However, the question remains. If today's atlases are not necessarily comprised of maps, but also by data sets and software that allow the creation of maps on-demand, just like a GIS, what is the real difference between GIS and digital atlases?

Kraak and Ormeling (1996, pp. 183-184) addressed this question as follows:

> "The maps in electronic atlas function as an interface with the atlas database. This combination of database and graphical user interface (GUI) and other software functions developed to access the information is different from a GIS: special care is taken to relate all data sets to each other, to allow them to be experienced as related, to let them tell, in conjunction, a specific story or narrative. There will usually be a central theme"

Bär and Sieber (1999) proposed three approaches for using GIS when developing digital atlases. The first they called Multimedia in GIS, which can be understood as the integration of multimedia functionalities within GIS systems in order to create cartographic products such as atlases and decision-support systems. The advantage of this approach, according to the authors, is that as multimedia is introduced into the GIS, all spatial and geometrical functions are predefined within the system and, hence, there is no need to develop them. The authors claimed, though, that this approach is "the fastest way of bringing full GIS functionality to multimedia" (Ibid., p. 236). However, the main drawbacks outlined were that: GIS systems provide limited multimedia functionalities, the design of the application can not be made independently and the user interface is not friendly, as GIS users are not necessarily multimedia developers.

The second approach is called GIS in Multimedia. According to Schneider (2001) this approach corresponds to the integration of GIS functionalities into multimedia authoring systems. On one hand, the advantage of this approach is that the user interface can be created with more flexibility, as the developer is not restricted to a GIS environment. Another advantage is that cartographic functions can be customised in order to meet a particular user's needs. Conversely, this approach had some disadvantages, among them the fact that this approach is hard to develop, because "even low level analytical functionality, data structures and GIS techniques have to be explicitly defined and implemented by the authors" (Bär and Sieber 1999, p. 236). Also this approach does not provide the same cartographic quality as well as analytical tools as the previous approach.

This approach was adopted by Ramos (2001) for developing a prototype digital atlas of agriculture in Sao Paulo state, Brazil. In her research the author summarised the disadvantages of the GIS in Multimedia approach, which were the painstaking work of development could dissuade people of working in this field, particularly professionals with little programming expertise, and the cost of the developer license of GIS packages is extremely high compared to the standard license. This second point is particularly important in developing countries such as Brazil, where research funds, most of time, are not sufficient to cover research expenses.

With the aim of overcoming the drawbacks associated with these first two methods, Bär and Sieber (1999) proposed a third approach, GIS Analysis for Multimedia Atlases. The authors claimed that none of the previous approaches were meant to specifically respect cartographic characteristics, thus, the proposed approach was intended not only to overcome the known limitations of GIS-based approaches, but to preserve its analytical potential as well. This approach is based on a multimedia atlas development environment, which is comprised of a GIS, the authoring system and a multimedia map extension that transforms GIS objects into cartographic objects. In this way GIS features would be preserved.

Schneider (1999) stated that the focus of digital atlases is on information presentation and summarised the main differences between GIS and digital atlases, or multimedia atlas information systems. These are shown in Table 6.5.

The differences between GIS and digital atlases shown in Table 6.3 are, sometimes, tricky or even subtle. For instance, the use of interfaces in digital atlases should be easy, but with these applications the interface is customised, but this does not necessarily result in an easy to use product. In addition, the control of digital atlas is said to be done by the author, however, as can be noticed by the many different kinds of atlas discussed, in

Table 6.5. Main differences between GIS and Multimedia Atlas Information Systems, after Schneider (1999).

	GIS	**Multimedia AIS**
Use of interface	Complex	Easy
Users	Experts	Non-experts
Computing time	Long	Short
Control by	Users	Authors
Main focus	Handling of data	Visualization of topics
Data	Unprepared	Edited
Output medium	Paper	Screen

some situations the author has full control over the atlas and in other situations they have little control.

The main focus of GIS, as claimed by Schneider (1999), is on data handling. However, it cannot be overlooked, though, that without proper visualization of the results of analysis, the analytical work in GIS environment would be compromised. It is believed that, considering this focus issue, the difference between GIS and digital atlases is that the first comprises the four classical functions: data capture, manipulation, analysis and presentation; and the second can comprise manipulation, analysis and presentation. These functions in digital atlases can be implemented at different levels of complexity, depending on the purpose of the atlas.

Additionally, the output medium for both, GIS and digital atlases, could be either paper or screen, therefore it is considered that this particular feature should not be used to distinguish between them.

Considering the comments above, it is believed that another view of the differences between GIS and digital atlases is necessary. The differences are shown in Table 6.6.

Finally, it is important to note that the subject of the similarities between atlases and GIS is not a new one. This question was raised by Bakker *et al* (1987); at that time, comparing national atlases on paper and GIS. The authors argued that:

> "The traditional national atlas can therefore be conceived as a non-characterised geographic information system in its own right: it was the first medium which enabled users to compare, overlay or otherwise combine data, for the first time presented at similar scales at a similar degree of generalization." (Bakker *et al.* 1987, p. 83)

Table 6.6. Differences between GIS and Digital Atlases, after Schneider (1999).

	GIS	Digital Atlases
Interface	Complex and complete	Customised according to the purpose and the target audience of the atlas
Users	Experts	Non-experts
Computing time	Depends on the project	Short
Controlled by	The user	The author, who allows different levels of control to the user.
Focus	Data capture, manipulation, analysis and presentation	Data manipulation, analysis and presentation. Not all these function must be provided, depending on the purpose of the atlas.
Data	Unprepared	Edited
Purpose	There is no purpose in GIS, the application is open for any kind of data input and analysis	The digital atlas has a purpose, and it was prepared to deliver the conveyed information.

Considering their point of view, as the first national atlas, the atlas of Finland, was published in 1899, one could assume that the question should be: is GIS a kind of atlas? The answer is, undoubtedly, not; but the converging point between atlases (in any medium) and GIS is that both are tools for spatial analysis and cognition.

6.8 The Evolving Medium: the Transition from Discrete Atlases to Internet Atlases

The study of contemporary atlases involves two distinctive transitions: from paper to digital and from discrete to networked atlases. Cartwright (1999) provided an extensive analysis on the evolving technology of optical storage media, from Videodiscs, in the early seventies to the modern DVDs. However, the rise of the Internet as a medium for cartography in the mid-nineties has changed the research focus of digital atlases developers: from discrete storage to the World Wide Web.

Peterson (2003) indicates the advantages of the Internet as a medium to cartography, for the author although the screen is still a major drawback for using computer as a medium for cartography " (...) the Internet makes it possible to distribute the map to many people. Therefore the sum total of map use/communication across all individuals is greater with the Internet" (Peterson 2003, p. 443). In this way, recent figures of Internet audience provide extra support for using Internet as a medium to digital atlases. According to the Computer Industry Almanac (http://www.clickz.com/stats/sectors/geographics/article.php/151151) the global online population corresponded to 934 million people in 2004, however the predicted figures for the following years show a steady increase of about 130-140 million people a year (1.07 billion in 2005, 1.21 billion in 2006 and 1.35 billion in 2007).

The expressive growth of the Internet audience makes it a very effective way of distributing cartography, and particularly atlases. There are many ways of publishing maps on the Internet. Kraak (2001a) provided an analysis on different Internet maps, identifying the following map types:

- *Static Maps* – Static maps are those maps that offer no more than a basic level of interactivity. In a way they are similar to paper maps and they are a very common way of publishing Web maps. The author identifies two kinds of static maps:
  - View only – Static view only maps are those maps where there is no interaction and/or dynamics whatsoever. Still raster maps are considered to fit in this category; and
  - Interactive interface and/or contents – This kind of Web map comprises clickable maps, maps where the user will get responses depending on the map element clicked. Maps where the user can switch layers on and off and use zoom and pan are examples in this category.
- *Dynamic Maps* – Dynamic maps are maps that show dynamic processes and/or contents. The author subdivides dynamic maps into two groups:
  - *View only* – Simple cartographic animations are examples of dynamic view only Web maps. Some examples are developed using animated *GIFs* or vector animations, the last can be developed, for example, using Flash (*Macromedia*) or SVG (Scalable Vector Graphics). These maps show the spatial dynamics of a feature usually over time. In this kind of Web map the user is generally provided with controls to pause, play, go forward or backwards, however, in some kinds of formats (such as animated *GIFs*) this kind of control is not viable and the animation is executed continuously; and
  - *Interactive interface and/or contents* – In this kind of Web map, the user can interact with the map in a more immersive way. Three-dimensional cartographic models are considered to fit in this category

as the user can literally freely navigate throughout the map. Other kind of interactive can be implemented as objects within the model can be linked to other Web pages.

However, the definition and classification of different Web maps is not a simple task because technology is constantly evolving and providing new capabilities and perspectives to Internet Cartography. For instance, it is fairly straightforward to implement Web links to any map object in vector maps, raster maps can also be linked to other pages, the whole map can be a hyperlink or the image can be subdivided into several different links. Therefore this kind of feature cannot be used to distinguish different kinds of Web maps.

Internet maps can be classified according to several aspects. Considering technology, they can be divided into maps based on proprietary technology or open standards. Considering architecture, they can be divided into client-side maps and server-side maps. Considering contents they could be divided into stand-alone maps and data-driven maps, and so on.

The GIS industry have early noticed the potential of the Internet as a medium for cartography and have developed a number of tools for Internet map publishing, therefore, Internet applications based on GIS tools are very common. A recent trend in Internet cartography is the use of open standards for publishing on the Web.

Recent developments in cartography for the Internet have been done towards using open standard technologies for Web publishing. *Open standards* can be understood as standards established by a public body (comprised by members of industry and/or public sectors) with the purpose of providing guidelines for their field of activity. By definition open standards are not proprietary, in other words, they do not belong to any individual or company. The World Wide Web Consortium (W3C) has focused on developing open standards for publishing on the Internet. The W3C developed the eXtensible Markup Language (XML), which is a standard for exchanging data via the Internet. Other standards were developed based on XML, amongst them the Scalable Vector Graphics (SVG) that corresponds to a standard for publishing vector graphics on the Internet.

SVG maps are basically text files that are rendered on the Web browser through a SVG viewer. It is expected, however, that as the format becomes popular further versions of Web browsers will be able to interpret SVG files with no need of a plug-in. Several examples of Internet cartography, and atlases, based on SVG can be found. Also, Newmann and Winter (2003) provided an extensive overview on the use of SVG and open standards for Internet cartography.

It is believed using SVG will allow publishing maps on the Internet in a more extensive and flexible way; however some programming knowledge is required in order to develop SVG maps. To minimize this drawback it is believed that further research is necessary to develop self-explanatory SVG map templates. Such templates would foster the use of SVG as a tool for Internet map publishing and provide means for developing interactive Internet maps at low cost.

6.9 Summary and Conclusions

Computers have become not only a tool for producing cartography, but a medium for it. It is difficult to find statistics about the number of atlases produced annually; as a result one could argue that any analysis regarding the importance of computers as a medium for cartography, and particularly for atlases, is nothing but conjecture. It is believed, though, that despite the lack of official figures, computers have emerged as a major medium for cartography, and the increasing number of digital atlases presented in the last international cartographic conferences corroborates it.

This chapter focused on understanding digital atlases by assuming that, as they are meant to be used in a different medium, they need a particular definition. However the search for a definition of digital atlases raises other questions for reflexion. The main point is the blurred distinction between digital atlases and GIS. As the technology for production of digital atlases evolves, it is harder to distinct between them. However, it is important to remember that digital atlases not necessarily include GIS.

The chapter also focused the Internet as a medium for digital atlases. Internet maps can be classified through different points of view and some definitions were discussed. Recently open standards technologies have emerged as an significant research trend in the field, it is believed that such technologies offer not only the opportunity of delivering interactive Internet atlases at low cost, but they can be used to involve more atlas developers in Internet publishing as well.

References

Aghion I, Barbillon C, Lissarrague F (1996) Gods and heroes of classical antiquity. Flammarion, Paris

Aguiar VTBd (1996) Atlas geográfico escolar. PhD. Thesis, Instituto de Geociências e Ciências Exatas, UNESP

Alonso PG (1968) The first atlases. The Canadian Cartographer, vol 5(2), pp 108-121

Aranaz F, Iguácel C, Romera C (2003) The national atlas of Spain: current state and new projects. In: Proceedings of the 21st International Cartographic Conference, Durban, ICA, pp 1351-1357

Bakker NJ, Elzakker CPJMv, Ormeling, F (1987) National atlases and development. ITC Journal, vol 1987 (1), pp 83-92

Bär HR, Sieber R (1999) Towards high standard interactive atlases: the GIS and multimedia cartography approach. In: Proceedings of the 19th International Cartographic Conference, Ottawa, ICA, pp 235-241

Biermann S, Smit L (2003) IDEA 2002: national housing spatial investment potential atlas. In: Proceedings of the 21st International Cartographic Conference, Durban, ICA, pp 1366-1374

Borchert A (1999) Multimedia atlas concepts. In: Multimedia Cartography, Springer-Verlag, Berlin, pp 75-86

Brewer CA, Suchan TA, Tait A (2003) Mapping census 2000 in the census atlas of the United States. In: Proceedings of the 21st International Cartographic Conference, Durban, ICA, pp 1375-1382

Buckley A, Meacham J, Steiner E (2003) Creating a state atlas as an integrated set of resources: book, CD-ROM and website. In: Proceedings of the 21st International Cartographic Conference, Durban, ICA, pp 1383-1397

Cartwright W (1999) Development of multimedia. In: Multimedia Cartography. Springer-Verlag, Berlin, pp 13-30

ClickZ Network (2004) Population explosion! http://www.clickz.com/stats/sectors/geographics/article.php/151151, Accessed: 15/11/2004.

Elzakker Cv (1993) The use of electronic atlases. In: Klinghammer I, Zentai L, Ormeling F (eds) Seminar on electronic atlases. Cartographic Institute of Eötvös Loránd University, Visegrád, Hungary, pp 145-155

Huber S, Schmid C (2003) 2nd atlas of Switzerland: interactive concepts, functionality and techniques. In: Proceedings of the 21st International Cartographic Conference, Durban, ICA, pp 1398-1405

Keates JS (1989) Cartographic design and production. 2nd edition, Longman Scientific & Technical, Essex

Koop O (1993) Tools for the electronic production of atlases. In: Klinghammer I, Zentai L, Ormeling F (eds) Seminar on electronic atlases. Cartographic Institute of Eötvös Loránd University, Visegrád, Hungary, pp 129-137.

Kraak M-J (2001a) Settings and needs for web cartography. In: Web cartography: developments and prospects. Taylor and Francis, London, pp 1-7

Kraak M-J (2001b) Web maps and atlases. In: Web cartography: developments and prospects. Taylor and Francis, London, pp 135-140

Kraak M-J, Ormeling F (1996) Cartography: visualization of spatial data. Longman, Essex

Kraak M-J, Ormeling F (2003) Cartography: visualization of geospatial data, 2nd edition. Pearson Education, Harlow

Newmann A, Winter AM (2003) Webmapping with scalable vector graphics (SVG): delivering the promise of high quality and interactive web maps In: Maps and the internet. Elsevier, Oxford, pp 197-220

Ormeling F (1995) Atlas information systems. In: Proceedings of the 17th International Cartographic Conference, Barcelona, ICA, pp 2127-2133

Ormeling F (1996) Functionality of electronic school atlases. In: Köbben B, Ormeling F, Trainor T (eds.) Seminar on electronic atlases II, Prague, ICA Commission on National and Regional Atlases, pp 33-39

Peterson MP (1995) Interactive and animated cartography. Prentice Hall, Englewood Cliffs

Peterson MP (1998) That interactive thing you do. Cartographic Perspectives, vol 29, pp 3-4

Peterson MP (1999) Elements of multimedia cartography. In: Multimedia cartography. Springer-Verlag, Berlin, pp. 31-40

Peterson MP (2003) Foundations of research in Internet cartography. In: Maps and the Internet. Pergamon Press, Oxford, p. 437-445

Pucher A, Kriz K, Hurni L, Tsoulos L, Hanewinkel C, Petrakos M (2003) STATLAS: statistical atlas of the European Union. In: Proceedings of the 21st International Cartographic Conference, Durban, ICA, pp 1411-1418

Ramos CdS (2001) Visualização cartográfica: possibilidades de desenvolvimento em meio digital. MSc Thesis, Instituto de Geociências e Ciências Exatas, UNESP

Richard D (1999) Web atlases - Internet atlas of Switzerland. In: Multimedia cartography. Springer-Verlag, Berlin, pp 111-118

Robinson AH, Morrison JL, Muehrcke PC, Kimerling AJ, Guptill SC (1995) Elements of Cartography. 6th edition, John Wiley & Sons, New York

Rystedt B (1996) Electronic atlases - a new way of presenting geographic information: the state of the art. In: Köbben B, Ormeling F, Trainor T (eds.) Seminar on electronic atlases II, Prague, ICA Commission on National and Regional Atlases, pp 1-2

Schneider B (1999) Integration of analytical GIS-functions in multimedia atlas information systems. In: Proceedings of the 19th International Cartographic Conference, Ottawa, ICA, pp. 243-250

Schneider B (2001) GIS functionality in multimedia atlases: spatial analysis for everyone. In: Proceedings of the 20th International Cartographic Conference, Beijing, ICA, pp. 829-840

Schneider B (2002) GIS-funktionen in atlas-informationssystemen. PhD Thesis, Institut für Kartographie, ETH

Siekierska E (1984) Towards an electronic atlas. Cartographica, vol 21(2,3), pp 110-120

Siekierska E, Williams D (1996) National atlas of Canada on the Internet and schoolnet. In: In: Köbben B, Ormeling F, Trainor T (eds.) Seminar on electronic atlases II, Prague, ICA Commission on National and Regional Atlases, pp. 19-23

Siekierska EM (1991) Personal communication at a meeting of the ICA Commission on national and regional atlases. ICA Commission on National and Regional Atlases, Bournemouth

Siekierska EM, Taylor DRF (1991) Electronic mapping and electronic atlases: new cartographic products for the information era - the electronic atlas of Canada. CISM Journal, ACSGC, vol 45(1), pp 11-21

Thrower NJW (1972) Maps & man: an examination of cartography in relation to culture and civilization. Prentice-Hall, Englewood Cliffs

Zhukovsky VE, Sveshnikov VV (2003) National atlas of Russia: progress in two recent years. In: Proceedings of the 21st International Cartographic Conference, Durban, ICA, pp 1425-1429

7 Hypermedia Maps and the Internet

Michael P. Peterson

Abstract. The highly-interactive, map-based multimedia presentation known as the hypermedia map has emerged along with the development of the interactive personal computer since the 1980s. Although combining maps with other forms of representation such as pictures is not new, this particular interactive type of map required an effective electronic form of distribution. The CD-ROM served that capacity briefly but was soon replaced by the Internet which emerged in the mid-1990s as the major form of information delivery. More maps are now distributed through the Internet than through any other medium. But, most of the maps that make their way to the Internet are simply static maps, often scanned from paper, not the highly-interactive hypermedia maps that were expected with this new medium. There are very few examples of hypermedia maps currently available through the Internet, and those that do exist are very difficult to find. The continued development of hypermedia is based on both the accessibility of hypermedia maps through the Internet and a system of remuneration so that the authors of these time-consuming products can be compensated.

7.1 Introduction

The Internet has changed the process of mapping and map use. The new medium has drastically increased the availability of maps and led to more interactive forms of mapping. The distribution of maps through the Internet is still relatively new and much work lies ahead in order to make the Internet an effective means of transmitting spatial information, particularly developing forms of mapping that use the potential of this new medium.

Based on the concept of hypermedia (Nielsen 1990), the hypermedia map is a map-based interactive multimedia presentation that combines some mix of text, pictures, video, graphic images, sound and other forms of media. For example, a hypermedia map presentation on Africa might include links to regions, countries, music, climate, or population—all linked together with maps that emphasize both location and thematic information.

The hypermedia map has evolved since the mid-1980s as methods of map delivery have evolved from the CD-ROM to the Internet. With the growing interest in interactive forms of communication, the integration of media with maps has become a major area of research and development. The goal is to transcend the static and sequential nature of information presentation and ultimately create a greater and broader understanding of the world in which we live (Peterson 1995, p. 127-128). The overall objective is nothing less than a revolution in how spatial information is communicated.

While we can envision and even implement highly interactive hypermedia maps that help guide users to a greater and broader understanding of the world, bringing such maps to a large audience through the Internet is an entirely different matter. If we are to be successful, the challenge is to not only conceive of useful hypermedia maps but also finding ways of marketing these maps, funding their creation, and making them available to people at a reasonable cost. This chapter examines the current status of Internet-delivered hypermedia maps, and ways of making these highly interactive maps more available. We first examine the Internet as a means of dissemination.

7.2 Maps and the Internet

7.2.1 Current Status

Concomitant with the increased use of other forms of media with maps has been the rise of the Internet as a major form of information delivery. According to the Computer Industry Almanac in 2005, there were 935 million Internet users or nearly 16% of the world's population (see Table 7.1). This is up from 533 million Internet users worldwide at year-end 2001 which at that time represented only 8.7% of the world's population. There were only 200 million Internet users at year-end 1998 and 61 million in 1996. It is expected that the number of Internet users will reach 1.46 billion by 2007.

With the growth of the Internet comes the expansion of map use through this medium. Internet map use was tracked at four major sites between 1997 and 2001. The growth in Internet map use was compared to the growth in the use of the Internet itself. It was found that while both growth rates are strongly exponential, the growth in the use of maps through the Internet exceeded the growth rate for the Internet itself (Peterson 2003). It can now be said that more maps are now distributed through Internet than through any other medium. Estimates put the number at over 200 million a

Table 7.1. Number of Internet Users

Year	Internet Users in Billions
2007	1.46
2005	1.00
2004	0.935
2001	0.533
2000	0.327
1996	0.061

Source: Computer Industry Almanac (2001, 2005)

day. This change in map distribution happened over only a decade. Never in the history of cartography has there been such a dramatic shift in how maps are delivered to the map user.

7.2.2 Experiments in Finding Online Maps

A variety of experiments have been conducted beginning in 2001 to examine the state of maps and the Internet, particularly in reference to their availability through search engines. Search engines are viewed here as a window to the Internet, a way of understanding what types of maps are easily available to Internet users. In order to improve the Internet as a medium for cartography, we need to understand the current state of map availability.

In an experiment conducted in 2001, a series of high school and college freshman were asked to find map resources through the Internet (Peterson 2001). Over 100 students were tested in five groups in a computer user room with 25 Pentium III 500Mhz computers. Every student had their own computer with Microsoft Internet Explorer and a relatively fast Internet connection. They were instructed to find three maps: 1) a map of Africa by country; 2) a map of Africa by country in PDF format; and 3) a map that places a star on a street map showing where they live.

Students could find these maps in remarkable speed. For example, it took only 16 seconds for some students to find a map of Africa. Almost all of the students found such a map within 45 seconds but some students required slightly more than a minute. The search strategies differed slightly between the high school and college students. The college students had learned to simply use the search field in the toolbar area of Microsoft Explorer that linked to Microsoft's MSN search engine. The high school students first went to Google to start the search and this slowed their search times.

It is interesting to note which map of Africa was most commonly found by students. It was not an interactive map. Rather, it was a scanned paper map in JPEG format (see Fig. 7.1). The map exhibits the typical JPEG compression artefacts that make this file format less than ideal for use with

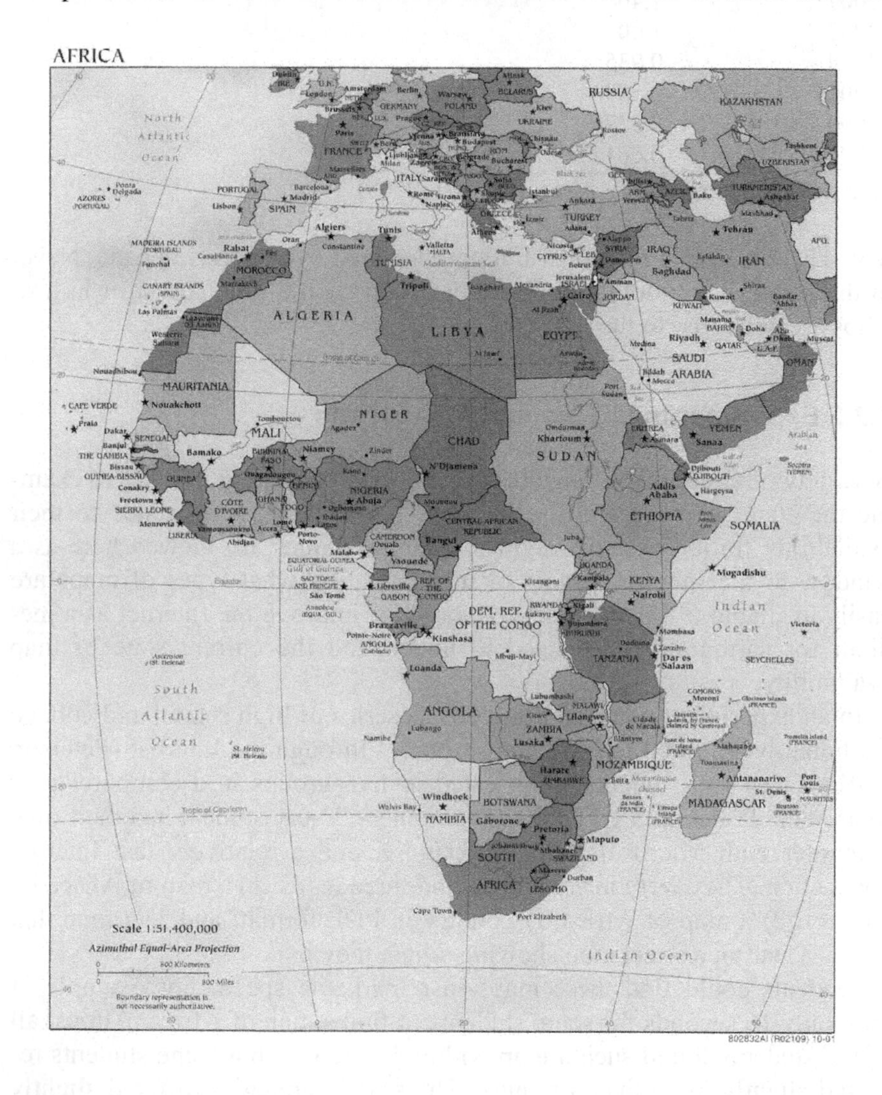

Fig. 7.1. High school and college students were asked to find a map of Africa. Most found this scanned JPEG map of Africa in under 30 seconds. The fastest search time was 16 seconds. Only a few had not found the map in 45 seconds. The map cannot be zoomed and most of the text is illegible because of the small size or JPEG compression artefacts.

maps. But, that didn't seem to matter to the students. It was the map that students were expecting to find of Africa and, although the map is barely legible and could not be zoomed or queried, they were quite happy with what they had found.

To find the PDF map of Africa, the Google search engine helped the high school students find this type of map faster than their college counterparts. Their fastest response time was 28 seconds while the fastest college student required 40 seconds. Loading the Acrobat plug-in took a good portion of this time. Some students required over a minute and were then provided help. Few could explain the advantage of the PDF file type and only a couple students knew that zooming was possible. They did know that PDF files took longer to load because the plug-in needed to start first, and they found this to be annoying and mentioned that they would often cancel the process when this happened with other documents. While they seemed impressed that one could zoom into the PDF map, in the end, many liked the JPEG map better because it loaded faster, although it was essentially illegible.

Making a map of home proved more difficult because the search engine was less useful for this task. Students tried to search for "map of home" but this did not yield links to any of the interactive street mapping programs. Some remembered MapQuest.com from a previous visit and went directly to that site. For the few students using the Yahoo search engine, they were provided a link directly to MapQuest.com. The main Yahoo page has a link to Yahoo's interactive mapping site but students did not look for this indicating that these students preferred using a search engine to a directory list. At that time, about 1/3 of the students had never used an online mapping site to find a particular location. In the end, the fastest map of home was produced in 30 seconds but most students could not complete the task in one minute and some had to ask for help. Once the map was made, students complained that the star indicating their home was not in the right location. This is a common complaint and results from the address matching process in which locations of addresses are estimated along street segments. This process often produces an incorrect location.

The most interesting observation from these experiments was that static maps are much easier to find than interactive maps. Search engines are oriented toward static pages because these are indexed more easily, and these pages usually have links to static files. Another observation was that students didn't understand how the interactive map was made. One student said it was "magic." Others said that the interactive street map was made through satellite imagery or used GPS. Although these may be involved, the process of bringing such a map to map user is much more complicated.

7.2.3 Finding Maps – A 2005 Update

The same experiment was conducted in early 2005, nearly four years after the first study in 2001. This time students in a third year college course were asked to find the same three maps but with faster computers (Pentium 4, 2.4GHz with 512 MB of memory). The results remained remarkably similar. Almost everyone found a similar map of Africa in less than 30 seconds. Some required less than 10 seconds. Finding the PDF map took slightly longer but most students were still able to locate the map in under 40 seconds.

The most interesting finding was that all of these students had been previously exposed to interactive street mapping sites and everyone completed this map in less than 45 seconds. About half completed this task in less than 30 seconds.

7.2.4 Finding Hypermedia Maps

To determine the level of difficulty in finding online hypermedia maps, an experiment was first conducted by the author with the Google search engine. The purpose of the experiment was to determine what type of hypermedia sites could be found within the top 20 links provided by Google's first two pages. Six different text strings, such as "Maps Hypermedia Africa," were entered into the search engine. Each of the top 20 links were examined and potential links were followed through two additional links to see if a desired page could be found. Further links may have been followed if a promising lead was found in the first two links.

The results of the search are presented in Table 7.2. As can be seen, very few hypermedia or multimedia maps could be found. No true hypermedia maps were found using a typical search string that most people would use. Many links were to commercial sites (.com), particularly to books on hypermedia. Other links were to libraries that would list library audio-visual collections but there were no links to hypermedia materials. The only hypermedia maps that were found during the entire search resulted from a search on "Africa Flash." In this case, Flash is referring to an online multimedia authoring program. Most people would not know to perform a search using the name of the proprietary file format.

A similar experiment was conducted with the same upper-division college students. In this experiment, the students were instructed to time themselves as they attempted to find a variety of maps of Africa that included pictures, sound, or video. An electronic stop watch implemented

Table 7.2. Summary of Google search for hypermedia maps of Africa. Although many maps and photographs of Africa were found, few sites integrated maps with other media. Those that did could only be found using the "Flash" file type keyword.

Google Search Text*	Sites with Maps in top 20	Sites with Pictures of Africa in top 20	Sites other Media of Africa in Top 20	Hypermedia Maps
Maps Hypermedia Africa	I	I		
Maps Multimedia Africa	~~IIII~~ IIII	III	II	
Maps Pictures Africa	II	II	I	
Maps Sound Africa	II	IIII	I**	
Maps Movies Africa	~~IIII~~		II	
Flash Africa	~~IIII~~	III	III	II

*Resource found within 2 clicks. Additional clicks may have been needed to activate resource.
**Only site with sound presented bird songs.

through a web page was used as the timer. Students were given the instructions as shown in Fig. 7.2.

The results largely paralleled what was found by the author. In many cases, students could not find any maps of Africa that included the desired media. Many students spent over five minutes trying to find a suitable map before giving up. This is far longer than most people would normally devote to such a task unless they were highly motivated.

Pictures were the most common media element found with maps. Maps with pictures were found with many of the search strings that were used. Even searching for "Maps Sound Africa" produced hits to maps of Africa that included pictures. "Flash Africa" led to the most sites with hypermedia maps. Searching for hypermedia content using the Flash keyword is problematic because it requires that one be aware of a program that is used to create hypermedia content.

A major advantage of the Internet is the ability to include pictures, sound and video with maps. Try finding maps that include one of these elements using the following text in the search engine. If you can't find a map with pictures, sound or video in the top 20 links provided by the search engine, enter "Not found." Do not proceed down more than 2 links in any page provided by the search engine.

Search Engine Text

1. Maps Hypermedia Africa
2. Maps Multimedia Africa
3. Maps Pictures Africa
4. Maps Sound Africa
5. Maps Movies Africa
6. Maps Flash Africa

Fig. 7.2. A portion of a questionnaire given to students in an upper-division geography course. Many students could not find examples of the desired maps on the Internet. Those that could find examples sometimes required over 8 minutes of search time. The search for Flash files produced the most results but required the knowledge that hypermedia maps would be presented using the Flash file format.

7.3 The Information Landscape

Landscapes come in many shapes and sizes. There are tropical valleys and windswept desert plains. There are landscapes formed by methods of farming and urban landscapes develop as a result of human patterns of habitation. O'Day and Jeffries (1993) use the term "information landscape" to refer to the landscape of information that we traverse when we venture into a hypermedia environment. They study the specific process used by individuals in finding information through hypertext. Researchers also examine what patterns of behaviours are exhibited when users browse information landscapes, how performance varies across the two distinctive tasks and how different navigational aids influence patterns of use Toms (1996). It has even been suggested that maps can serve as a guide to the information landscape (Block 2002).

The information landscape that most people experience is the one presented by search engines. It is the only effective tool at present to traverse the information landscape. These automated systems explore this ever growing landscape and collect keywords that are entered into a database. Keyword searches are subsequently made by the search engine user. Of course, there is more to this landscape than can be found by the search engine. Some estimate that search engines have only indexed a third of all web pages. Before we proceed, it is useful to examine how these systems work.

7.3.1 Search Engines

Search engines are based on a program called a web robot or a spider that traverses web pages in an automated fashion. There are two steps in the traversing process: 1) words are entered into a special index; and 2) other links are found for later traversing. Web robots only need to be told to search the main page and they find all linked pages automatically. If a page has links to 100 pages, and each of these has links to another 100 pages, a web robot, in theory, will create an index for all of these of pages – and this can soon number into the millions. The index created by the web robot cannot be too large, no more than about 2 KB. If the index were larger, it could essentially be making a copy of the entire page.

After making a collection of these small index files of billions of pages, the web robot begins to prioritize the links by keyword. This is where problems start because the process is inevitably arbitrary and based on inadequate information. Each search engine also prioritizes in a different manner as well. Companies do not provide the exact algorithms because this would lead to abuse by those people who trying to get their web pages to show up near the top. There is somewhat of a Catch-22 situation in which we need to know how the web robots work so that they provide better matches, but as soon as we know how the robots work, they get abused by those that want to manipulate them and they no longer provide good matches.

The ranking system used by Google, called PageRank, relies on a system that counts the number of times pages are linked from each other. It also incorporates a weighting factor that analyzes the "importance" of the page that is making the link. Important, high-quality sites receive a higher ranking, which Google remembers each time it conducts a search. Google states that it goes beyond the number of times a term appears on a page and examines all aspects of the page's content (and the content of the pages linking to it) to determine if it's a good match for a query.

Some basic search engine rules seem to be shared by all search engines. Smaller pages are given preference over longer pages with a similar set of keywords. The text in the title of the page is given a higher priority than text in the page. The META tag, listed near the top of the page, is another way of making the web robot assign a high priority. Before the META tag was available, a high rank for a page was only possible if a phrase was used repeatedly throughout the text. Now, the phrase just has to appear in the META tag. Popular pages that have more links directed to them are listed higher. This tends to cause new pages to fall into an "anonymity trap" since they are not linked from other pages, and they can't be linked because they can't be found.

Search engine rankings are extremely competitive and the operators of websites are under pressure to make sure that their sites are listed near the top. Webmasters will spend a considerable amount of time making sure that their sites are highly placed on the major search engines. Most search engine companies now accept cash in return for a high placement. Search engines have become a necessary if frustrating part of people's lives.

Sherman and Price (2001) identify five major problems with search engines:

1. Cost of crawling: Crawling the web is very expensive and time-consuming, requiring a major investment in computer and human capital.
2. Crawlers are dumb: Crawlers are simple programs that have little ability to determine the quality or appropriateness of a web page, or whether it is a page that changes frequently and should be re-crawled on a timely basis.
3. Poor user skills: Most searchers rarely take advantage of the advanced limiting and control functions that all search engines offer.
4. Quick and dirty results: "Internet Time" requires a fast if not always thorough, response. A slower, more deliberate, search engine would not gain user acceptance although it might lead to better results.
5. Bias towards text: Search engines are highly optimized to index text. For non-text pages, such as images, audio, or streaming media files, the search engine can do little more than record filename and location details.

7.3.2 Media Search Engines

Media search engines are designed to index images, sound, or video files, sometimes collectively referred to as "multimedia." Image searching is the most common and is offered by major search engines like Google and Yahoo. WAV, MIDI and MPEG sound files can be found through specialized search engines like MusicRobot.com. Music sharing programs like Kazaa and Limewire locate and download music and other types of media through a point-to-point protocol. Yahoo includes a search for video files on its search page. Media search engines represent an alternative to the text-based search engine and represent a more promising technology for finding hypermedia maps.

Image search engines generally index images in the GIF, JPEG, or PNG format and represent a major way that Internet users find maps. Google reports an index to more than 880 million images. Of these, approximately 3,750,000 are images associated with the "map" keyword. These dominate in any type of place name search. For example, an image search for "Africa" results in 12 maps in the top 20 links, and 23 in the top 40 links. Of these 23 maps, 13 are in the JPEG format. The Yahoo image search finds 10 maps in the top 20 images and 14 maps in the top 40 image links. Google finds a total of 145,000 hits for "Africa map" while Yahoo has 30,400 hits. Maps, in the form of images, represent a major component of Internet image content.

The primary problem with image search engines is that the indexes are created from words associated with the images, and not the images themselves. This is because the image file cannot be searched for keywords. Rather, the indexing of the image is done by examining the title or other words associated with the image. It might also take into account any accompanying 'ALT' picture tags coded into the HTML page or look for clues from the image's context – for example, the words or phrases that are close to the image, or the 'META' tags found at the top of the HTML coding. The nature of the Web site and its provider may also be taken into account. The image search indexing process can easily be fooled with a non-descriptive filename or associated text that does not relate to the image. For example, a search engine would likely be fooled by a page on South America with a mislabeled link to a map of Africa.

An alternative to the automated approach are collection-based search engines that index a single or small number of image collections. Commercial stock photo collections, like Corbis, that offer images for sale, will implement their own search engine-like procedure to find images from within their collections. These search engines are augmented with "human indexers" that assign keywords for each picture.

All of these types of image search engines are text-based in that their indexes are created from words associated with the images. There have been attempts to create *content-based search engines* that 'index' visual characteristics of an image, such as its shape and colour. However, these attempts are still largely experimental and are often limited to single, proprietary image collections (TASI 2004).

Yahoo is the only major search engine that provides a video search option. The search engine returns links to movies in five different formats (AVI, MPEG, QuickTime, Windows Media, and RealPlayer) but few movies of these movies incorporate maps. For example, "Africa map" returns only one movie that includes a map in the 96 movies that it returns for this search. The 7,800 video files that are listed for "Africa" contain very few map examples. Even a search for "animated map" that should be represented in a video database leads to only about 100 hits. Clearly, the current video search option is not an effective way to search for hypermedia maps.

7.3.3 Map Search Engine

A search engine that is designed specifically for maps would serve many purposes. Most of the maps that are currently available through the Internet are classified as pictures. As a result, a place name image search results in a confusing mix of maps and pictures. The image search engine would benefit from a clear distinction between pictures and maps, something that could be determined in an automated way by looking at the level of pixel color variability in a file. Presumably, the map would have large areas that have the same color or shading and this would not be true of ordinary pictures that are characterized by gradations of pixel values.

Determining map content in other types of multimedia files would be more difficult. These files have a more complicated structure and performing similar automated inspections of files may not be sufficient to separate map and non-map content. An author could self-define the map content of a hypermedia product and insert this into an associated header or metadata file but enforcing such a self-classification of content would invariably lead to problems. A workable, automated map search would provide a better alternative and could categorize content by both file type and media content.

7.4 Online Commerce

7.4.1 Internet Marketing

The major problem with finding hypermedia maps on the Internet may not be that search engines can't find them. The problem may be that they just don't exist in great numbers. The ones that do exist have been produced by large print organizations like National Geographic that are promoting their publications through their online hypermedia products. The production of hypermedia map products requires a considerable capital in human and computer resources. Without a system of remuneration, these types of maps are simply not profitable to make.

Like the map houses in ancient port cities that served sailors, one could envision hypermedia map houses, where groups of cartographers would craft hypermedia maps for sale. The maps would embody the latest in mapping and multimedia technology and the products would be eagerly anticipated by a devoted group of students and scholars. As an independent map house, the cartographers would not need to orient their maps toward the sale of a physical product like a magazine, CD-ROM or DVD. Instead, the maps could take full advantage of the interactive and multimedia elements embodied in the new medium. No such map houses exist and the reason is not related to a lack of appropriate technology. Rather, it is because the marketing of such maps is problematic.

Marketing has been the bane of the Internet since the appearance of the Web. The lack of a workable model to make money through the Internet was one factor in the downfall of the Internet speculation bubble beginning in 2000. While a great many purchases are now made through the Internet, most of these are for physical goods or travel. The purchase of Internet-delivered content has not fared well. The one exception is the online music industry and this deserves a closer look.

7.4.2 Apple's iTunes

Apple's iTunes store was so successful that in first few weeks online, the company sold over two million songs. This was particularly remarkable because only individuals with Macintosh computers running later versions of their operating system could initially buy songs from their store. Apple has since added iTunes for Windows. The iTunes store is such a success story that it might be worth taking a closer look at what Apple managed to get right and consider if the approach can be applied to market other products.

Prior to iTunes, attempts to sell music through the Internet were based on a subscription service model that used proprietary formats. Copy protection schemes were incorporated into the music files that prevented copying. One service not only made it impossible for consumers to burn CD's of music they downloaded, it also required consumers to pay a monthly fee in order to keep listening to music they had already downloaded. Essentially, consumers were not purchasing music, they were renting it.

Apple was the first business to understand that consumers want music purchased on the Internet to have the same properties as music they buy on a CD. So, Apple allowed consumers to buy individual songs for 99 cents each, or albums for ten dollars. Once downloaded, consumers can easily burn a CD or transfer the music to a portable MP3 player.

Apple's success with iTunes marked a milestone in the history of electronic commerce and sent shockwaves through the Internet marketing business (Morris 2003). It provided a model for marketing something besides a physical product. iTunes was able to sell music in the form of bits and bytes, rather than a physical product like a CD.

7.4.3 The 99¢ Map

For centuries, printing was the only practical method of transferring large amounts of information. Publishers and bookstores sold the artefacts and libraries developed to store them. As Libicki, et.al. (2000, p. 75) point out, words differ from books: "Their pricing, the property rights inherent in their expression, and the challenges of their distribution follow from the physical form they take." But, implicit in the idea of intellectual property rights in information is that authors need to "be compensated with money lest their creative incentive whither" (Libicki, et.al, 2000, p. 76). As they point out, it is very easy to copy information electronically and without a model that makes such copying difficult, there is little reason to "liberate the content of information" from its form.

The model chosen by Apple's iTunes is to make material available at a low enough price so that copying becomes more of a hassle than it is worth. Although one can still obtain music files for free through point-to-point protocols, this illegal method of file exchange has drastically declined. If 99¢ can help stop the illegal copying of music, it may represent the price point of individual maps if marketing can create a demand.

7.5 Summary

The hypermedia map offers the potential of more engaging forms of interactive maps, and the Internet is a functional method of delivery. While examples of hypermedia maps are available through the Internet, they are difficult to find with current search engines – even with the newer media search engines that have recently appeared. Hypermedia maps are not well represented on the Internet. As a result, people don't see the advantages of combining maps with other types of media.

Search engines are based on creating indexes of words. The specific words that are used in labelling material is extremely important in the indexing process. Not only does the word need to be unique so that it is distinctive from other resources but it needs to have a meaning that is generally understood by most people. While hypermedia is a unique concept, the word is not generally understood and most people do not use it in referring to interactive material. Other terms like "interactive multimedia" are recognized by more people, and more people would likely use this in a search. While hypermedia maps would have a broad appeal, at present, there are few online examples, and these can only be found with great difficulty. In order for this type of map to be more mainstream, academic terms like *hypermedia* will either need to become more broadly understood, or new terms will need to be introduced that more effectively describe this type of map to the public.

The real challenge for hypermedia maps will be to create a demand and market that would fairly compensate hypermedia cartographers for their efforts. The 99¢ solution may represent a possible method of marketing. But, a demand needs to be generated for such maps and it seems that most people are perfectly satisfied with the "static" maps that are currently available for free.

References

Block M (2002) Mapping the Information Landscape. Searcher, vol 10(4), http://www.infotoday.com/searcher/apr02/block.htm

Computer Industry Almanac (2001) 15 Leading Countries in Internet Users Per Capita, http://www.c-i-a.com/pr0904.htm

Computer Industry Almanac (2005) Worldwide Internet Users will Top 1 Billion in 2005, http://www.c-i-a.com/200010iuc.htm

Libicki M, Schneider J, Frelinger DR, Slomovic A (2000) Scaffolding the new web: standards and standards policy for the digital economy. Rand Corporation, Santa Monica, CA

Morris GE (2003) The Apple iTunes Store: how Apple got it right. Advertising and Marketing Review. http://www.ad-mkt-review.com/public_html/air/ai200308.html

Nielsen J (1990) Hypertext and Hypermedia. Academic Press, Boston

O'Day VL, Jeffries R (1993) Orienteering in an information landscape: how information seekers get from here to there, In: Proceedings of the SIGCHI conference on human factors in computing systems, Amsterdam, The Netherlands, pp 438-445

Peterson MP (1995) Interactive and animated cartography. Prentice-Hall, Englewood Cliffs, NJ

Peterson MP (2001) Finding maps through the Internet: an investigation of high school and college freshman. Presented at the Annual Meeting of the Association of American Geographers, Los Angeles, CA

Peterson MP (ed) (2003) Maps and the Internet. Elsevier Press, Amsterdam

Sherman C, Price G (2001) The invisible web: uncovering information sources search engines can't see. Cyberage Books, Medford, NJ

TASI - Technical Advisory Service for Images (2004) A review of image search engines. http://www.tasi.ac.uk/resources/searchengines.html

Toms E (1996) Exploring the information landscape. In: Proceedings of the ACM CHI 96 Human Factors in Computing Systems Conference, Vancouver, Canada, pp 63-64

8 In Pursuit of Usefulness: Resetting the Design Focus for Mobile Geographic Hypermedia Systems

Karen Wealands

Abstract. Mobile geographic hypermedia provides a means whereby geospatial data can be delivered to users as 'rich' information via highly portable devices and wireless telecommunication networks. To operate successfully, systems based on mobile geographic hypermedia combine geospatial information perception, knowledge generation and communication. Each of these aspects may be largely ineffective, however, without in-depth consideration of the usefulness (utility + usability) of the representations and the methods of interaction involved. It is argued here that rather than being driven by the underlying technology, the design of mobile geographic hypermedia systems should be approached from a usefulness perspective. Not only will this ensure their use, but ultimately their commercial success.

8.1 Introduction: Mobile Geographic Hypermedia

Geographic hypermedia is a form of cartographic representation based on the combination of interactive computing tools with advanced distribution techniques (e.g. the Internet and the World Wide Web), that is intended to address the growing need for more intuitive geospatial information presentation (Cartwright & Peterson 1999). More specifically, it is the use of multimedia – e.g. maps, text, speech, sounds, images, video, animation, and so on – within computer-based cartographic systems, allowing for double encoding of information, interactivity and the use of complementary information, in order to produce more realistic representations of geographic space, support knowledge construction and ultimately ensure efficient communication of the underlying geospatial data (Buziek 1999).

Whilst the majority of geographic hypermedia systems are designed for and accessed using stationary (e.g. desktop, laptop) computing devices and wired distribution networks, a subset are based around highly mobile devices (e.g. mobile phones, SmartPhones and handheld computers) and wireless telecommunication networks. Such infrastructure provides a greater degree of physical freedom and enables an array of mobile geographic hypermedia systems that are potentially available anywhere, at any time. By way of example, mobile Location-Based Services (mLBS) are a specific class of mobile geographic hypermedia systems, which additionally incorporate device positioning and thus enable applications that can utilise a user's location as a filter for data querying and representation. A useful working definition for mLBS is thus: *wireless services which use the mobile Internet, along with the location of a portable, handheld device, to deliver applications that exploit pertinent geospatial information about a user's surrounding environment, their proximity to other entities in space (eg. people, places), and/or distant entities (eg. future destinations), in real-time* (Urquhart *et al.* 2004).

MLBS incorporate diverse sets of geospatial (and other) data and appeal to a wide variety of users, ranging from everyday consumers to private industry, governments and the military (Niedzwiadek 2002). As a result, numerous applications are possible, with Table 8.1 providing some common examples. In general, mLBS rely on: (i) the accurate determination of a device's location (e.g. via automatic positioning techniques such as mobile cell-based triangulation or GPS); (ii) connectivity to the mobile Internet over a wireless network; (iii) access to geospatial search engines incorporating relevant data; and (iv) useful applications allowing the user to access the geographically-related information they require to achieve their goals. It is in this last point that the role of mobile geographic hypermedia becomes most evident, since it provides the interface to the underlying data.

8.2 Challenges to Mobile Systems Design

There are numerous benefits that can be attributed to mLBS and mobile geographic hypermedia systems in general. Firstly, the high mobility of the devices and networks used to deliver the services makes them equally as convenient to carry and use as paper maps. Secondly, mobile Internet connections ensure that the geospatial and other information being represented are always as current and as accurate as the underlying data. And finally, where real-time positioning of the user is available, context can be incorporated into representations in order to increase their relevance to the user's current situation. Countering these advantages, however, there are new challenges for cartographers brought about by the combination of technologies and dynamic environments characterising mobile geographic hypermedia systems. Indeed, just as the introduction of interactive computing required new developments in cartographic theory and methodology

Table 8.1. Examples of mLBS-driven applications (based on Niedzwiadek 2002).

Types of Location Information	**Applications**		
	Consumer	*Business*	*Government*
Positions	Where am I?	Contact nearest field service personnel	Location-sensitive reporting
Events	Nearest theatre playing the movie I want to see?	Local training announcements	Local public announcements
Distributions	House hunting in low density area	Sales patterns	Per capita open space
Assets	Where is my mobile phone?	Status of utility field devices	Where are the street sweepers?
Service Points	Where are the sales?	Targeted advertising	New zoning
Routes	How do I get there?	Taxi dispatch	Emergency dispatch
Context	Show me the nearest…	Nearby competitors	Local commerce
Directories	Where can I buy …?	Best supplier within 50km	Public services
Transactions	Must purchase in a specific location	Location-sensitive billing	Location-sensitive tolls
Sites	Tourist attractions to visit	Candidate store sites	Environmental monitoring stations

(Dransch 2001), the fields of *Mobile Cartography* (Reichenbacher 2001) and *TeleCartography* (Gartner & Uhlirz 2002) have emerged to respond to the constraints of the mobile medium. Table 8.2 divides the limitations into two major categories – technical infrastructure and context of use – which are discussed in detail below.

8.2.1 Infrastructure Constraints

Despite continuing improvements in the mobile devices and wireless networks used to deliver mobile geographic hypermedia systems (Hassin 2003), they will never truly equal the capabilities offered by desktop computers and wired networks in terms of display, interaction and performance

(Weiss 2002). Perhaps the most apparent limitation in this respect is the restricted screen size and resolution of most mobile devices: for example, the user interface of a typical handheld computer measures 240 × 320 pixels and supports 16-bit colour, which is much lower than that of a desktop computer (commonly larger than 800 × 600 pixels, with 32-bit colour). Faring even worse are mobile phones, with screen resolutions around 100 × 80 pixels and 256 colours (Weiss 2002). These screen real estate limitations pose implications concerning how much information can be visually represented to a mobile user at any one time (which varies according to device type), and the need to ensure that the information displayed is pertinent and useful to the user (Holtzblatt 2005). Related to this are issues concerning wireless networks, which have comparatively slower connections, higher latencies, lower bandwidth and less coverage than wired networks (Hassin 2003), limiting the amount and timeliness of data that can be reasonably delivered to a mobile device. Further compounding the problem is: (a) the limited storage and processing power of many mobile devices – with the data and processing that underlies most systems having to reside on the networks rather than the devices (Weiss 2002; Lee & Benbasat 2003) – and (b) their high rates of power consumption, combined with short battery lives – diminishing device operation and requiring frequent recharging, which is unsuitable for, and difficult during, prolonged mobile use (Hassin 2003).

Positioning accuracy is another concern, with various techniques in existence. These range from network based solutions (with accuracies anywhere from 50m to upwards of 2km) to satellite-based, handset-centric positioning (e.g. GPS – accurate to 10m or better) and hybrid techniques such

Table 8.2. Limitations related to mobile geographic hypermedia systems.

Technical Infrastructure	Context of Use
Display	*Situation*
• Small screen size / limited resolution [D]	• Location
• Low colour range [D]	• Time
Interaction	*Environment / surroundings*
• Limited input opportunities [D]	• Dynamic settings
• Restricted output capabilities [D]	
Performance	*User characteristics*
• Slow connections / high latencies [N]	• Goals
• Limited storage / restricted processing [D,N]	• Tasks
• High power usage / short battery life [D]	• Interests
• Variable positioning accuracy [D, N]	• Abilities

[D] Device related   [N] Network related

as network-assisted GPS (A-GPS), which provides location accuracies up to 1m (Zeimpekis *et al.* 2002; Mountain & Raper 2000). Unfortunately, the inequitable availability of the technologies involved, and their widely varying accuracies, impacts on the representation of the user's location within any given mobile geographic hypermedia system. In terms of user interaction with the service, and ultimately the data, a further issue involves the limited input capabilities inherent with most mobile devices. Keyboards are generally small or absent, often replaced by touch screens, predictive text, handwriting recognition and/or voice input – each of which lacks the speed and accuracy of a desktop computer's input techniques (Weiss 2002; Sandnes 2005). Output capabilities are again restricted (introduced above with respect to screen displays), with early mobile geographic hypermedia consisting of simplistic text and/or low-level graphics (e.g. Fig. 8.1a). Fortunately, today's devices are becoming more sophisticated in terms of providing true multimedia output – i.e. audio (voice, sound), visual (text, graphics, animation) and/or haptic (vibration) techniques, e.g. Fig. 8.1b – however device size and network performance remain limiting factors.

8.2.2 Contextual Constraints

Schmidt (2000) describes context as a combination of location, environment, situation, state, surroundings and task, among other things. For the purposes of this chapter, however, a simplified definition is adopted, focussing on the user aspects of *location*, *time*, *goals*, *tasks* and *environment*. With this in mind, it is generally assumed that different users in different

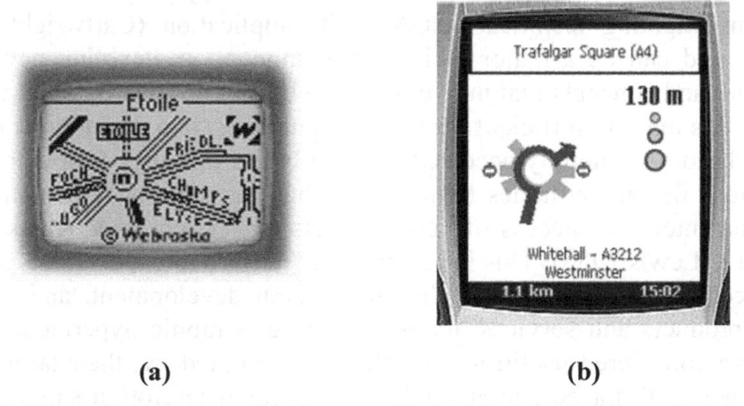

(a) (b)

Fig. 8.1. Examples of geographic hypermedia in (a) early mobile systems and (b) more recent mobile offerings (images reproduced with permission of Webraska Mobile Technologies, SA).

contexts have differing goals, thus they require different sets of geospatial (and other) information, as well as alternative representations of, and methods of interaction with, the data. This makes it both important and difficult for mobile geographic hypermedia systems to present the user with only the information that is genuinely of interest to them in their current context (Coschurba *et al.* 2001). Moreover, the dynamic nature of the user's location is an integral aspect, requiring the underlying content to be both up-to-date and accurate and introducing the idea of 'time-critical' information. Finally, the environmental aspect of context emphasises the importance of mobile geographic hypermedia techniques being sensitive to the dynamic settings within which they are accessed. Lumsden & Brewster (2003) contend that when a mobile device is used in the context of mobility, the user's visual attention must remain with the surrounding environment, rather than the device. Kjeldskov (2002) sees this as a significant consideration affecting the amount of information that should be presented to the user and the level of interaction required, since small and cluttered displays place high demands on a user's attention.

Despite identifying these issues, it can be argued that mobile geographic hypermedia systems has thus far been largely driven by the underlying technology, as opposed to the needs of the end users. The next section discusses this, along with the limitations of such an approach, by introducing a different focus for mobile hypermedia research and development.

8.3 Usefulness and Geographic Hypermedia

Technology-driven development concentrates on applying new technology and investigating technical issues in its application (Cartwright *et al.* 2001), and can be a major facilitator of innovation, enabling new technologies and concepts that may never have been arrived at by other means. While this may be sufficient for research purposes, in a development sense it has become generally accepted that the 'build and they will come' and 'one tool fits all' attitudes typical of technology-driven development do not guarantee the success of new products (MacEachren & Kraak 2001; Gould & Lewis 1985). This is because user needs and expectations are seldom considered as part of technology-driven development, and as with most products and services, unless mobile geographic hypermedia products are considered useful to, and therefore adopted by, their target markets, they will not be commercially successful. Therefore a supplemental approach is advocated, focused on the optimisation of system usefulness. Note that this perspective does not seek to discount the value of taking a technology-driven approach early on, when cutting-edge technologies are still in their infancy. Rather it describes a complementary, user-driven methodology for further development, capitalising on the technology whilst ensuring the usefulness of products.

8.3.1 The Concept of Usefulness

Whilst there are numerous characterisations of the term *usefulness*, and seemingly no general consensus on its definition, this chapter takes the lead of Nielsen (1993), viewing it as the degree to which a system can be used to achieve a given goal, and acknowledging its component concepts of utility and usability (after Grudin 1992). According to Nielsen (1993), *utility* concerns whether the system can (at least in theory) do what the user requires it to do, whilst the International Organization for Standards formally defines *usability*, in the context of human-computer interfaces, as "the extent to which a product can be used by specified users to achieve specified goals with effectiveness, efficiency and satisfaction in a specified context of use" (TC 159/SC 4 1998), where:

- Effectiveness is the accuracy and completeness with which users achieve specified goals;
- Efficiency concerns the resources expended in relation to the accuracy and completeness with which users achieve goals; and
- Satisfaction relates to freedom from discomfort, and positive attitudes toward the use of the product (Jokela *et al.* 2003).

These concepts, and inherently the users themselves, are of paramount importance to the acceptance of mobile geographic hypermedia products. If users cannot use the services to (a) achieve their goals, and (b) do so with efficiency, effectiveness and satisfaction, they will be unlikely to adopt them.

8.3.2 Designing for Usefulness

As Dransch (2001) identifies, cartographic research has long been dominated by issues of technique (e.g. the possibilities for interaction) and data (e.g. what can be done with it), rather than user needs. She goes on to recommend that future studies in this area should support both map-making and map-using and should target the specific tasks the maps are used for as well as how they are used. MacEachren & Kraak (2001) support this view, highlighting a need to integrate work on human spatial cognition with technological development, and address user differences. In more general terms, Slocum *et al.* (2001) propose a research agenda dealing with the development of a methodology based on a user-centred approach for evaluating the usability of alternative geovisualization techniques.

Usefulness of Geospatial Information

In 2002, a workshop was held to develop a research agenda related to improving the usability, and thus accessibility, of geospatial data in general (Wachowicz *et al.* 2002). Whilst outwardly focused on usability, a large number of concepts and issues arose from the discussions which were equally applicable to usefulness as a whole, including:

- Geospatial data has the potential to support decision-making that is faster and more informed, however poor data *usefulness* can counteract this and ultimately inhibit geospatial data usage.
- The effects of geospatial data (e.g. information acquisition, time saving and satisfaction measures) use can help to characterise its *usefulness*.
- Geospatial data usage is no longer confined to the realm of expert users, with benefits to be gained from targeting non-expert users with a view to improving *usefulness*.
- Representation of existing geospatial data sets (i.e. the 'human interface') is an area for constant improvement.
- Important research priorities include: the development of formal rules for ensuring geospatial data *usefulness*; linking *usefulness* to user tasks; and the differences between geospatial and non-geospatial data *usefulness*.

Even more recently, the International Cartographic Association established a new ICA Working Group on Use and User Issues, concerned with addressing the subject of usefulness within cartographic data, systems, products, interfaces and research. In particular, the group aims to focus on (i) the user – their characteristics, use contexts, goals, tasks and requirements; (ii) usability – including User-Centred Design and methods of evaluation; and (iii) improving user abilities – from user training to collaborative mapping to map use in education. In light of both of these initiatives it is clear that the cartographic community has become increasingly concerned with the user and their experiences with geospatial information.

Research in Pursuit of Useful Mobile Geographic Hypermedia

In the realm of generic mobile Internet services, developers and researchers have begun to realise the benefits of making utility and usability a focus of development, in order to design products that meet users' needs and expectations. Wireless Application Protocol (WAP) services are a prime example of the failure of technology-driven development in this realm, with studies uncovering numerous usability problems leading to user rejection of the earlier services (Ramsay and Nielsen 2000). Building on such experiences, others have endeavoured to improve new mobile Internet services by conducting research early in the product development lifecycle, in order to understand user requirements and thus design for them from the outset (Helyar 2002).

Specific to the area of mLBS, the focus has also turned to usefulness. Much activity in this respect comes from the field of Human-Computer Interaction (HCI), which categorises mLBS under the broader topic of 'context-aware' computing (Cheverst *et al.* 2000). A number of studies have sought to define user needs for location-aware services (Kaasinen 2003), with projects such as GUIDE (Cheverst *et al.* 2000) and HIPS (Broadbent and Marti 1997) having employed techniques for ensuring systems that meet users' goals and are easy to use.

These projects have, however, been mainly concerned with issues of overall system appearance, functionality, information content and methods of interaction, with little or no emphasis on the appropriateness of the component geographic hypermedia, with one notable exception (Chincholle *et al.* 2002). Taking a different approach, researchers within Cartography have been working to specifically develop geographic hypermedia for mLBS that will support users' geospatial tasks. Particular attention has been paid to map design (Uhlirz 2001, Wintges 2002), with some treatment of non-map geospatial information presentation (Brunner-Friedrich and Nothegger 2002).

Whilst these studies are undeniably revealing, it may be argued that the resulting representations and services are bound by the same constraints as technology-centred development, and thus their usefulness cannot be assured (a review of the related literature suggests that the design work is not based on an assessment of user needs, nor have the results been evaluated).

This chapter therefore recommends an approach to designing mobile geographic hypermedia systems that is more in line with that of the HCI researchers, namely User-Centred Design.

8.3.3 A User-Centred Approach

User-Centred Design (UCD) is an approach commonly employed in computer systems design, having developed under the premise that in order to ensure the usefulness and commercial success of a system, all design activities should position the end user as their focus so that the final product is easy to use and ultimately meets their needs (Gould and Lewis 1985). Essentially a UCD approach addresses the questions: '*How do I understand the user?*' and '*How do I ensure this understanding is reflected in my system?*' (Holtzblatt and Beyer 1993).

The ideals and techniques of UCD originated in the early works of researchers Gould and Lewis (1985) and Norman and Draper (1986), with the former proposing three basic principles: (1) an early focus on understanding users and their tasks; (2) empirical measurement of product usage by *representative users*; and (3) an iterative cycle of design, test and measure and redesign. Since that time, UCD has become the subject of much research and literature, having come to be viewed by many as an integral factor in the development of successful commercial software products (Bias and Mayhew 1994, Butler 1996, Mayhew 1999, Nielsen 1993, Rubin 1994). In 1999 an international standard was established – 'ISO 13407 Human-Centred Design Processes for Interactive Systems' – providing guidance for UCD by way of describing the rationale, planning, principles and activities of UCD practice (Jokela *et al.* 2003). Developed by a board of international researchers and practitioners in the field, the standard discusses the four main activities of UCD which are carried out iteratively until the defined objectives (in terms of usefulness) have been met. These are presented in Fig. 8.2 and described below (Jokela *et al.* 2003, Maguire 2001).

A. *Understand and specify the context of use* incorporating user characteristics (e.g. knowledge, skills, experience, education, training, attributes, habits, preferences, capabilities), user goals/tasks (including system use) and the environment of use (technical, physical and social); in order to support user requirements specification and provide a basis for later evaluation activities.
B. *Specify the user and organisational requirements* using the previously defined context of use (particularly user goals and tasks), in order to evolve measurable criteria against which the usefulness of the product will be evaluated, and to define user-centred design goals and constraints.
C. *Produce design solutions* based on the established design goals, guidelines and constraints and incorporating HCI knowledge (relating to visual design, interaction design, usability, etc.). An iterative process of design production to support evaluation at different stages of the system development lifecycle.
D. *Evaluate designs against requirements* throughout development, employing appropriate prototypes and applying the goal-/task-based crite-

ria developed previously. Important for determining the degree to which the user/organisational objectives have been met and in obtaining feedback relating to design refinements.

As Jokela *et al.* (2003) identify, ISO 13407 does not aim to outline detailed methods for completing the activities, with such descriptions already contained within numerous methodology publications. There are in fact a multitude of methods available for conducting UCD, with Table 8.3 providing an illustrative selection, grouped by the main activities identified above.

The ideals of UCD are grounded in a number of disciplines, most notably cognitive psychology (the study of human perception and cognition), experimental psychology (the use of empirical methods to measure and study human behaviour) and ethnography (the study, analysis, interpretation and description of unfamiliar cultures) (Mayhew 1999). Furthermore, it aligns closely with the social sciences – concerned with the study of peoples' beliefs, behaviours, interactions and institutions (Neuman 1997) – which is particularly noticeable in the correlation between specific UCD techniques and social research data collection methods, both quantitative and qualitative (Urquhart *et al.* 2004).

Comprehensive models exist for implementing a UCD methodology for the development of a system, complete with detailed discussions of optimal activities and methods (see for example Nielsen 1993, Mayhew 1999). It is not the purpose of this chapter (nor is it desirable), to provide a definitive UCD methodology for developing mobile geographic hypermedia sys-

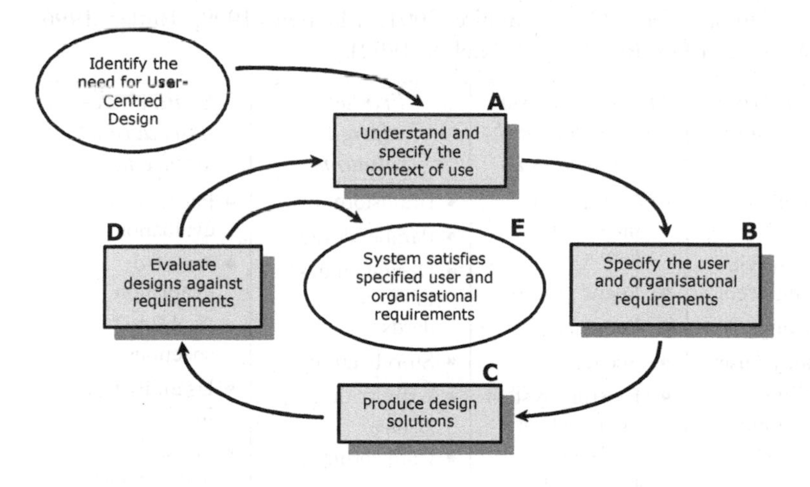

Fig. 8.2. The main activities of UCD (ISO 13407 in Jokela *et al.* 2003).

tems. In fact, the final set of methods used in any one undertaking must be carefully selected and tailored to meet the unique requirements of the system. Moreover, factors such as time, budget and available resources will dictate the feasibility of using individual UCD techniques. Ultimately however, provided that Gould & Lewis' (1985) base principles are adhered to and a rigorous process is followed, any combination of UCD methods maybe employed to achieve the required aims (Nielsen 1992).

8.4 Discussion

It is the intention of this discussion to build on the ideas presented above, providing further justification for a user-driven approach to the design of mobile geographic hypermedia systems. In doing so, it is fundamental to highlight the role of the user in the process. Undeniably, the user is the central figure in the success of any mobile geographic hypermedia system, which is directly related to their adoption or rejection of the product. Adoption is in turn contingent on usefulness which, when optimised, should benefit the user through increased productivity (via effective operation), reduced errors (caused by inconsistencies, ambiguities, etc.), and less need for training and support (via reinforced learning), thus leading to improved acceptance (through satisfaction and ease of use) (Maguire 2001). As discussed previously, the usefulness of a system is measured by its util-

Table 8.3. Methods for UCD (Maguire 2001; Mayhew 1999; Butler 1996; Holtzblatt & Beyer 1993; Rubin 1994; Nielsen 1992).

Understand and specify the context of use	*Specify the user-organisational requirements*	*Produce design solutions*	*Evaluate designs against requirements*
• Context of use analysis • Survey/interview of existing users • Focus groups • Field study / user observation • Diary keeping • Task analysis	• User requirements interviews • Focus groups • Personas • Scenarios • Existing system / Competitor analysis • User / task / domain models	• Brainstorming • Parallel design • Design guidelines and standards • Storyboarding • Affinity diagrams • Card sorting • Prototyping	• Participatory evaluation • Assisted evaluation • Usability Inspections • Usability testing • Satisfaction questionnaires • Post-experience interviews

ity and usability in the eyes of the user. Herein lies the major limitation of technology-driven development, which rarely involves consultation with the user – least of all during the design process. In contrast, a UCD methodology requires that the user is actively involved *at all stages* of the project lifecycle, from gathering and specifying requirements through to evaluating design solutions. Only in this way can the final product truly be claimed to have been designed for the user.

A UCD approach to the design of mobile geographic hypermedia systems may be considered especially crucial in light of the characteristics of the users themselves and their use contexts. Returning to the example of mLBS, arguably the most profitable market segment for these systems are consumers, a large proportion of whom have little or no formal training in the interpretation and analysis of geospatial information. Such people are considered 'non-expert' users who are generally less adept at understanding cartographic representation forms than 'expert' users of traditional geospatial systems – who are often formally trained in spatial thinking and reasoning (Golledge 2003). Cartwright *et al.* (2001) emphasise the importance of understanding and designing for a range of user abilities stating that "access to geospatial information and the interfaces that provide the 'gateways' to this information, need to be designed in sympathy to all users, so as to ensure equity of access and use" (p.48). Looking to the actual use of mobile geographic hypermedia systems, this takes place in dynamic, changing environments which will naturally take the primary focus of the user's attention. Hence the operation of these systems are relegated to lesser levels of user focus, and thus attention, making the optimisation of their usefulness (and in particular their usability) under distracting conditions especially important.

So what does the new approach to system design advocated by this chapter really mean for mobile geographic hypermedia systems and geographic hypermedia in general? To deal first with the latter, the concepts and processes related to usefulness and UCD are indeed generic and have been employed in computer systems design for many years (as the references attest). Therefore they can be easily applied to the design of all types of geographic hypermedia systems with the following benefits also universally relevant. Of major benefit, adhering to the guidelines of UCD compels creators of geographic hypermedia systems to find out who their end users are and exactly what types of geospatial information they need at the outset, which they can then use to drive design, rather than making assumptions and designing for 'what is possible'. This is particularly pertinent for mobile systems, which combine innovative technology with new and inexperienced users, and arguably results in a greater likelihood that the final product will be adopted. Furthermore the involvement of end users, together with an associated expenditure of time and effort, throughout the design and development lifecycle helps to identify and eliminate problems of usability with the component geographic hypermedia as they arise, thus avoiding the time and expense associated with resolving usability issues within a completed product. This equates to 'getting it right the first time'.

Another benefit involves the consideration of specific constraints within the delivery medium, including both technological and context of use, which is especially important for mobile systems. This ensures that inappropriate geographic hypermedia forms are discarded early on (e.g. detailed visual media designed for the desktop environment that will likely be inappropriate for use within display-limited mobile systems), and thus enables the design to focus on more suitable representations. Complementary to this, speaking with the end users may even prompt novel forms of geographic hypermedia which would not otherwise have been revealed through the application of pure theory and/or the technological possibilities of the medium. In closing, a final benefit comes from the iterative nature of the UCD process. By employing a cycle of design evaluation and refinement, alternative forms of (mobile) geographic hypermedia can be trialled and compared through feedback from real users before being committed to development, again optimising the efficiency of the development process whilst ensuring the most appropriate result.

8.5 Conclusion

Mobile geographic hypermedia systems have the potential to deliver geospatial information to (often new) users where and when they want it. To date, however, user needs have been largely assumed while developers race to produce novel applications based on cutting-edge mobile technology. While the resulting systems are often highly desirable (e.g. for their innovation and attractiveness), they do not necessarily meet real users' needs or account for the unique cartographic constraints of the medium, and therefore they run the risk of rejection in the marketplace. This chapter has argued that in order to ensure the commercial success of mobile geographic hypermedia products (and indeed geographic hypermedia products in general), usefulness should be the key driver of system design. In this way, the user should be made the focus of all design and development activities so that the final product not only meets their requirements but is effective, efficient and satisfactory to use. A focus on usefulness does not only benefit commercial products, however. It also has the potential to drive research resulting in new forms of geographic hypermedia for the mobile medium, which will in turn benefit the mobile geographic hypermedia systems of the future.

Acknowledgements

The author would like to acknowledge the following people for their intellectual input into the concepts presented in this chapter and/or their review of its content: Assoc Prof William Cartwright and Dr Suzette Miller of RMIT University (Australia), Prof Dr Doris Dransch of GeoForschungsZentrum (Germany) and Mr Peter Benda of Telstra (Australia).

References

Bias RG, Mayhew DJ (eds) (1994), Cost-justifying usability. Academic Press, Inc, Boston, USA

Broadbent J, Marti P (1997) Location aware mobile interactive guides: usability issues. In: Proceedings of the Fourth International Conference on Hypermedia and Interactivity in Museums (ICHIM97), Paris, France

Brunner-Friedrich B, Nothegger C (2002) Concepts for user-oriented cartographic presentation on mobile devices - a pedestrian guidance service for the TU Vienna. In: Gartner G (ed) Proceedings of the ICA Commission on Maps and the Internet Annual Meeting, Institute of Cartography and Geomedia Technique, Karlsruhe, Germany, pp 189-196

Butler KA (1996) Usability engineering turns 10. Interactions, vol 3(1), pp. 58-75

Buziek G (1999) Dynamic elements of multimedia cartography. In: Cartwright *et al.* (eds) Multimedia cartography, Springer, Berlin, pp. 231-244

Cartwright W, Crampton J, Gartner G, Miller S, Mitchell K, Siekierska E, Wood J (2001) Geospatial information visualization user interface issues. Cartography and Geographic Information Science, vol 28 (1), pp 45-60

Cartwright W, Peterson MP (1999) Multimedia cartography. In: Cartwright *et al.* (eds) Multimedia cartography, Springer, Berlin, pp 1-10

Cheverst K, Davies N, Mitchell K, Friday A, Efstratiou C (2000) Developing a context-aware electronic tourist guide: some issues and experiences. In: Proceedings of CHI 2000, ACM, The Hague, pp 17-24

Chincholle D, Goldstein M, Nyberg M, Eriksson M (2002) Lost or found? a usability evaluation of a mobile navigation and location-based service. In: Paterno F (ed) Proceedings of Mobile HCI 2002, Springer, Sept, Pisa, pp 211-224

Coschurba P, Baumann J, Kubach U, Leonhardi A (2001) Metaphors and context-aware information access. Personal and Ubiquitous Computing, vol 5(1), pp 16-19

Dransch D (2001) User-centred human-computer interaction in cartographic information processing. In: Proceedings of 20th International Cartographic Conference, Beijing, China, pp 1767-1774

Gartner G, Uhlirz S (2002) Maps, multimedia and the mobile Internet. In: Gartner G (ed) Proceedings of the ICA Commission on Maps and the Internet Annual Meeting, Institute of Cartography and Geomedia Technique, Karlsruhe, Germany, pp 143-150

Golledge R (2003) Thinking spatially, http://www.directionsmag.com/article.php?article_id=277

Gould JD, Lewis C (1985) Designing for usability: key principles and what designers think. Communications of the ACM, vol 28 (3), pp 300-311

Grudin J (1992) Utility and usability: research issues and development contexts. Interacting with Computers, vol 4(2), pp 209-217

Hassin BG (2003) Mobile GIS: how to get from there to here. http://www.jlocationservices.com/LBS-Articles/Mobile%20GIS.pdf

Helyar V (2002) Assessing user requirements for the mobile Internet. Appliance Design (3).

Holtzblatt K (2005) Customer-centred design for mobile applications. Personal and Ubiquitous Computing, vol 9, pp 227-237

Holtzblatt K, Beyer H (1993) Making customer-centred design work for teams. Communications of the ACM, vol 36(10), pp 93-103

Jokela T, Iivari N, Matero J, Karukka M (2003) The standard of user-centred design and the standard definition of usability: analyzing ISO 13407 against ISO 9241-11. In: Proceedings of Latin American Conference on Human-Computer Interaction, Rio de Janeiro, Brazil, pp 53-60

Kaasinen E (2003) User needs for location-aware mobile services. Personal and Ubiquitous Computing, vol 7(1), pp 70-79

Kjeldskov J (2002) Just-in-place – Information for mobile device interfaces. In: Paternò F (ed) Proceedings of the 4th International Symposium, Mobile HCI 2002, Springer, Pisa, Italy, pp 271-275

Lee YE, Benbasat I (2003) Interfact design for mobile commerce. Communications of the ACM, vol 46 (12), pp 48-52

Lumsden J, Brewster S (2003) A paradigm shift: alternative interaction techniques for use with mobile & wearable devices. In: Proceedings of the Conference of the Centre for Advanced Studies on Collaborative Research, Toronto, pp 197-210

MacEachren AM, Kraak MJ (2001) Research challenges in geovisualization. Cartography and Geographic Information Science, vol 28(1), pp 3-12

Maguire M (2001) Methods to support human-centred design. International Journal of Human-Computer Studies, vol 55(4), pp 587-634

Mayhew DJ (1999) The usability engineering lifecycle: a practitioner's handbook for user interface design. Morgan Kaufmann Publishers, San Francisco

Mountain D, Raper J (2000) Designing geolocation services for next generation mobile phone systems. In: Proceedings of the Association for Geographic Information Conference, AGI 2000, London, UK

Neuman WL (1997) Social research methods: qualitative and quantitative approaches. 3rd Edition, Allyn & Bacon, Needham Heights, MA
Niedzwiadek H (2002) Where's the value in location services? http://jlocation-services.com/EducationalResources/WhatareLocationServices.htm#
Nielsen J (1992) The usability engineering life cycle. IEEE Computer, vol 25(3), pp 12-22
Nielsen J (1993) Usability engineering. Academic Press, San Diego, USA
Norman DA, Draper SW (1986) User-centered system design: new perspectives on human-computer interaction, Erlbaum, Hillsdale, NJ
Ramsay M, Nielsen J (2000) WAP usability deja vu: 1994 all over again. Report from a field study. Nielsen Norman Group, Fremont, CA, USA
Reichenbacher T (2001) The world in your pocket, towards a mobile cartography. In: Proceedings of the 20th International Cartographic Conference, Beijing, China, pp 2514-2521
Rubin J (1994) Handbook of usability testing: how to plan, design, and conduct effective tests, John Wiley & Sons, Inc., New York, USA
Sandnes FE (2005) Evaluating mobile text entry strategies with finite state automata. In: Proceedings of MobileHCI'05, ACM, 19-22 September, Salzburg, pp 115-121
Schmidt A (2000) Implicit human computer interaction through context, personal and ubiquitous computing, vol 4 (2,3)
Slocum TA, Blok C, Jiang B, Koussoulakou A, Montello DR, Fuhrmann, S, Hedley NR (2001) Cognitive and usability issues in geovisualization. Cartography and Geographic Information Science, vol 28 (1), pp. 61-75
TC 159/SC 4 (1998) Ergonomic requirements for office work with visual display terminals (VDTs). Guidance on usability, International Organization for Standardization, ISO 9241-11
Uhlirz S (2001) Cartographic Concepts for UMTS-location based services. In: Proceedings of 3rd International Symposium on Mobile Mapping Technology, Cairo
Urquhart K, Miller S, Cartwright W (2004) A user-centred research approach to designing useful geospatial representations for LBS. In: Proceedings of 2nd Symposium on LBS and TeleCartography, Vienna, Austria.
Wachowicz M, Riedemann C, Vullings W, Suarez J, Cromvoets J (2002) Workshop report on spatial data usability. In: Proceedings of the 5th AGILE Conference on Geographic Information Science, Palma, Mallorca, Spain
Weiss S (2002) Handheld usability. John Wiley & Sons Ltd., West Sussex, England
Wintges T (2002) Geo-data visualization on personal digital assistants (PDA). In: Gartner G (ed) Proceedings of the ICA Commission on Maps and the Internet Annual Meeting, Institute of Cartography and Geomedia Technique, Karlsruhe, Germany, pp. 177-183
Zeimpekis V, Giaglis GM, Lekakos G (2002) A taxonomy of indoor and outdoor positioning techniques for mobile location services. ACM SIGecom Exchanges, vol 3(4), pp. 19-27

9 Cruiser: A Web in Space

Manolis Koutlis, George Tsironis, George Vassiliou, Thanassis Mantes, Augustine Gryllakis, and Kriton Kyrimis

Contribution: Nectaria Tryfona

Abstract. This chapter presents Cruiser, a geographically-oriented content management and delivery system, providing a platform for developing rich internet applications and services that make use of spatial data, maps and location-aware information. The chapter addresses the technological and business need behind it, and discusses Cruiser's unique feature, namely the *channel*, that allows for tuning, navigation, browsing, searching, enriching and sharing of geographic space, at a personal level. Finally, the target application domain as well as, the benefits of adopting this platform are presented.

9.1 Introduction

In the many facets of everyday communication activities that take place over the internet, there is a need to refer to issues of spatial nature and geographical concepts. This is typically accomplished through the use of verbal descriptions or through maps. Nevertheless, the use of maps over the internet for conveying and manipulating geographic positions, regions, itineraries, distributions, and networks is a difficult task, while the typical practices used for displaying image-based maps in web-pages result in a very poor user experience. What's more, advanced uses, such as, for example, the combination of textual references with their map representations (e.g., displaying the multimedia description of an entity whenever it is selected on the map, or vice versa), searching with spatial criteria, or dynamically simulating physical phenomena, require expertise and technical infrastructure that are beyond reach in most cases.

Cruiser, a new medium for publishing, searching and communicating on the web, based on maps and geographic space, was conceived and developed to address the aforementioned business and technological need. It provides a geographically-oriented content management and delivery platform for developing internet applications that make use of spatial data and geo-coded information.

By adopting modern software architectures and state-of-the-art technologies, Cruiser provides unique quality of user experience by offering the ability for:

- Real-time fly-over 3D terrains and seamless map browsing over the internet. Using Cruiser, users can freely navigate and pan-around in terrains of great detail and maps of any scale, zoom-in/out at will to any place of interest, measure distances, and print personalized maps with suitably chosen symbols of interest (e.g., eliminating places or marks of non interest). Flying and map browsing can be achieved even over dialup internet connections.
- Access to any type of geo-coded information based on its spatial distribution. As users browse maps and navigate through landscapes, they can selectively display or hide descriptive information of the various land features and geographic entities (e.g. cities, road network, companies, products, etc) by applying various filters and criteria (e.g., "show only 3-star hotels", or "cities with population greater than 3000"). Information is displayed on the maps and landscapes with various symbols which can be selected to display linked multimedia information (html links, videos, animations, photographs, etc).
- Criteria based searches. In addition to accessing information while browsing, users can search for in-formation by issuing structured queries and then selectively jump to the places conveying the re-quested information. Queries can be based either on pure lexicographic criteria (e.g., by specifying the name of a place, a city, a company, etc), or on spatial criteria, like "distance from point", or "inclusion in shape" which are graphically determined on the maps, or both.
- Embedding of personal content. Cruiser offers users the ability to attach various types of objects and personal information like points (flags, notes, photos, etc), itineraries with commentaries and polygon shaped regions of interest, again, accompanied by information. In that way users can create their "personal excursion albums" or "personal cadastral", and share this information with friends who can dis-play this information on their own Cruisers.

It important to note that the Cruiser user interface experience avoids the pitfalls of fragmented web-page based applications which suffer from interruptions, and slow responsiveness, as web pages are reloaded after every user interaction. Wait times are eliminated, thanks to effective data compression techniques which are deployed for the geographic data, while data caching allows the off-line use of Cruiser, in cases where an internet connection is not available.

The rest of the chapter is organized as follows: Section 9.2 presents the unique feature of Cruiser, namely the 'channel' and discusses the way it works. Section 9.3 gives the technical description of the platform, focusing on the Cruiser server, Cruiser clients, and Cruiser applications. Section 9.4 discusses the benefits of adopting and using the Cruiser environment. Section 9.5 presents the application domain and finally, Section 9.6 concludes and gives related work in this field.

9.2 The "Channel" Feature

Cruiser-based applications are offered as "channels" through a game-like interface which allows seamless navigation within the 3D geographic space and 2D maps. Fig. 9.1 depicts a snapshot of the navigation in 3D Greece.

Users can just "tune in" to a desired channel (see below), and visit places, browse available information, inspect objects and geographical features, access web sites that are linked to the objects of concern, conduct searches using spatial criteria, add notes and place-marks, share geographic content with others, personalize the views and print customized maps. Fig. 9.2 shows the use of the Barcelona area channel.

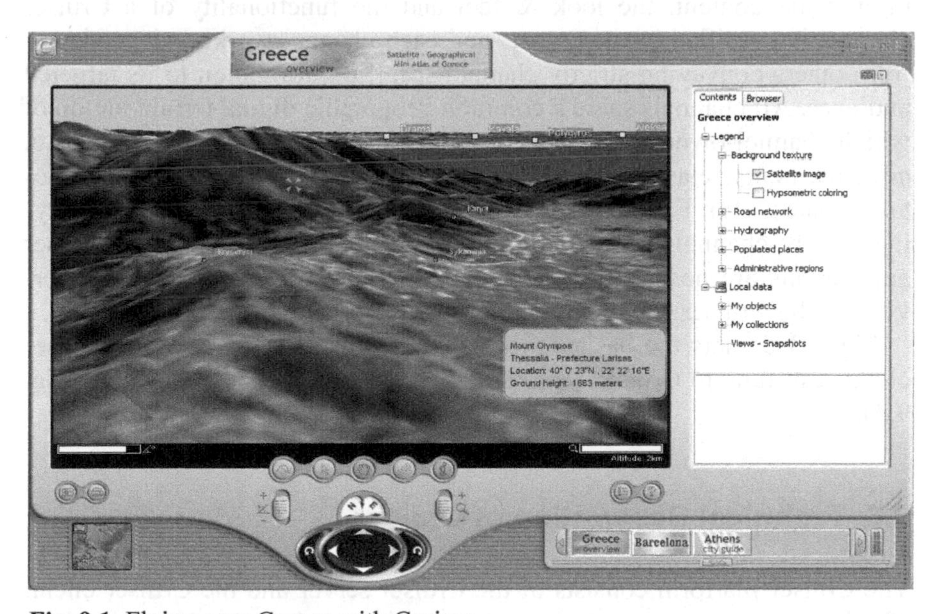

Fig. 9.1. Flying over Greece with Cruiser

Fig. 9.2. A Cruiser channel for Barcelona

Cruiser channels can be setup on demand, managed and branded by interested customers (service or content providers, governmental organizations, companies for enterprise uses, individuals, etc). The geographic and multimedia content, the look & feel and the functionality of a Cruiser channel is specified by its owner, and each channel works independently from others conveying strictly channel-specific information to its targeted audience. All channels share a common geographic digital terrain on top of which channel-owners add their own multimedia geo-coded content (e.g. advertisements, classified-ads, real-estate property, photos, notes, annotated routes, etc). Each channel works independently from others conveying strictly channel-specific information to its targeted audience. Cruiser applications are hosted by Cruiser servers which stream content and services to the Cruiser clients through channels over the internet or intranets.

Fig. 9.3 depicts the basic concepts of the channel architecture. Cruiser clients can tune to Cruiser servers and browse all available channels (Fig. 9.4).

9.3. Technical Description of Cruiser

The Cruiser platform consists of the Cruiser server and the Cruiser client. The latter is a rich internet client that provides a shell that hosts and exe-

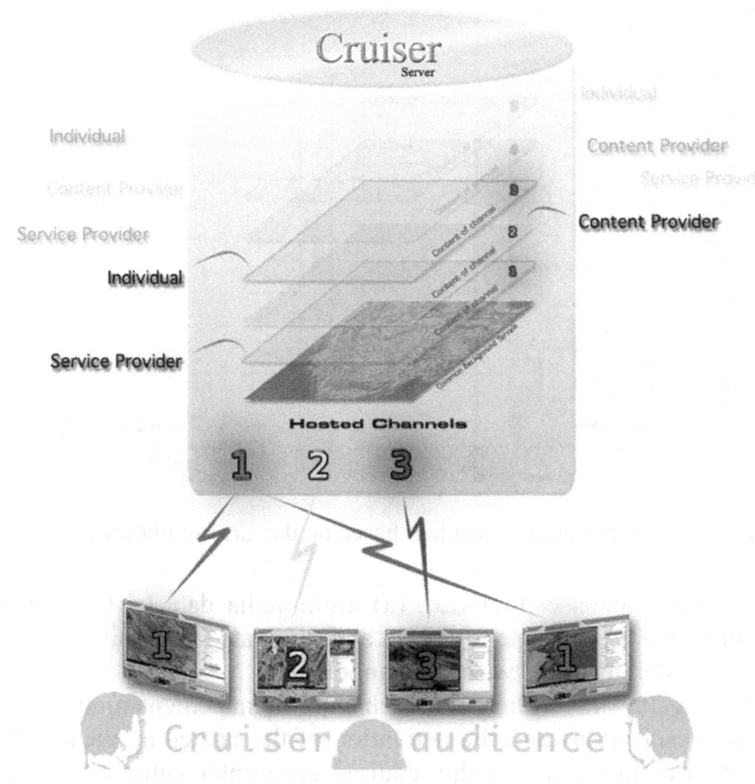

Fig. 9.3. Architecture of Cruiser Channels

cutes applications composed by *software components* in various combinations of user interface and functionality. This gives Cruiser the ability to be "*polymorphic*", i.e., to adapt dynamically to the requirements of many different applications, which implement, each time, the targeted services, by changing user interface, look & feel (skins) and functionality, according to each case.

Fig. 9.5 presents the platform architecture. Next, the parts comprising the platform are described.

9.3.1 Cruiser-Server

The Cruiser-server consists of the following parts (modules):

1. *Geographic Database*. It includes the set of geographic data which are used by all the hosted channels-applications: three-dimensional terrain models, satellite images, and vector maps in various scales.

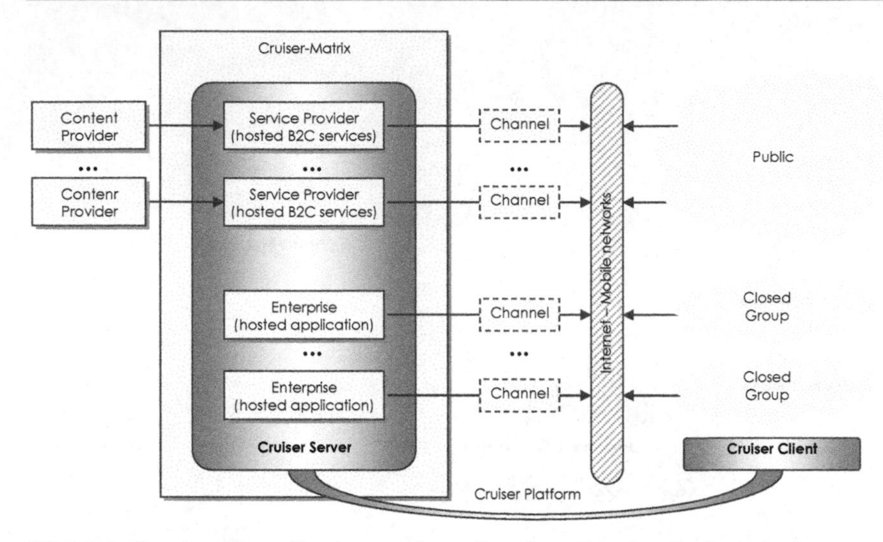

Fig. 9.4. Hosting of applications and services based on the Cruiser platform

2. Content & User Database. Includes: (a) multimedia data (rich media) accompanying the geographic objects; and (b) data about the users-subscribers registered for the various channels.
3. *Channel Manager*. This is the central channel management module. It manages the descriptions and the activation of the hosted applications, as well as their connection with the required geographic substratum and multimedia content.
4. *Geo-spatial data streaming module*: It manages the data requests by the Cruiser clients and handles the real-time transmission of the geographic data..
5. *Query-Lookup processing module:* It manages the requests by the Cruiser channels for searches for geographic objects and descriptive data through structured queries.
6. *Geo-spatial processing module.* All maintenance, extension and processing (translation to internal formats, compression, etc.) operations for the total of the data, used by each service, take place through this module.
7. *Data entry & geocoding module*: Through this module are entered new geographic objects. Either by the service support personnel or by remote users of the system, using the Cruiser client and web-forms (e.g., input of the position of a real estate property and its descriptive data by the seller).
8. *Real-time feeds module*: It manages real-time geographic data sources that are used for display by the channels (e.g., weather satellite images, GPS streams of moving vehicles, etc.).

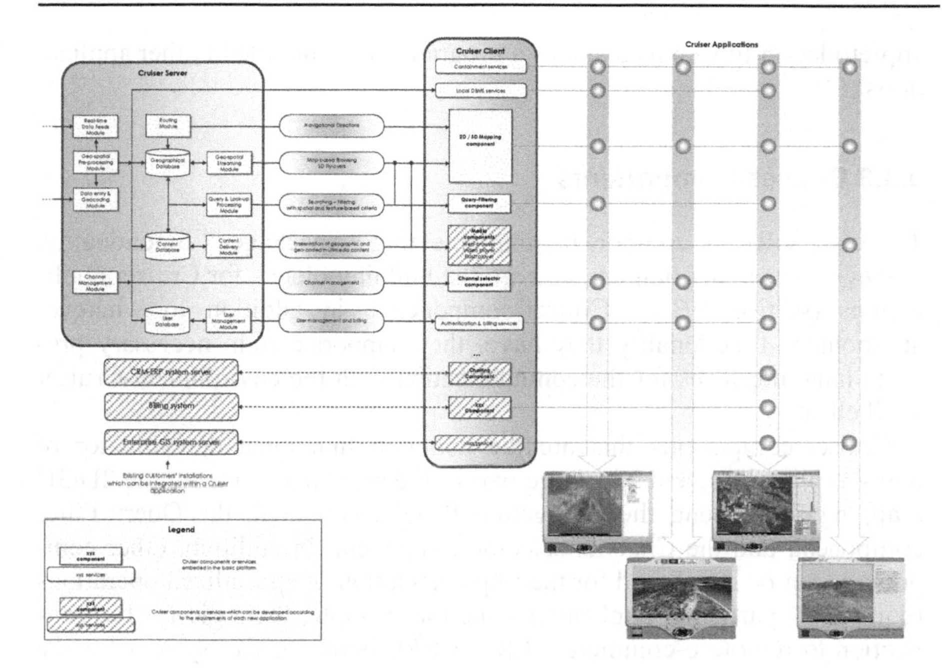

Fig. 9.5. Block diagram of the architecture of the Cruiser platform

9. *Subscription management module*: It implements the various subscription charging schemas for the information providers (e.g., the time an entry will be kept, according to the charging package) and for the consumer, per service, producing the necessary operation and usage statistics of the services.

9.3.2 Cruiser-Client (Cruiser)

This is the specially designed, rich client application, through which all provided services are used. Cruiser is installed as a desktop application using a simple procedure, and is updated continuously, with any updated versions, in a user-transparent manner. Cruiser can be used either of-line or on-line. In the latter case, it is connected to the Cruiser server (through a proprietary API) and has access to services and content provided by the server.

From a technical viewpoint, Cruiser is a shell that hosts application (Cruiser-applications, section 9.3.4) which are composed by software components (Cruiser-components, section 9.3.3), providing them with a series of services regarding their appearance and operation. Furthermore, Cruiser is itself a component that can be incorporated in other applications (as an applet/plug-in in web pages or as a programmable module in other applications).

9.3.3 Cruiser-Components

These are software components implementing a "packaged" functionality, however simple or complex, and are the building blocks for Cruiser applications (section 9.3.4). Cruiser-components, in addition to whatever autonomous functionality they have, they support certain necessary programming interfaces for the communication with the environment-Cruiser shell client.

Cruiser components that are the common functional denominator of many applications, and which are part of the basic platform, are the 2D/3D mapping component, the Navigation Panel component, the Query-Filter component, and the Channel selector component. In addition, other components can be developed for the implementation of specialized operations required by particular applications, as, for example, the logic for the connection to remote e-commerce, ERP, CRM, booking, etc. system for the implementation of buying or selling products, room reservations, etc.

9.3.4 Cruiser -Applications

A Cruiser-application is composed, regarding its user interface layout and its functionality, by Cruiser-components, and is distributed by the Channel manager of the Cruiser server. This architecture of composing applications using components gives the ability to implement a wide range of applications even with a small number of Cruiser-components.

A typical Cruiser-application consists of a client and a server part. The client part is installed in each Cruiser-client, while the server part may be distributed in many servers, including the Cruiser-server. In addition, a Cruiser-application can be only a local (desktop) application.

All Cruiser-applications are available to the Cruiser-client through the Cruiser environment. Each application comprises a different channel. The Cruiser platform offers the following services to its applications:

- Automatic deployment wherever there is a Cruiser-client. The client has a list of all available applications. When the user selects an application, it is automatically downloaded and installed in the Cruiser-client. In addition, there is completely transparent support for the distribution of newer versions of an application, centrally by the Cruiser server. Apart from the application code, the deployment refers to its database information, as well as the parts of the information about the geographic substratum, related to the application.

- The geographic substratum in various levels of detail. The navigation and management, in general, of the substratum is done through the 2D/3D navigation component, as well as the geographic information management component. The former provides visual map management services, while the latter provides services accessing the geographic information and incorporating it in the application logic.
- Geographic services, as, for example, distances between points, the points of interest that are within a certain distance from some other point, and many others. These services are provided to the application either in the form of a local API, by the Cruiser-client, or in the form of remote web services, by the Cruiser-server.
- The local relational DBMS services, which the application can use for the deployment of database information used locally. Through the same DBMS, access is provided to the descriptive information of the geographic substratum.
- Local caching capabilities of the remote database information of the application, and access of this information via the same API with which the local database information of the application is accessed.
- Uniform user management system, with profiles, access rights to operations and data, identification. Using the same id, the user can subscribe to different services of Cruiser-Matrix. In every case, user management is a service provided centrally by the platform to its users, and, through this service, subscribers to services/applications are managed.
- Support for various charging schemas for the use of the services of an application. An application will be able to charge for parts of its functionality or for its full use, in various ways. Some of the basic ways that must be examined are time-based charging (for services whose use is priced based on time), volume-based charging (for services priced based on data volume), charging based on how many times a function was used, with support for different charging for each function.
- Complete control of the user interface by the application. An application can use diverse technologies (Flash, .Net, Java) and different components, to compose both its user interface and functionality. For example, the platform offers applications the ability to handle a wide spectrum of multimedia content types.
- A Cruiser-application (stand-alone of rich internet client) can run as:
  - An application inside the Cruiser client. The application UI is surrounded by the UI of the Cruiser client.

- An autonomous application. In this case, the application runs inside the Cruiser client again, since in every case it uses its services. In this mode, however, the Cruiser UI is the minimum that is required to host the UI of the application. In this case, Cruiser operates in single-application mode.
- A component, part of more complex applications. The intent is for a Cruiser-application to be able to become an ActiveX / .NET and Java component of other applications. In this case, as well, the application runs inside Cruiser, which is in single-application mode.
- As a browser plug-in, comprising part of the web interface of a web-based application, i.e., an application used inside a web browser.

9.4 Benefits

End-users enjoy a wealth of innovative characteristics offered by Cruiser which resembles more of video-games than of classical «applications». Unique features such as combination of 2D and 3D navigation, configurable level of detail, graphical querying and embedding of personalized content bring advanced GIS technologies to the non-expert large public.

From a business perspective, for service and content providers, Cruiser offers an innovative service/application deployment model with very fast development cycles, low costs and small time-to-market. Cruiser's application-hosting model, in combination with its component-oriented architecture, ensure:

- Flexibility in developing new applications with rich functionality supplied to order.
- Flexible charging model: Customers pay only for what they use (from a single to thousands of geo-coded items) and for just the amount of time they use it (from a single day to many years).
- Turn-key solutions which eliminate the need to invest in infrastructure, geographic data, and techno-logical know-how (as is typically required by other existing solutions) in order to provide services which make use of maps and geo-coded data.
- Low costs thanks to the economies of scale in development and time resources that are made possible through the re-use of software modules and geographic data.

9.5 Application Fields

Cruiser has a potentially huge application domain. Some of the more prominent sectors include: Advertising, Directory Services, Tourism, Real Estate, Navigational services, Portal / informational services, Education, On-line Gaming, Location Based Services, Geo-Marketing, CRM Applications, Fleet Management, Property Management, Risk Management Applications, User Communities, Governmental applications (Fig. 9.6). Both the private sector and the public can use Cruiser. An example of private channel for enterprise use by authorized users can bc, the Fire Department hiring a channel, and using it to coordinate its fire-fighting work by monitoring the development of a fire and the movement of its vehicles.

On the other hand, examples of channels providing services to the broad public are travel channels (to plan excursions, vacations, etc.), real estate channels (for both sellers and buyers), directory services channels(Yellow Pages) and many others.

9.6. Conclusions and Related Systems

This chapter presents Cruiser, a new medium for publishing, searching and communicating on the web, based on maps and geographic space, was conceived and developed to address the aforementioned business and tech-

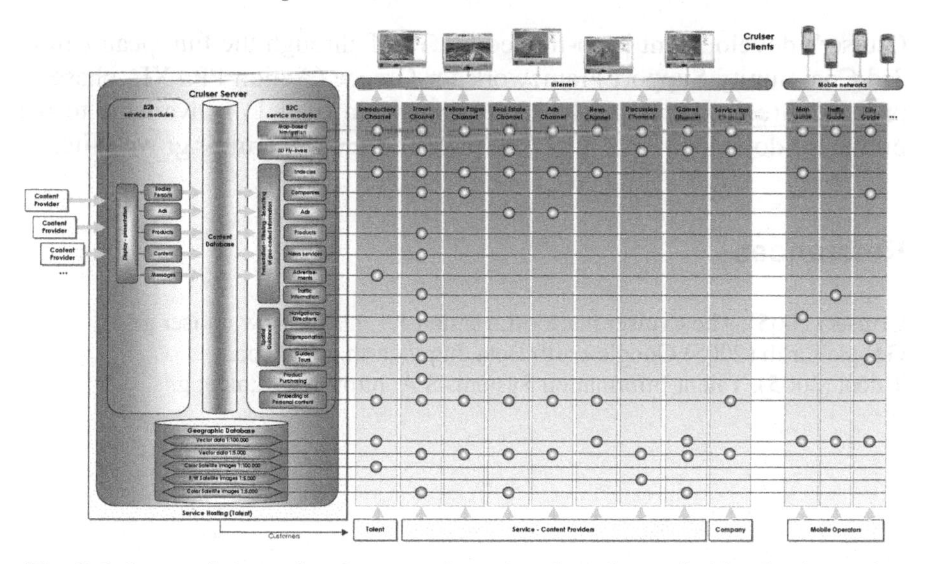

Fig. 9.6. Setup of channels – integrated services based on primitive basic services of the Cruiser platform

nological need. Cruiser provides a geographically-oriented content management and delivery platform for developing internet applications that make use of spatial data and geo-coded information.

In the spectrum of publishing, searching and communicating geographic content on the web. Cruiser can be compared with Google Earth (beta version). It has two main advantages to the latter public domain software:

- One user client – Many service providers.

While Google Earth is based on the 1-1 relation between provider and client (i.e., the client can only see what the one (only) provider has to offer), Cruiser follows the 'TV principle': *one* client sees what *many* providers are offering (the concept of channels). Users can 'tune' to their favorite channels or zap through them at any time via the Cruiser controls.

- One user client – Many applications.

As described in Section 9.3, Cruiser is not *one* 'application', but rather a "shell" capable for hosting *many* geo-spatial and mapping internet applications. This allows for multi-working on the user side.

The vision is, for the digital geographic space, Cruiser to become what a web-browser is for the web and eventually create a «web in space».

Acknowledgements

Cruiser's development is co-funded by GSRT through the European Union 3rd Community Support Framework for Greece (Action PPAXE, phase 2, project "Greek Matrix"). The Cruiser platform (v. 1.1) is available to the public for download as of 12.12.2005 from the www.cruiser.gr web site.

References

Cruiser (2005). The Cruiser platform, version 1.1, http://www.cruiser.gr
Google Earth (2005) Google Earth Beta. http://earth.google.com
Talent (2005). Talent Information Systems S.A, http://www.talent.gr

PART III

TECHNOLOGIES FOR CONTENT INTEGRATION

10 Merging Hypermedia GIS with Spatial On-Line Analytical Processing: Towards Hypermedia SOLAP

Yvan Bédard, Marie-Josée Proulx, Sonia Rivest, and Thierry Badard

Abstract. Geographic Knowledge Discovery (GKD) requires systems that support interactive exploration of data without being slowed down by the intricacies of a SQL-type query language and cryptic data structures. GKD requires to compare maps of different phenomena or epochs, to dig into these maps to obtain detailed information, to roll-up data for more global information and to synchronize maps with tables and charts. This can be done by combining the capabilities of GIS with those of OLAP, leading to SOLAP (Spatial OLAP). To enrich the GKD process, we added hypermedia documents to SOLAP. Hypermedia SOLAP provides a more global perception of the situation without requiring the advanced knowledge typically required by Hypermedia GIS. This chapter presents hypermedia SOLAP concepts and gives an example dealing with the erosion of lands and roads.

10.1 Introduction: The Power of Coupling SOLAP with Hypermedia

According to Wikipedia (2005), the first hypermedia system was the MIT Aspen Moviemap. From a Hypermedia GIS point of view, it is interesting to note that this project merged a digital map of the city of Aspen (Colorado, USA) with hypermedia information, namely digital movies (Naimark 1997). Built in 1978 for military purposes as an ARPA (Advanced Research Projects Agency) project, this system allowed the user to select any point in the city of Aspen to start a virtual visit by viewing, on-demand, a film in the selected direction. This freed the users from following a given route that was predefined by somebody else.

Hypermedia is an extension of hypertext that uses intertwined hyperlinks. It is essentially used to access multimedia information in a random manner at the will of the user. Such flexibility in the search for new information replicates to the way human brains think, that is in a non-linear manner. The type of information provided by such systems uses various digital media to present documents such as texts (descriptive text, narrative and labels), graphics (drawings, diagrams, charts or photographs), videos, sounds (music and oral narration) and animations (changing maps, objects and images) (Bill 1994). Having such information at hand enriches the perception of the users as more senses are actively recruited.

An interactive manner of navigating through information bears many similarities with the manner used by OLAP technology. Both Hypermedia and OLAP systems allow the user to see the desired information by clicking on a word, on a spreadsheet cell, on a piece of a pie chart and so on without forcing a given sequence to navigate into the database. However, two major differences exist between these two families of technologies: the types of data that are used and the media that are supported. OLAP technology uses multidimensional datacubes to manage numeric data called "measures" that are aggregated and cross-referenced for several axes of analysis, each having several levels of details and that are called "dimensions" (see next section). Typically, such measures are derived from detailed data imported from heterogeneous transactional sources and restructured for OLAP navigation through a data warehouse architecture and ETL (Extract, Transform, Load) processes. These measures can then be visualized as values through two media: spreadsheets and statistical diagrams (pie chart, bar chart, histogram, etc.). However, as opposed to a conventional spreadsheet, data can be drilled down, rolled up, drilled across, sliced and diced easily and rapidly. This allows the user to interactively explore the data from one level of detail to another and to see how the different axes of analysis interact together (see next section for examples and Bedard *et al*., 2002). The capability to interactively explore multi-resolution cross-referenced aggregated numerical data is not an objective of hypermedia technology. However, as opposed to OLAP, hypermedia technology supports a large variety of digital media (photos, videos, sound tracks, etc.). Consequently, hypermedia supports both the reproduction of raw signals (e.g. audio waves in record tracks, emitted light in pictures) as well as the presentation of explicit data that have been interpreted *a priori* (e.g. words, numbers), while only the latter is possible with OLAP.

In spite of such capabilities, these technologies do not harness the full power of map data. Map data are the raw materials that produce the geographic information that leads to knowledge regarding the position, extent and distribution of phenomena over our territories. Visualizing geographic phenomena on maps facilitates the extraction of insights that help to understand these phenomena. Such insights include spatial characteristics (position, shape, size, orientation, etc.), spatial relationships (adjacency, connectivity, inclusion, proximity, exclusion, overlay, etc.) and spatial distribution (concentrated, scattered, grouped, regular, etc.). Like other media, maps do more than present phenomena, they help to place them in context and they support the thinking process. Integrating maps with other digital media naturally leads to applications that allow the user to investigate cartographic elements in an interactive mode and to link these elements with a hypermedia database. Such applications are called Hypermedia GIS and typically emphasize the exploration of a hypermedia database using spatial data as the starting point (e.g. a position). Hypermedia GIS provide the essential concepts and techniques for many new GIS applications in visualization, spatial decision support systems and spatial database management and exploration (Hu 2004). Numerous studies in cognitive sciences have shown the superiority of images over numbers and words to stimulate understanding and memory (Buzan and Buzan 2003, Fortin and Rousseau 1989, Standing 1973). Similarly, SOLAP provides new capabilities to explore cartographic data interactively in ways not available in today's GIS. Thus, properly combining maps, spatial analysis, hypermedia navigation and OLAP capabilities should help users to compare different regions, create global views, investigate details, get new hints, discover correlations between phenomena in the same region, see the evolution of the phenomena over different epochs, understand spatial relationships between phenomena, better sense a phenomenon, better communicate findings, etc. In other words, combining Hypermedia GIS with SOLAP (Spatial OLAP) should offer a very promising environment for the user.

Since the fundamental concepts as well as examples of Hypermedia GIS are provided throughout this book, we focus on introducing the new technology that is SOLAP and on showing the potential of Hypermedia SOLAP for spatially-referenced hypermedia applications. Using the first Hypermedia SOLAP application ever developed, we illustrate the inherent capability of SOLAP to support the non-linear approach for accessing information at different levels of abstraction as well as its "natural fit" with hypermedia. Finally, we describe the potential evolution of Hypermedia SOLAP from the traditional architecture to a web service architecture.

10.2 SOLAP Concepts

Geographic Information Systems (GIS) have successfully provided users with new capabilities not available 25 years ago. They have been developed for gathering, storing, manipulating, and displaying spatial data (see Longley *et al.* 2001) and to put the power of digital maps in the hands of users. However, these systems are transaction-oriented, and like every transactional system (e.g. database management systems), they do not efficiently address summarized information, cross-referenced information, and interactive exploration of data. Furthermore, GIS do not efficiently deal with temporal data or with multiple levels of data granularity. Finally, GIS are still difficult to use for non-expert users while the demand for geographical information by many different professions or the general public is increasing (Viatis and Tzagarakis 2005).

Besides transactional technologies, there exist specific tools to easily query and navigate datasets with optimized data structures specifically built for knowledge discovery and decision-support. This category of tools, called OLAP (on-line analytical processing), is designed for rapid and easy exploration of different levels of aggregated data that are cross-referenced among themselves. This approach is intuitive and allows users to construct their analysis by clicking and navigating directly on the data (Yougworth 1995), in a way that is similar to the use of hyperlinks. The general architecture of an OLAP application is composed of a multidimensional database, an OLAP server, and OLAP clients (different implementation strategies are described by Thomsen *et al.* (1999).

OLAP allows users to navigate within data without knowing query languages such as SQL. With simple mouse clicks, a user may "drill down" within different levels of data and "drill up" to a more general level of data. Several books and papers address the fundamental concepts of OLAP and their underlying multidimensional databases (e.g. Berson and Smith 1997, Gouarné 1999, Thomsen 2002, Vit *et al.* 2002). An OLAP user interface displays data into statistical charts and spreadsheet tables that are dynamic and that directly support OLAP operations such as the different types of drills. Although OLAP does not address spatial data, one can combine OLAP and GIS software to create SOLAP applications. Three combination approaches were described by Bédard *et al.* (2005): GIS-centric, OLAP-centric and hybrid.

We define SOLAP as a category of software that allows rapid and easy navigation within spatial databases and that offers many levels of information granularity, many themes, many epochs and many display modes synchronized or not: maps, tables and diagrams. SOLAP relies on the multidimensional paradigms and on an enriched data exploration process made available by explicit spatial references and maps. SOLAP adds dynamic navigation into spatial data, maps, and symbols on maps. SOLAP maps are created dynamically from multidimensional data combined in the spatial datacube. This differs from some visualization and multimedia tools where navigation operations are associated to a sequence of predefined maps stored on the server. In comparison to the latter and to some commercial OLAP with minimal graphic capabilities, SOLAP facilitates the update of cartographic data. In comparison to most geo-visualization solutions (e.g.

as presented in Dykes, MacEachren and Kraak 2005), SOLAP adds new visualization challenges in order to maintain a continuous train-of-thought in an interactive analysis process. These challenges have been described by Rivest *et al.* (2005).

SOLAP fills the "analysis gap" between spatial data and geographic knowledge discovery since response times remain below Newell's cognitive band of 10 seconds (Newell 1990) and therefore can support this process in an interactive manner at whatever level of abstraction. It has been demonstrated that it is possible to achieve such performance for large datasets (Marchand 2004). In spite of a short history, SOLAP already reaches a first level of maturity with its own concepts, technologies and applications. Commercial solutions have recently become available and have been implemented in Canada, the United-States, France, Portugal and other countries. The number of scientific papers and book chapters on the subject has also begun to multiply (for more details, see Bedard *et al.* 2005).

The datacube, or multidimensional database approach introduces new concepts such as "dimensions", "members", "granularity", "measures", "facts" and "datacubes". For example, in a digital library where we need to calculate the number of documents for different types (e.g. digital maps, raster images, photographs), different publication dates and different regions, the *dimensions* are the themes to be analyzed (i.e. "documents", "time", "territory"). Each dimension contains *members* (e.g. "map", "2001", "Quebec City") that are organized hierarchically into levels of *granularity*, or levels of details (e.g. "province" ->"county" -> "city" for the "territory" dimension). The members at one level (e.g. "municipality") can be aggregated to give the members of the next higher level (e.g. "county") within a dimension. Three categories of dimension can be defined: descriptive (thematic), temporal and spatial ("non-cartographic" in the case of a conventional OLAP tool, "cartographic" in the case of a SOLAP tool). The *measures* (e.g. the number of documents) are numerical values calculated from the cross-reference of all the dimensions, for every combination of members from all levels of abstraction. A measure can be considered as the dependent variable while members of the dimensions are the independent variables (e.g. the measure "number of document" depends on the members of each dimension, such as "2003" for "time", "map" for "document" and "Quebec City" for "territory"). Each unique combination of dimension members and measures represents a *fact* (e.g. "3000" documents of type "map" were produced in "2001" for "Quebec City"). Those facts are aggregated according to the hierarchies of each dimension to finally obtain the total number of documents of all types for all publication dates and the entire country. The resulting set of facts is called a *datacube* (or *hypercube* if there are more than 3 dimensions). An impor-

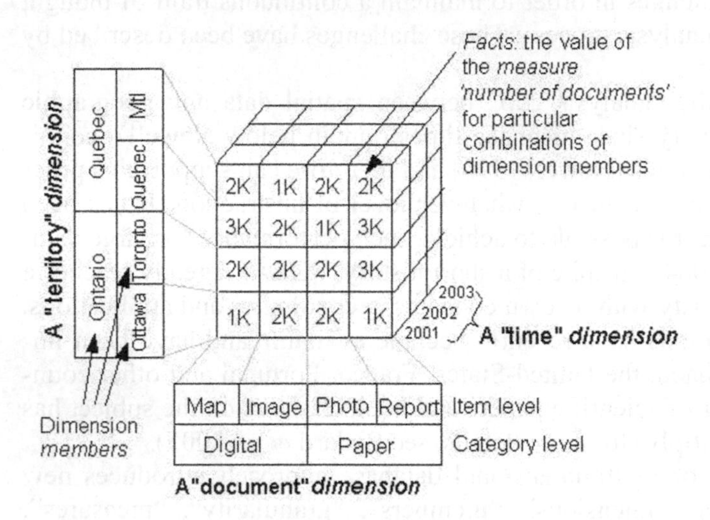

Fig. 10.1. A datacube showing the multidimensional database concepts

tant part of the datacube is typically pre-computed to increase query performance. This performance depends on the number of hierarchical levels for each dimension and on the number of members that compose the hypercube. Fig. 10.1 presents the multidimensional database concepts of the explained datacube.

This datacube can be described in UML (Unified Modeling Language, Blaha and Rumbaugh 2005) using packages for dimensions and cubes, and of classes for dimension levels and cube measures (Fig. 10.2).

Regarding spatial dimensions, three types can be defined: non-

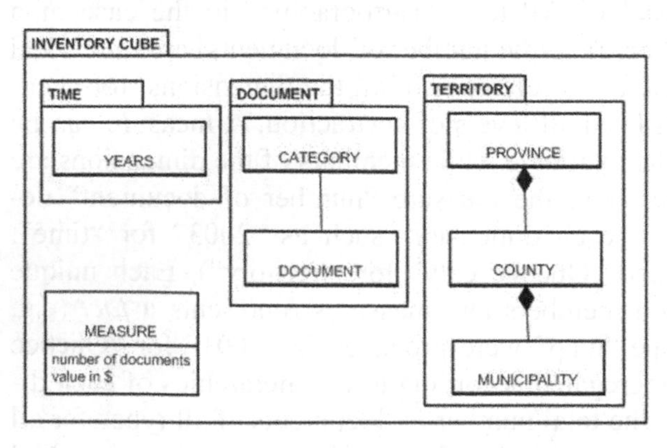

Fig. 10.2. A datacube expressed using the UML formalism where packages are cubes and dimensions and classes are dimension levels and measures

geometric, geometric, and mixed (Bédard *et al.* 2001). For the first type, spatial reference uses nominal data only (e.g. place names), i.e. no geometry or cartographic representation for the dimension members. It is the only type of spatial dimension supported by non-spatial OLAP and it is treated like other descriptive dimensions, leading to poorer spatio-temporal analysis and the non-discovery of certain correlations between phenomena that do not follow predefined boundaries. On the other hand, geometric spatial dimensions comprise, for every level of detail, the geometric shape(s) of every member (e.g. polygons to represent the boundaries of each municipality). This allows the dimension members (e.g. Quebec City) to be displayed and drilled on maps. The mixed spatial dimensions comprise geometric shapes for every member of only a subset of the levels of details. The members of the geometric and mixed spatial dimensions can be displayed on maps using visual variables (e.g. color) that relate to the values of the measures contained in the datacube. Fig. 10.3 shows an example of the three types of spatial dimensions expressed with spatially-extended UML.

The same classification can be made with the hypermedia aspect of a dimension. First, a non-hypermedia dimension exists when no dimension members are associated to hypermedia documents. Every SOLAP project seen so far fits in this category. Second, a dimension is said to be hypermedia if all dimension members, at all levels of detail, are associated with multimedia documents. The mixed hypermedia dimension is comprised of hypermedia documents for every member of a subset of the levels of the dimension. The hypermedia aspect can be applied to the highest level of aggregation only, to the finest level of detail only, or to intermediate levels. Furthermore, hypermedia documents can enrich non-spatial, mixed or

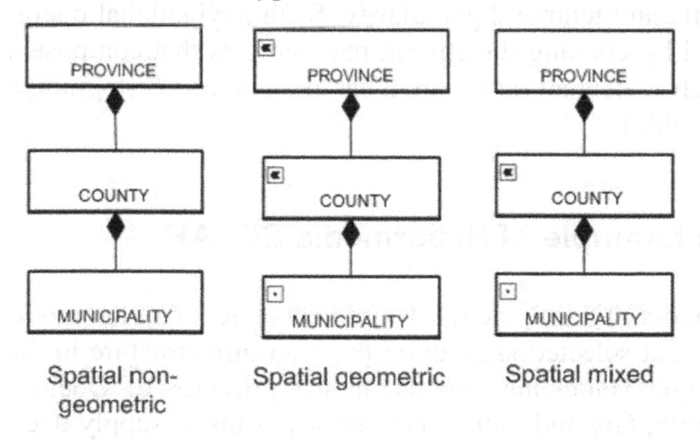

Fig. 10.3. Three types of spatial dimensions from left to right: non-geometric, geometric (2 polygonal levels, 1 punctual level) and mixed (1 polygonal level and 1 punctual level)

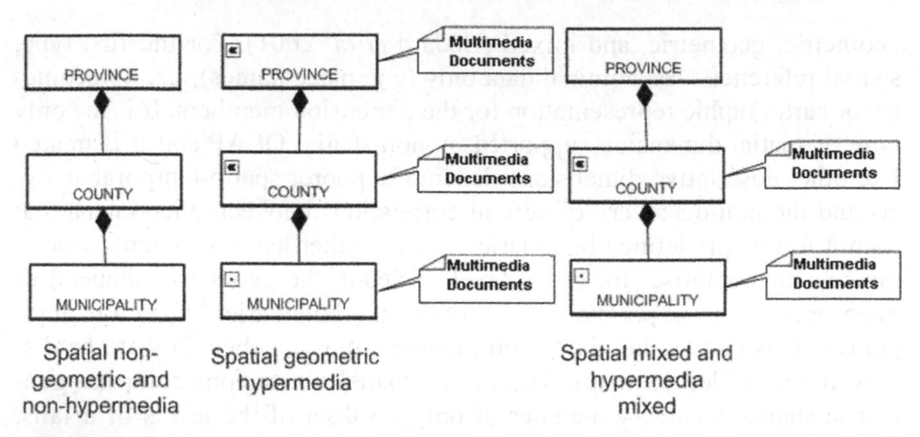

Fig. 10.4. Three different combinations of hypermedia and spatial aspects of a dimension

spatial dimensions. Fig. 10.4 illustrates an example of the three types of hypermedia dimensions, applied to a spatial dimension.

In a SOLAP user interface, navigation operators are similar to those used in OLAP but they take advantage of the spatial datacube structure. For example, two basic operators are drill-down and drill-up. Drill-down allows digging into lower levels of detail of a dimension and drill-up allows climbing into higher levels of detail. These operations are now available in the different types of displays, including maps, and can be specialized according to the type of dimension they manipulate (Rivest *et al.* 2003). One can drill on a dimension independently of the other dimensions. For example, a spatial drill allows the user to alter the level of granularity of a spatial dimension (e.g. territory) while maintaining the same level of thematic and temporal granularity. Such navigational operations can be executed by clicking directly on the elements that compose a display (e.g. a geometric element on a map, a bar element of a histogram, a cell of a spreadsheet table).

10.3 Application Example of Hypermedia SOLAP

To illustrate Hypermedia SOLAP, we show a project concerning the monitoring of erosion rates at selected sites along the road infrastructure in the Gaspésie and Îles-de-la-Madeleine regions, in the province of Quebec, Canada (Bilodeau 2005, Grelaud 2005). This project aims to supply decision-support tools to the Quebec Ministry of Transportation (MTQ) to help insure a better follow-up of erosion problems along roads in these regions.

The tool that was developed allows users to synthesize, to centralize, to aggregate, and to represent, in a simple way, all the pieces of information required to facilitate visualization, analysis and decision-making. In other words, the Quebec Ministry of Transportation (MTQ) faces erosion problems for road infrastructures in coastal areas and wants to improve the monitoring, control, and planning of the protection of coastal banks. Erosion is a natural process that has been progressing for millenniums as the coasts are subject to winds, waves, currents, tides, storms and to human activities. The MTQ identified 60 sites where the erosion of the coast is or will be a problem along highways 132 and 199. To solve these problems, the department builds infrastructures such as bridges, shoulders, and walls.

To support their work, the MTQ produces a large amount of documents (maps, plans, aerial photographs, LIDAR profiles, softcopy photogrammetry, research and technical reports, etc.) when inspections, measurement sessions, or research projects are conducted on the erosion sites. The first part of the project consisted in the inventory of all collections of documents related to erosion sites. To query this data inventory, a datacube was produced with 3 dimensions: time (production date), erosion site and document type. For this first datacube of the project, we had to allow easy visualization of the number, location and types of documents for each individual site or group of sites. In particular, we had to consider the cases where one specific document, for example a digital map of the coast of Bonaventure, is linked to more than one erosion site, leading to an aggregation formula that counts the documents only once at aggregated levels (i.e. municipality and county). This simple datacube is the example used to introduce hereafter the Hypermedia SOLAP. First, Fig. 10.5 presents the navigation that will be used in Figs. 10.6 to 10.9 and where all multimedia document types are linked at the lowest level of detail.

For this datacube, the user interface starts with two panels. First, in the left panel, the user selects the dimensions and measures from the upper section to drag and drop them into the lower section boxes called "columns" and "rows". This allows the display of the selected dimensions and measures (e.g. "type of documents" per "region") using only select, drag and drop (no SQL-type query to select the desired elements). Then, for the right panel (the display panel), the user selects from the horizontal toolbar the desired display type (maps, tables or graphics). Now, the user can "drill down" or "drill up" into the different levels of details of the selected dimensions using the navigation tools of the horizontal toolbar. For example, it is possible to create a map presenting the "number of documents" by "document type" for "counties" (general level of information) simply by dragging the measure and two dimensions into columns and rows and by clicking on the map display button (3 drags, 1 click, 3 seconds to produce

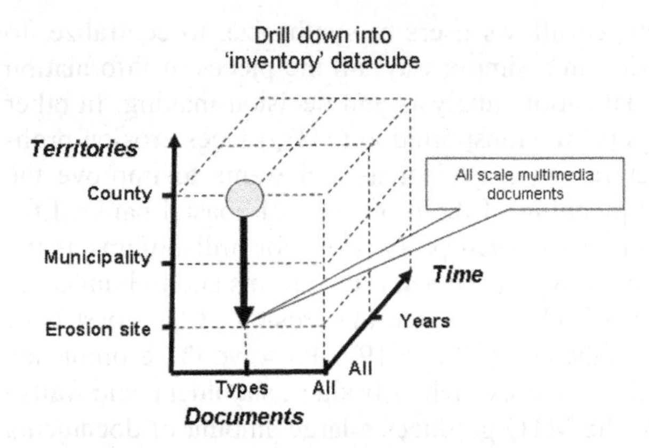

Fig. 10.5. Navigation into the territory dimension where all multimedia document types are linked to the lowest level of detail

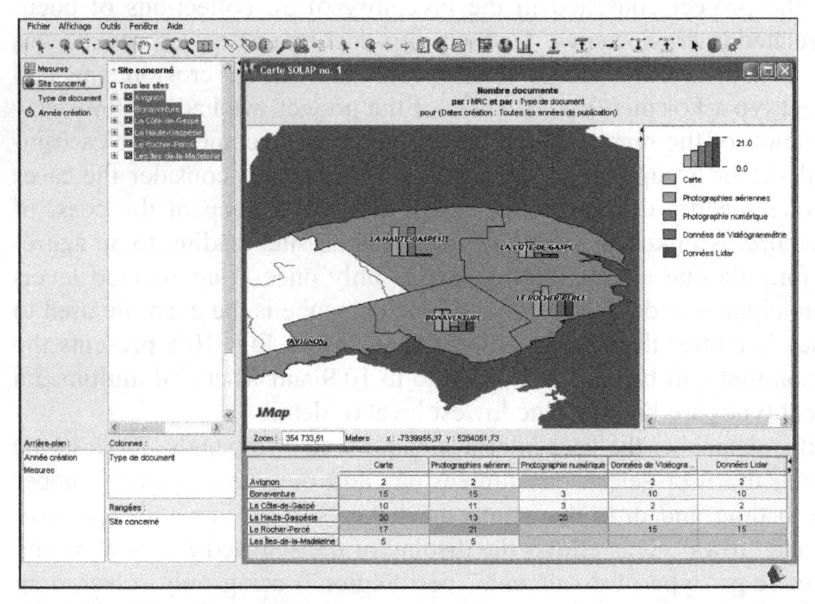

Fig. 10.6. Map and table displays presenting the "number of documents" for some "document types" and "county"

the map). Different displays for the same information are just one click away (e.g. table, histogram, pie chart) and they can be synchronized or not according to the demands of users. For example, Fig. 10.6 shows the number of documents, for several document types, per county (in both map and table).

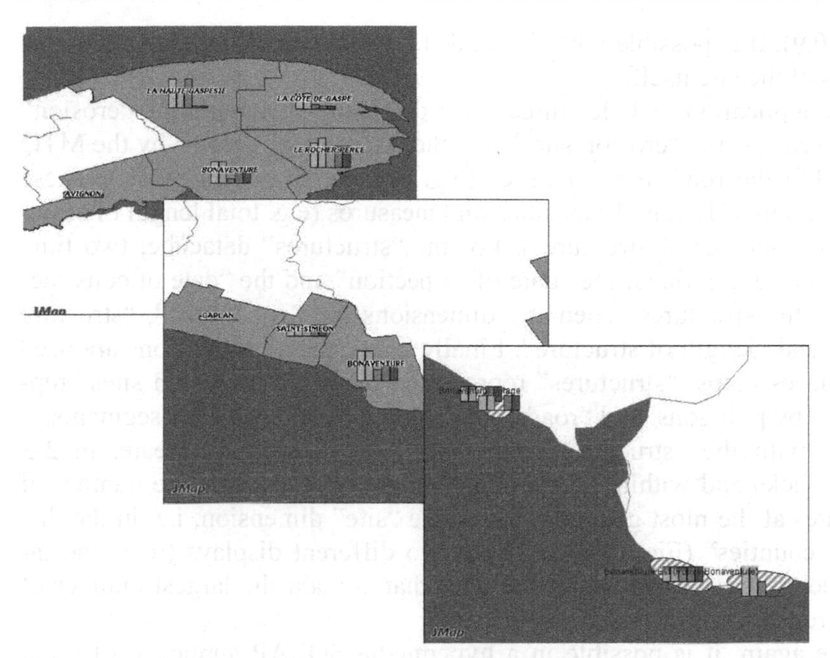

Fig. 10.7. Number of document by "county" (first map), by "municipality" (second map) and by "erosion site" (third map)

The advantage of SOLAP truly starts here, i.e. when one wants to explore the data from the first display. To navigate within this datacube in order to visualize the "number of documents" by "erosion site" (more detailed level of information), the user can simply navigate interactively at the various levels of details of the data. For example, to get information about a particular erosion site, the user can drill down spatially through the spatial dimension, i.e. from "county", to "municipality" and then to "erosion site". With 2-3 mouse clicks, users navigate in those map displays, tables or charts within seconds (Fig. 10.7).

In a SOLAP tool, different display types (map and table in this example) can be synchronized together to always reflect the actual analysis results. Thus, if the user desires such behaviour, a manipulation (drill down or drill up, for example) in one of these displays propagates the navigation operation in all active displays (Fig. 10.8).

Therefore, when users identify a site of interest (e.g. Bonaventure Est) in a natural way that fits with hypermedia navigation (i.e. simply clicking on the item of interest, no SQL-type query language or cryptic database structure), they can continue working this way and use the multimedia tool in the horizontal toolbar. They can then open the form related to the element of interest where hyperlinks provide access to multimedia documents

(Fig. 10.9). It is possible with this added information to better evaluate the erosion of the site itself.

This application includes three other datacubes managing the "erosion" characteristics, the "erosion sites" and the "structures" erected by the MTQ to solidify the road infrastructures (e.g. ripraps, shoulders, walls). These cubes contain different dimensions and measures (e.g. total length of structures and number of structures). For the "structures" datacube, two time dimensions are defined: the "date of inspection" and the "date of construction" of the structures. Thematic dimensions are "coast type", "structure status" and "length of structure". Finally three spatial dimensions are used to produces maps: "structures" represented by points, "erosion sites" represented by polygons, and "roads" represented by lines of road segments.

Now with the "structures" datacube, it is possible to create, in 2-3 mouse clicks and within 3-5 seconds, a map and a table of the number of structures at the most general level of the "site" dimension, i.e. in the different "counties" (Fig. 10.10). These two different displays (map and table) allow the user to identify the sites that contain the largest number of structures.

Once again, it is possible in a hypermedia SOLAP application to link multimedia documents to all levels of detail of information. Fig. 10.12 presents the navigation example that will be used in Figs. 10.13 to 10.15 and where different document are linked to different levels of detail.

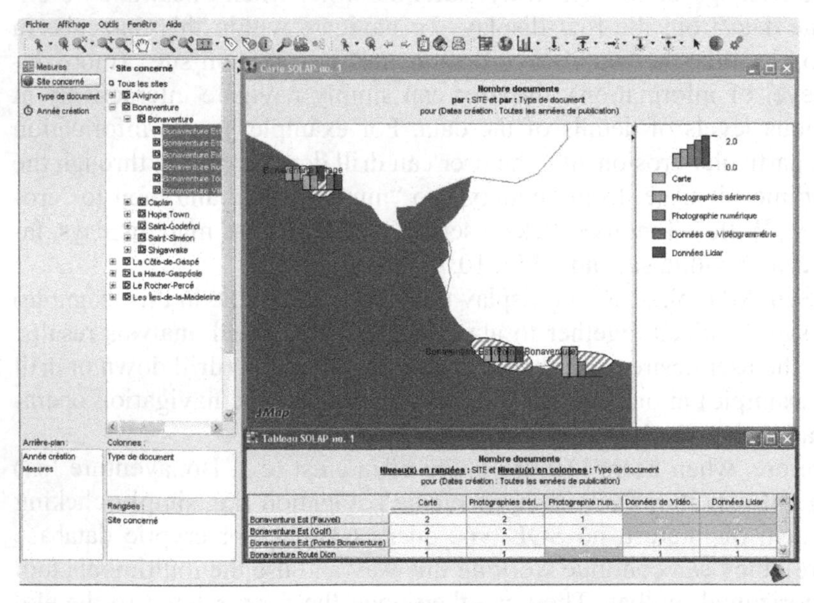

	Carte	Photographies aéri...	Photographie num...	Données de Vidéo...	Données Lidar
Bonaventure Est (Fauvel)	2	2	1		
Bonaventure Est (Golf)	2	2	1		
Bonaventure Est (Pointe Bonaventure)	1	1	1	1	1
Bonaventure Route Dion	1	1		1	1

Fig. 10.8. Erosion sites map and table where documents are counted by "document type"

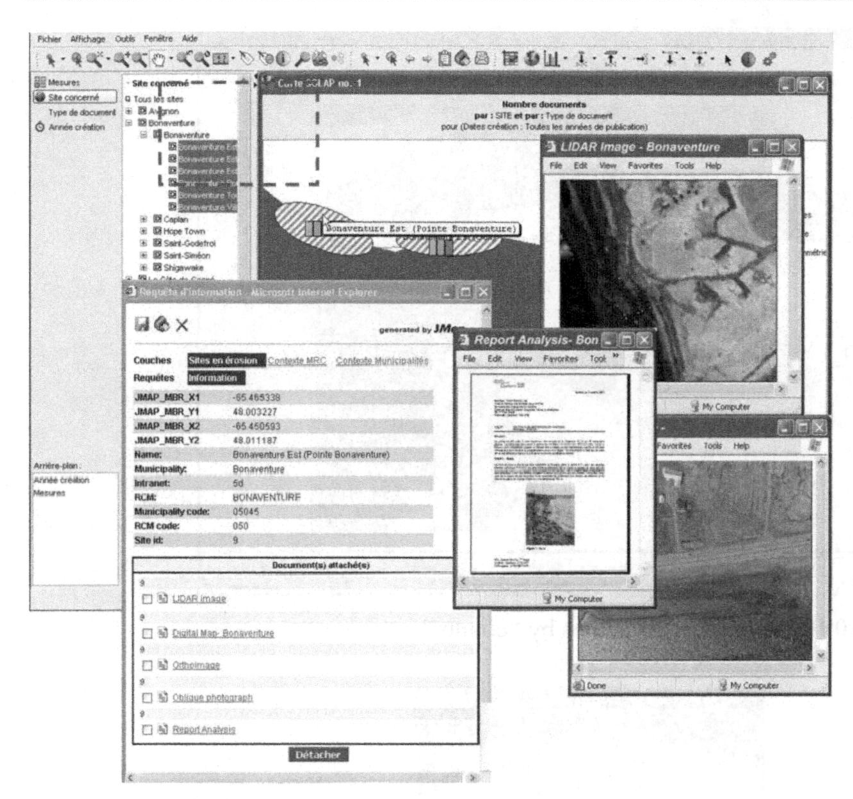

Fig. 10.9. Access to multimedia documents (e.g. report, digital map, LIDAR data and Orthoimage) of the Bonaventure "erosion site"[1]

As explained before, from this point, the user can drill down spatially to see more details through the spatial dimension (Fig. 10.11).

At the municipal level (second level of detail), it is possible to visualize multimedia documents linked to a municipality by opening the municipal "form"; here the user finds a list of hyperlinks to the municipal multimedia documents. At this level of detail, only documents that show the municipality level of information are linked, i.e. small-scale digital map. For example, for Bonaventure, there exists a digital map of the coast that localizes all the structures built on this littoral (Fig. 10.13). In our application, documents exist for the three most detailed levels ("municipality", "erosion site" and "structure"). Eventually, it would be interesting to find more general documents to link to the county level (e.g. very small-scale remote sensing images).

[1] For this figure and the following ones, SOLAP environment and toolbars were removed to increase picture size and improve readability.

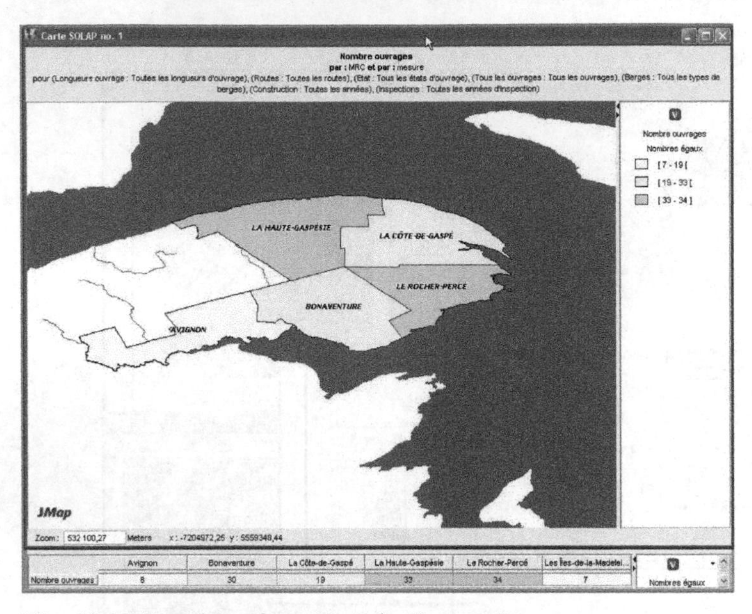

Fig. 10.10. Number of structures by "county"

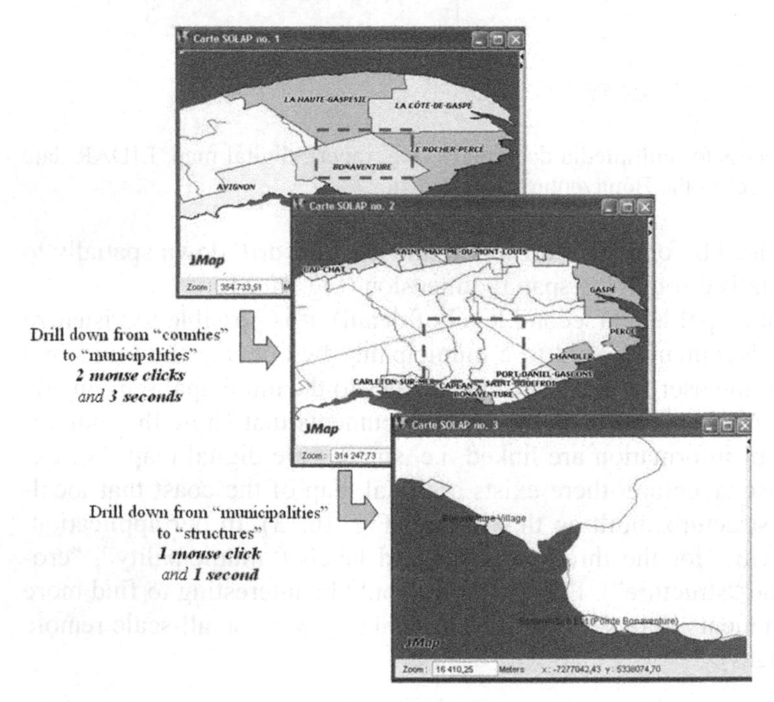

Fig. 10.11. Number of structures by "county" (first map), by "municipality" (second map) and by "erosion site" (third map)

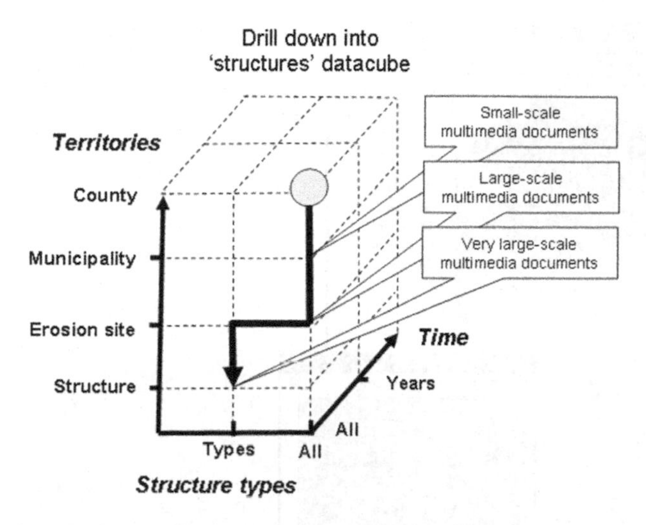

Fig. 10.12. Navigation into the territory and structure types dimensions where different document scales are linked to different levels of detail

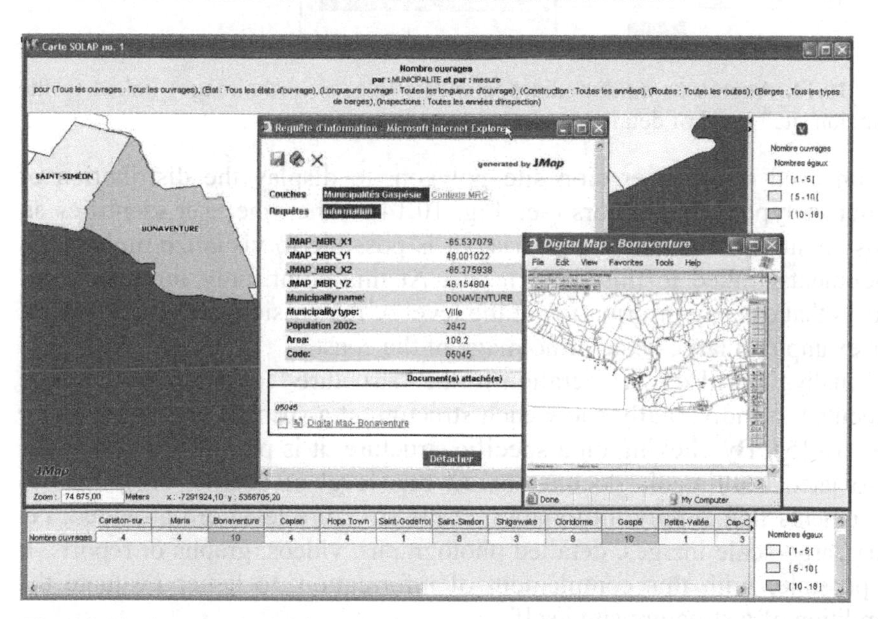

Fig. 10.13. Access to multimedia documents (e.g. a digital map) liked to the "municipality" level of detail

The user can then drill down "spatially" to produce a map of "erosion sites". In this view, the user can drill down "thematically" in the structure dimension (in the left panel) to show, on the map display, the number of structures per "structure type". In this case, the tool uses histograms, su-

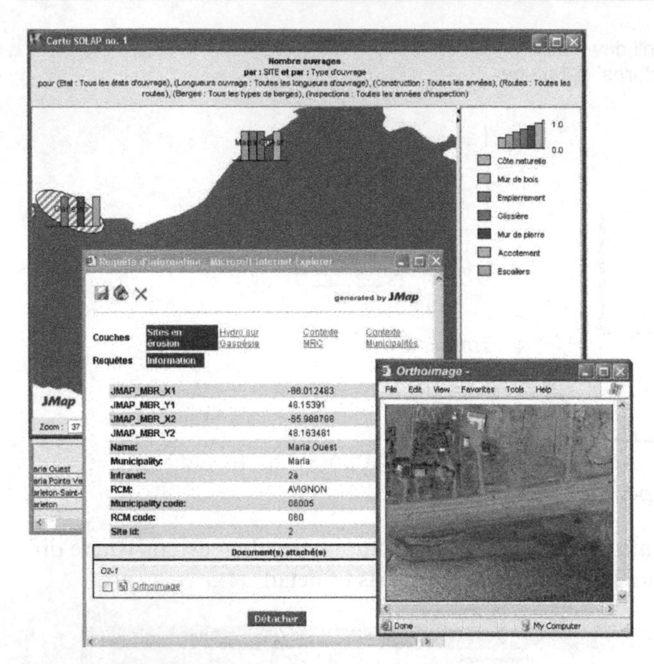

Fig. 10.14. Access to multimedia documents (e.g. an orthoimage) linked to the "erosion site" level of detail

perimposed on each erosion site polygon, to display the distribution of structure types using colors (see Fig. 10.14). When the user identifies an erosion site of interest (e.g. Maria), it is possible to visualize multimedia documents linked to this erosion site. At this point, only links to documents that display information at the level of the erosion site are provided, for example, a large-scale orthoimage of the coast.

Finally, a drill down operation to the "structure" level of detail can be executed to individually show each structure at a specific erosion site (see Fig. 10.15). By clicking on a specific structure, it is possible to access the associated multimedia documents. At this level of details, only links to documents that supply information at the structure level are provided, i.e. very large-scale images, detailed photographs, videos, graphs or reports. It is possible, with this complement of information, to better evaluate the condition of a structure site itself.

10.4 Evolution of SOLAP Hypermedia

Future work in this field will focus on the definition of location based (web) services, and the design of Service Oriented Architectures (SOA) to

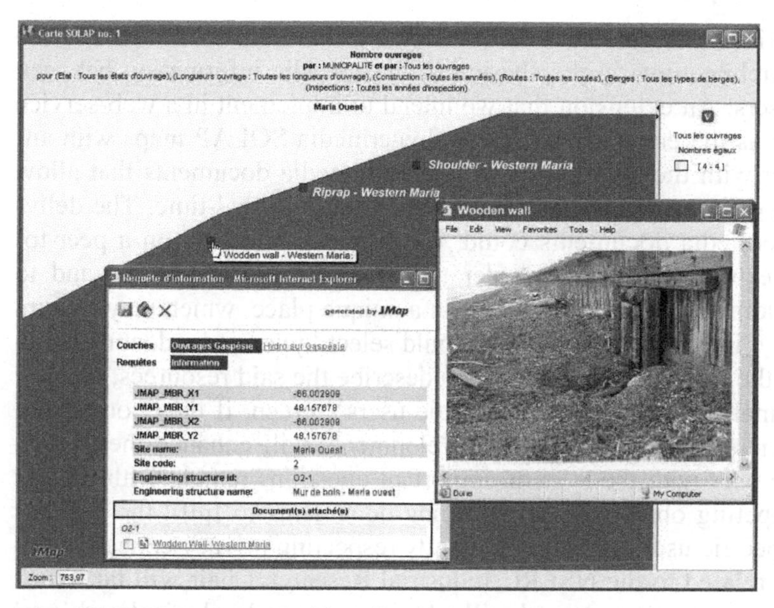

Fig. 10.15. Access to multimedia documents (e.g. a photograph) liked to the "structure" level of detail

support these services, in order to enable the real time delivery of contextual hypermedia information to the end user. By contextual we mean that the information provided depends on the "context of the user". For a mobile user, who accesses the system through a wireless network, on a device like a PDA or a smartphone, the context could be defined by his location and his surrounding. A context aware hypermedia SOLAP application (or service) could thus enhance the map and all the information already provided to the user through the real time delivery of hypermedia information in relation with the location of the user. Such services could be of great importance in emergency or crisis situations. For instance, different groups of firemen at the same operating theatre could have real-time access to specific information (e.g. reports of difficulties encountered by other groups; the position and the movements of other units, video images stemming from cameras in the surroundings or filmed by other units, detailed maps of a specific building sent from headquarters, etc.). This could significantly influence the way they will perform their work. In addition, they could cross-reference data on-the-fly and they could obtain global pictures on-demand through the aggregation of information, with the simplicity and speed of a SOLAP user interface. In this example, not only the absolute location of a specific unit, but also the relative positions of the different groups, have an impact on the information delivered to users.

The previous example also illustrates the possible role of such Service Oriented Architectures in the sharing of hypermedia information between different users. An extension that we intend to implement in a web-service architecture is the enhancement of the hypermedia SOLAP maps with annotations or with the addition of ad hoc multimedia documents that allow the sharing of information between different users in real-time. The delivery of hypermedia documents could, for instance, be based on a peer-to-peer architecture approach in order to increase the transfer rate and to avoid the storage of shared resources in a unique place, which may require large storage infrastructures. A user could select hypermedia documents to display on the basis of metadata which describe the said resources, or, they could automatically be displayed on the users' screen, if they correspond to the "user profile". While such developments will enhance the experience of the user with the system, important questions remain with respect to the competing objectives of providing documents to fulfil the requirements of specific users and concomitantly respecting privacy concerns. Future works related to the NSERC Industrial Research Chair will take such considerations into account and will address some technological and legal issues raised by the distribution and the sharing of such Hypermedia and SOLAP information over networks. As we have designed and developed the SOLAP technology used for the project presented in this chapter (now commercially available as JMap Spatial OLAP Extension, see Kheops 2005), we will continue improving this technology, including its enrichment with other types of data such as hypermedia.

10.5 Conclusion

Although it was beyond the goal of this chapter to present in details the new technology called SOLAP, we have provided an overview of this technology and attempted to demonstrate, using examples, that SOLAP technology naturally fits with hypermedia GIS. The "document inventory" datacube shows interesting ways to query inventory results such as statistics on documents types, areas covered, publication dates and so on. The "structures" datacube shows interesting ways to get statistical data at the appropriate level of analysis along with the corresponding hypermedia documents. The management of multimedia documents into a spatial datacube is innovative. Such Hypermedia SOLAP uniquely combines explicit spatio-temporal aggregated data (to get the global views) with statistics resulting from the cross-referencing of all dimensions at all levels of analysis. Furthermore, Hypermedia SOLAP also combines the above with implicit information from hypermedia documents at the corresponding levels of analysis (to get more insights from raw information). In addition, the inherent capability of SOLAP to support a non-linear approach for navigating within databases is also a novel way to provide access to hypermedia documents at different levels of detail. As we have seen, it is possible, within a Hypermedia SOLAP datacube, to organize documents by spatial level of detail, i.e. by the region of interest shown or studied by the document. However, other classification alternatives could have been implemented, such as creating groups of documents by level of detail (e.g. digital vs non digital document types) or groups of epochs to produce temporal levels of detail (e.g. publication epochs 1995-99, 2000-04). Also, we have illustrated that users can access multimedia documents by clicking on the geometric elements shown on a map, as is done with Hypermedia GIS; however, all the other types of SOLAP displays can also provide access to hypermedia documents (e.g. by clicking into a table cell or a histogram bar). Finally, we have described the potential evolution of Hypermedia SOLAP from the traditional architecture to a web service architecture.

Aknowledgements

We acknowledge the financial support of Canada NSERC Industrial Research Chair in Geospatial Databases for Decision-Support and its partners (http://mdspatialdb.chair.scg.ulaval.ca/). We also acknowledge the collaboration of the Ministère des Transports du Québec for the application dataset and the contribution of undergraduate students who worked on the development of the application.

References

Bédard Y, Merrett T, Han J (2001) Fundamentals of spatial data warehousing for geographic knowledge discovery. In: Miller H, Han J (eds) Geographic Data Mining and Knowledge Discovery, Taylor & Francis., London, pp 53-73

Bédard Y, Gosselin P, Rivest S, Proulx M-J, Nadeau M, Lebel G, Gagnon M-F (2002) Integrating GIS components with knowledge discovery technology for environmental health decision support. International Journal of Medical Informatics, vol 70(1), pp 79-94

Bédard Y, Rivest S, and Proulx M-J (2005) Spatial on-line analytical processing (SOLAP): concepts, architectures and solutions from a geomatics engineering perspective. In: Wrembel R, Koncillia C (eds) Data warehouses and OLAP: concepts, architectures and solutions. Accepted for publication

Bédard Y, Proulx M-J, Rivest S (2005) Enrichissement du OLAP pour l'analyse géographique : exemples de réalisation et différentes possibilités technologiques. Revue des Nouvelles Technologies de l'Information - Entrepôts de données et l'Analyse en ligne, Cépaduès-Éditions, France, pp 1-20

Berson A, Smith SJ (1997) Data warehousing, data mining and OLAP. McGraw-Hill, pp 612

Bill R (1994) Multimedia GIS - definition, requirements, and applications. In: Shand PJ, Ireland PJ (eds) European GIS Yearbook. NCC Blackwell and Hastings Hilton Publishers Ltd, Oxford, pp 151-154

Bilodeau F (2005) Érosion des berges en Gaspésie. Research report, Centre for Research in Geomatics, Laval University, pp 44

Blaha M, Rumbaugh J (2005) Object-oriented modeling and design with UML. 2nd Edition, Pearson Prentice-Hall, pp 477

Buzan T, Buzan B (2003) Mind map, dessine-moi l'intelligence. 2e édition, Éditions de l'Organisation, Paris, pp 328

Dykes J, MacEachren AM, Kraak MJ (2005) Exploring geovisualization, Elsevier, Amsterdam, pp 730

Fortin C, Rousseau R (1989) Psychologie cognitive: une approche de traitement de l'information, Presses de l'Université du Québec, pp 434

Gouarné JM (1999) Le projet décisionnel - enjeux, modèles et architectures du data warehouse. 3e édition, Eyrolles, pp 240

Grelaud B (2005) Développement d'un outil d'aide à la décision pour la réfection et l'entretien des routes érodées sur les côtes de la Gaspésie et des Îles-de-la-Madeleine. Research report, Université Paris 1, pp 91

Hu S (2004) Design issues associated with discrete and distributed hypermedia GIS. GIScience & Remote Sensing, vol 41(4), pp 371-383

KHEOPS (2005) JMap spatial OLAP. http://www.kheops-tech.com/en/jmap/solap.jsp (visited 10/15/05)

Longley PA, Goodchild MF, Maguire DJ, Rhind D (2001) Geographic information systems and science, Wiley, pp 472

Naimark M (1997) A 3D moviemap and a 3D panorama. SPIE Proceedings, vol 3012, San Josc

Newell A (1990) Unified theories of cognition. Harvard University Press, Cambridge, pp 576

Marchand P (2004) The spatio-temporal topological operator dimension, a hyperstructure for multidimensional spatio-temporal exploration and analysis, Unpublished doctoral dissertation, Laval University, Canada, pp 126

Rivest S, Bédard Y, Proulx M-J, Nadeau M (2003) SOLAP: a new type of user interface to support spatio-temporal multidimensional data exploration and analysis. In: Proceedings of the ISPRS Joint Workshop of WG II/5, II/6, IV/1 and IV/2 on Spatial, Temporal and Multi-Dimensional Data Modelling and Analysis, Quebec, Canada

Rivest S, Bédard Y, Proulx M-J, Nadeau M, Hubert F, Pastor J (2005) SOLAP: merging business intelligence with geospatial technology for interactive spatio-temporal exploration and analysis of data. In: Peuquet D, Laurini R (eds)

Journal of International Society for Photogrammetry and Remote Sensing (ISPRS), Advances in spatio-temporal analysis and representation, In press
Thomsen E, Spofford G, Chase D (1999) Microsoft OLAP solutions. John Wiley & Sons, pp 495
Thomsen E (2002) OLAP Solutions. 2nd Edition, John Wiley & Sons, pp 661
Standing L (1973) Learning 10000 pictures. Quaterly Journal of Experimental Psychology, vol 25, pp 207-222
Vaitis M, Tzagarakis M (2005) Towards the development of open cartographic hypermedia systems. In: Proceedings of the 1st Int. Workshop on Geographic Hypermedia: Concepts and Systems (GEO-Hypermedia'05), Denver, USA.
Vitt E, Luckevich M, Misner S (2002) Business intelligence, making better decisions faster. Microsoft Press, pp 202
Wikipedia (2005) Aspen movie map. http://www.wikipedia.org/ (visited 10/10/05)
Yougworth P (1995) OLAP Spells Success for Users and Developers. Data Based Advisor, pp 38-49

11 A Hypermedia Afghan Sites and Monuments Database

Ralf Klamma, Marc Spaniol, Matthias Jarke, Yiwei Cao, Michael Jansen, and Georgios Toubekis

Abstract. Cultural heritage management is an excellent application domain for geographical hypermedia information systems. Many people with different tasks and levels of profession like fieldworkers, researchers, project and campaign officers, cultural bureaucrats etc. collaboratively producing and consuming different media like photographs, video, drawings, books, etc. must deal with exact geographic information about moveable or unmovable objects of interest. Implemented information systems must obey all standards in the different domains to overcome classical failures of isolated solutions which do not scale beyond the scope of a single project. We present a conceptual approach which integrates geographic information, multimedia information, cultural heritage information and collaborative aspects in a single information model. This conceptual approach was used to design and implement a web-based information system on top of a single commercial database covering all mentioned aspects. This information system was deployed for a project in the conservation of cultural heritage in Afghanistan to prove the validity of the concepts.

11.1 Introduction

Due to coverage by global media people all over the world have become heightened aware of disasters taking place even at remote spots around the globe. This reaches from natural catastrophes like the Asian Tsunami and draught catastrophes in Africa to (civil) wars as in Afghanistan. Similarly, an increased social responsibility can be recognized by e.g. big worldwide spending campaigns. However, these actions are very often of short term effect only. The reason is clear. The shift in the focus of global media after the strike of a disaster commonly leads to decline in perception and thus in a decreasing support for the affected. In this aspect, this chapter proposes a trust building community approach that provides an option to users in whatever a country all over the world for a sustainable relief work by means of information systems. The aim is to provide communities with a cheap, long-lasting, and flexible mobile environment of geographic information system with hypermedia. It allows them to build up the disaster struck area more or less in a self-organizing way. Additionally, networked experts from all over the world may contribute to the overall process without requiring them to be physically present on site.

11.1.1 Motivations and Problems

Afghanistan is a country with a long history and rich cultural heritage. Cultural heritage worldwide faces damages resulted from the nature and human. This problem is especially severe in Afghanistan during the civil war and Taliban regime in the past 20 years. Since Afghanistan was on the way to democracy, many organizations around the world have made great effort to make up for the break of cultural heritage management there. Under the appeals and guidance of UNESCO, the International Council on Monuments and Sites (ICOMOS) Germany[1] cooperates with Department of Urban History (Prof. Dr. Jansen), RWTH Aachen University to recover the cultural heritage.

Cultural heritage management includes documentation, evaluation of conservation measures and execution of measures etc. How to preserve the cultural heritage effectively with the modern technologies is a central question for the archaeologists, historians, architects, and computer scientists. Therefore, along with the practical conservation work, the Department of Urban History had developed an MS Access-based database application for documentation. So far, the system has been used in the department. However, the limitations of MS Access, which is used as a personal database management system, cause some problems during the use:

1. It doesn't support multiple users in a network environment.
2. It manages the spatial information of a site or monument in a usual relational database. So it doesn't support the special spatial aspects.
3. It stores multimedia data in the file system, independent of the MS Access database.

To solve these problems, Department of Urban History cooperates with Department of Information Systems and Database Technology (Prof. Dr. Jarke), RWTH Aachen University to find a solution by means of modern information technologies.

[1] see http://www.icomos.org/germany/german.html

11.1.2 Scenarios

The revised application could be applied in various scenarios. First, a great amount of information including spatial information, text documents, pictures, and audio-visual data need to be managed in databases for information search, retrieval and exchange. Next, many researchers and scientists working in this field are resided all over the world. They need a channel to communicate and cooperate among them. This channel also enables an intergenerational cooperation, because no cultural heritage management work was done over two decades on site during the civil war in Afghanistan. There is a gap between the experiences accumulated by the prior generations and those collected newly. Finally, the information can be shared by various organizations and individuals in the fields of tourism, museums, and e-learning etc., when there are channels to bridge those diverse user communities.

As illustrated in Fig. 11.1, different kinds of fieldwork are carried out by diverse communities in Afghanistan. Researchers rehabilitate a quarter with several mosques and other historic buildings destroyed by the war. They guard the historic site to prevent more damages. They look for several artifacts missing from the national museum in wars. They take photos, record videos and draw sketches of monuments and sites. They position the historic buildings and monuments by using GPS handsets. And they record all related information in documents as research material. There is a scenario especially worth mentioning: due to the two decades' break-up, many former researchers and fieldworkers look forward to possibilities to impart their experiences and skills to the young generation.

Meanwhile, researchers and administrative employees worldwide manage and study the information collected in fieldtrips. They input the collected information into a central database. They exchange the information from various sources. They browse and download the available information in internet. They use the information for their presentations, research papers and books. Moreover, all the communities communicate and cooperate with each other.

11.2 Concepts of ACIS and Requirements Analysis

With regard to the existed problems in the MS Access-based application and the analysis of use scenarios, a new community information system *ACIS* has been proposed which stands for *A*fghan *C*ommunity *I*nformation *S*ystem for Cultural Heritage Management.

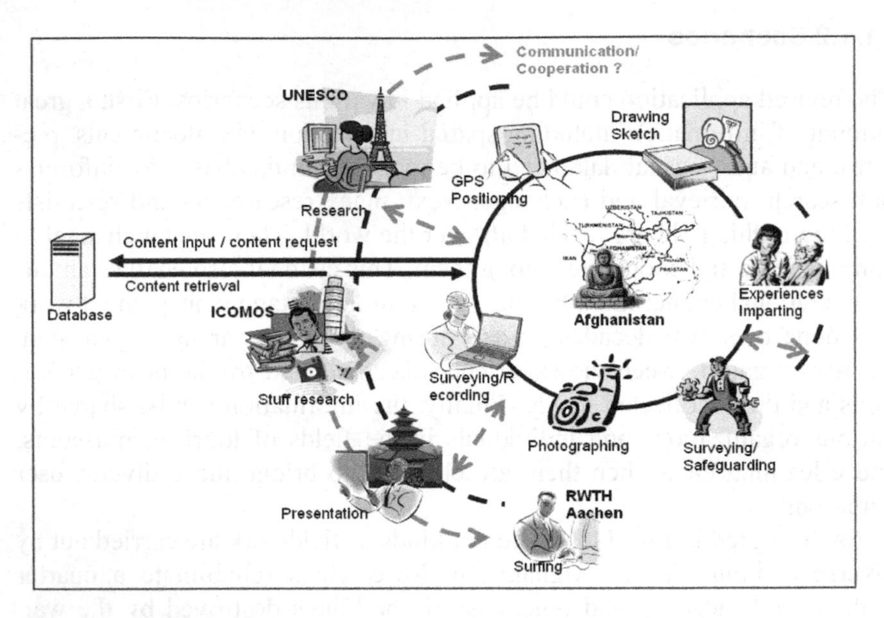

Fig. 11.1. Use Scenarios

11.2.1 Basic Concepts

The following concepts have been proposed to realize ACIS, namely a hypermedia-based geographic community information system for cultural heritage management.

- Web Community
  The concept of a *Community of Practice* could provide diverse user communities communication channels for intra-generational and inter-generational, as well as intra- and interdisciplinary cooperation. The potential users come from three sectors: *Government and administration sector* such as members of UNESCO, *Research sector* such as students and lecturers of different majors, and *Preservation sector* such as engineers and scientists in the cultural heritage conservation field. All user communities can cooperate together well in a web community environment.

- Geographic Information System (GIS)
  A great amount of information stored in the database pertains to sites and monuments in Afghanistan. Each site or monument has its geographic location information. Textual information alone is not able to represent the spatial information properly and efficiently. Thus, GIS

technologies including cartography and spatial queries processing are the central concept of ACIS.

- Hypermedia with multimedia standards
  A great number of photos and audiovisual files are research stuff as important as the textual information for researchers in this field. They can belong to cultural objects as sites or monuments. They might also be multimedia files that record a campaign in detail. It has confused researchers for a long time, how to manage and search and retrieve the multimedia information efficiently. Hence, the modern multimedia database technology and metadata standards such as MPEG-7 (ISO 2003) could be the solution to the enhancement of multimedia information retrieval and exchange. Moreover, the multimedia can be well embedded in a hypertext link and become so-called hypermedia in a web network environment.

- Cultural heritage management
  The cultural heritage object should be represented and described in detail precisely for easy management. Several eminent metadata standards in this field such as standards defined by Getty Institute (Getty 2000) are keys to describe the cultural heritage objects properly.

These four sub concepts together potentially facilitate a community information system to realize the various use scenarios and to meet the requirements that will be discussed in the next section.

11.2.2 Requirements Analysis

After workshops with architects from Department of Urban History at RWTH Aachen University and art historians from Seminar of Oriental Art History at Bonn University, the requirements of ACIS concerned with the aforementioned concepts respectively can be concluded as follows:

In the aspect of *community*, the input user interface should be as simple as possible. Multi-language-interface and multi-user-interface are supported for users in different countries and for users working in different disciplines. Users can communicate with each other via email and forum service. And the community activities are warranted by users' rights management. In the aspect of *GIS*, certain search catalogues should be defined for sites and monuments. The query results can also be displayed in maps which are generated dynamically. Graphic spatial query tools need to be developed to support users' interactive queries on the maps. Spatial data should be input into the database with simple mechanism. In the aspect of *hypermedia*, suitable metadata standards will be used to enhance multimedia information search and retrieval. In the aspect of *cultural heritage management*, thesaurus mediation service could be launched to enhance the interoperability among users who work in different disciplines and use different terminologies.

In addition, offline work should be synchronized into the central database, in case that internet is not available on site. The system complexity should be possibly lowered. The system should keep its extensibility and openness for further development.

Certainly, it is hard to fulfill all the requirements in the prototype of ACIS. Therefore, they are implemented according to certain priorities. The functionalities can be easily extended in such an open system in future.

11.3 Related Work and Theories

An in-depth survey of the state-of-the-art technologies related to the proposed concepts has been made and some results are listed as follows.

11.3.1 GIS Technologies

A widespread definition of GIS is provided by the National Centre of Geographic Information and Analysis in NCGIA (1990):

> *A GIS is a system of hardware, software and procedures to facilitate the management, manipulation, analysis, modeling, representation and display of geo-referenced data to solve complex problems regarding planning and management of resources.*

Although GIS has been developed for decades, its progress has not been so eminently accelerated like recent years due to the rapid development of hardware. Moreover, the development of GIS benefits from the Geographic Positioning System (GPS) technology, which as a fundamental data source provider for GIS can measure geographic information precisely. Certainly, there are still many factors that influence the accuracy of location calculation mentioned in Jones and Jones (2004). However, GPS has been contributing to the development of GIS greatly.

Furthermore, many GIS products and services have been developed to process geographic data and to provide user interfaces. They can be mainly grouped in two categories: backend geographic database technology and middle tier web services that process geospatial data from the database and present them to clients through networks. The most widespread geospatial

Table 11.1. A Comparison of GIS Web Servers

Standards	Map Image Format	Implementation	Geo-database	Comments
ArcGIS Server	SVG, GIF, PNG, JPEG etc.	--	Oracle, PostGIS, ArcSDE etc.	Substantial, well developed
Oracle MapViewer	GIF, PNG, JPEG	Java API	Oracle, ArcSDE	A new technology for Oracle Spatial
UMN MapServer	TIFF, GIF, PNG, ERDAS, JPEG, EPPL7	Support Python, Perl to access C API	Oracle, Sybase etc. through Perl DBI module	Open source providing out-of-the-box services
GeoServer	SVG, TIFF, GIF, PNG, JPEG etc.	Using GeoTools[1] Java API	Oracle, ArcSDE, PostGIS	Open source in development
GRASS	PPM, PostScript, JPEG, GIF etc.	C API	Oracle, ArcSDE, PostGIS, etc.	Well developed open source; widely used
Deegree	TIFF, JPEG, GIF, BMP	Java API	Oracle, ArcSDE, PostGIS, etc.	Well developed open source

data infrastructures include ESRI GIS Infrastructure (ESRI 2002), Oracle Spatial technologies (Oracle 2002), PostGIS (Momjian 2001), UMN MapServer[2], the Geoserver Project[3], GRASS GIS[4] and Deegree (Fitzke *et al.* 2003). A comparison of the middle tier web servers is listed in Table 11.1.

At the same time, the most significant open geospatial standards are implemented by OGC[5]. Its core metadata describes dataset title, spatial representation type, geographic location of the dataset etc. (OGC 2000) and (OGC 2003). Metadata for GIS can be categorized in two groups: metadata to describe a geometric object for visualization of a single object, and metadata to represent a map for map rendering with various geometric objects.

[2] See http://mapserver.gis.umn.edu/

[3] See http://geoserver.sourceforge.net/

[4] See http://grass.itc.it/

[5] Open Geospatial Consorsium: http://www.opengeospatial.org/

11.3.2 Hypermedia with Multimedia Standards

Multimedia is important research material in cultural heritage management. Before, research material has been kept in form of books, films, photos, drawings and so forth. Nowadays, more and more multimedia in material has been digitalized and managed with modern hardware and software technologies. In the meantime, the management, search and retrieval of a great amount of multimedia information become key problems for the researchers. So a lot of multimedia standards have been launched to try to solve these problems.

Four spontaneous questions are brought together in the development of multimedia technologies. They also imply a common model to cope with multimedia information (Fig. 11.2).

1. In which form will multimedia be coded, so that the multimedia can be readable by various multimedia processors?
2. How can the multimedia content be interacted by users?
3. How can the multimedia content be searched and retrieved efficiently?
4. How can the multimedia content be delivered, accessed and consumed?

To answer the first and the second questions, many multimedia encoding technologies have been developed such as Real-Time Protocol (RTP) and Real-Time Control Protocol (RTCP) & Resource Reservation Protocol (RSVP) by the Internet Engineering Task Force (IETF)[6], QuickTime by Apple Computer, Inc., the Resource Interchange File Format (RIFF)[7] by Microsoft and IBM, and MPEG-1, MPEG-2 and MPEG-4 by the Moving Picture Expert Group (MPEG)[8], and H. and G. and T. series[9] of standards by the International Telecommunication Union (ITU)[10] specified for video, audio, data and control, and so forth. These standards provide multimedia product manufacturers the guidelines to develop their products to ensure compatibility of various platforms on a worldwide basis.

Referred to the third and the fourth questions, several significant standards for multimedia description and consumption have been defined:

Dublin Core metadata defines a core set of 15 elements, namely Dublin Core Metadata Element Set (DCMES) (Weibel *et al.* 1998). They include *Title, Subject, Description, Type, Source, Creator, Publisher, Contributor, Rights, Date, Format, and Language* and so on. Due to the simplicity of DCMES,

[6] See http://www.ietf.org/

[7] See http://www.saettler.com/RIFFMCI/riffmci.html

[8] See http://www.chiariglione.org/mpeg/

[9] For example, the standards include G.723.1 for audio, H.264 for video, T.120 for data, and H.225 for control and so on.

[10] See http://www.itu.int/home/index.html

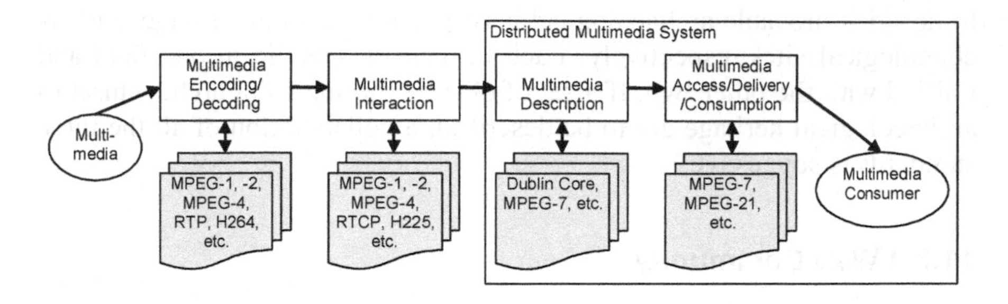

Fig. 11.2. Multimedia Process Model

Dublin Core enables a good compatibility to the other metadata standards. However, the use fields of Dublin Core metadata are mainly within libraries and museums. GIS standards can not be used straightforwardly in Dublin Core document, until a mapping between the element sets of both standards is established.

MPEG-7 is regarded as one of the most complete multimedia metadata standards in order to enhance multimedia information interoperability and retrieval (ISO 2003, Kosch 2003). The MPEG-7 Multimedia Description Schemes (MDS) are defined especially for systematical description of multimedia. They define variety of aspects which are divided in 5 element groups: Basic Elements, Content Management and Description, Navigation and Access, Content Organization, and User Interaction.

MPEG-21 is defined to provide a multimedia framework for multimedia access, delivery and consumption, in order to enable interoperability, content protection and content adaptation in a distributed multimedia system (Kosch 2003). It aims at enabling the transparent and augmented use of multimedia resources across a wide range of networks and devices (Bormans and Hill 2002).

11.3.3 Cultural Heritage Standards

It is hard to obtain an overview of metadata standards in the field of cultural heritage management even to get the total number of those standards. Among them, several eminent metadata standards in this field are MIDAS (2003), CIDOC (2004) and ObjectID and Core Data Standard by Getty Institute (Getty 2000). All these standards can be implemented in XML.

However, there are several features of those standards. First of all, many standards in domain cultural heritage have not been launched standalone. For example, Object ID, Core Data Index and Core Data standard are used to describe movable archaeological heritage, architectural heritage, and archaeological sites respectively. Each of them is specific in one field and unified with the other two. If a set of objects varying from architectural to archaeological heritage are to be described, a collaboration of all the three standards is suggestive.

11.3.4 Web Community

Kumar *et al.* (2002) considers a web community as a collection of web pages that deal with a common topic, presumably created by people with overlapping interests. Another term *virtual communities* can refer to a wider range than web communities.

A widely-applied definition of *Communities of Practice* was given by Wenger (1998):

> *Communities of practice are groups of people who share a concern or a passion for something they do, and who interact regularly to learn how to do it better.*

And ACIS aims at building up such a community of practice, providing its diverse user communities a communication channel and enabling them an intra-generational and intergenerational learning. Regarding to the research work in cultural heritage, community of practice is the best solution to a long-lasting cultural heritage conservation network. It supports more and more researchers to be involved, to share resources and to carry out diverse fieldwork.

There are several related existed applications that can be compared closely.

- The Historic Buildings, Sites and Monuments Record Database (HBSMR) developed by exeGesIS SDM in partnership with English Heritage's National Monuments Record (NMR) and the Association of Local Government Archaeological Officers (ALGAO) is an MS Access database application (HBSM 2002).
- The project of Semi-Automatic MPEG-7 Metadata Generation (Yasuda-Aoki 2003) was developed at Center for Collaborative Research, University of Tokyo. It aims at reducing the input work for multimedia annotation in a content-based image retrieval system.
- The Afghanistan map project is one of the projects developed by ALOV Map Free Java GIS[11]. It generates the map of Afghanistan dynamically.

[11] See http://alov.org/Afgan/afgan.html

Users can check the map theme selections such as cities, airports, and roads etc. on the left hand side.

- *Afghan map viewer* provided by Telemorphic, Inc. provides digital map of Afghanistan[12]. This application is implemented with Java Applets. The map covers airfields, coastline, country boundaries, roads, rivers, lakes, population centers, colorized digital elevation model, etc.

Generally speaking, more and more applications for cultural heritage management have switched to employ GIS technologies recently. However, they focus only on one or two points of the related technologies. A community information system concerned with all the four concepts has not been developed with regard to the related work.

11.4 System Design of ACIS

Aiming at a community information system with open architectures and standards, ACIS uses the common three-tier web architecture. The front-end are the user interfaces running on the client side. They enable the build-up of a virtual community. The backend are joint databases that combine a conventional relational database, a relational object-oriented geospatial database, and an XML metadata database. The intermediate layer, also called the business logic, connects the application on the client side with the databases. This middle tier is composed of server applications. The main functionality of map producing and representation, multimedia search and retrieval, user's profile management, and metadata management is carried out there. Besides the usual web server side application, a web map server is needed to process clients' map requests and responses. It can access data stored in geospatial database and the related map metadata.

11.4.1 Data Model

The data model is the starting point of the whole system design, and it is especially important for the database design that follows an entity relationship diagram. As illustrated in Fig. 11.3, the main entities are *Object* including *Geo-object* and cultural *Item*, *Source* including *Document* and *Event* which consists of *Fieldwork* and *Snapshot*, and *Media* that represents different multimedia such as image, video etc. The strongest relation-

[12] See http://www.telemorphic.com/maplicity/afghan.htm

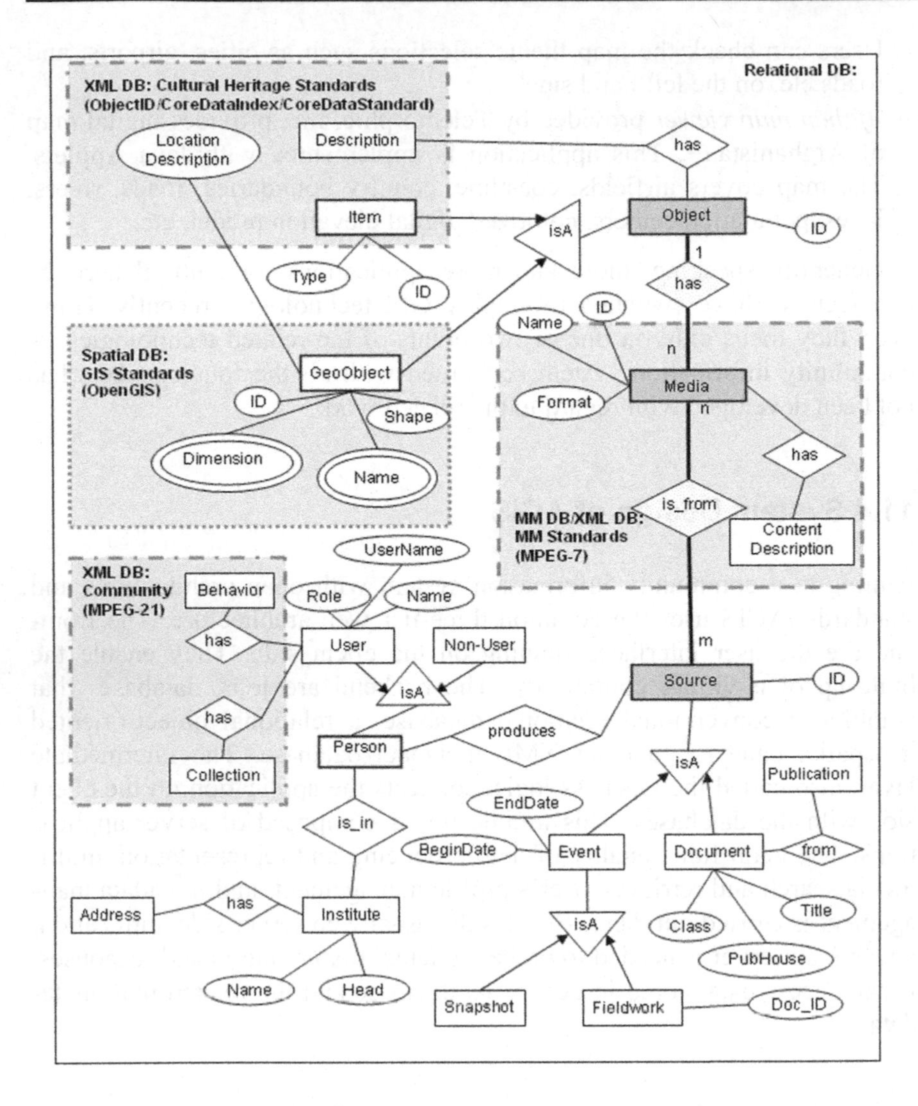

Fig. 11.3. Main Entity Relationship Diagram

ship in the data model is: each *Object* could have many *Media* which comes from certain *Source*. In addition, many other entities are aggregated with the three main entities. Especially, the *Person* entity composed of *User* and *Non-user* is the actuator of *Source*, who could be involved in a fieldwork, or the author of a written document.

The data model can also be seen as a composition of different entity parts that use the ACIS concepts respectively. For example, *Geo-object* represents GIS and uses GIS metadata standards and is stored in a spatial

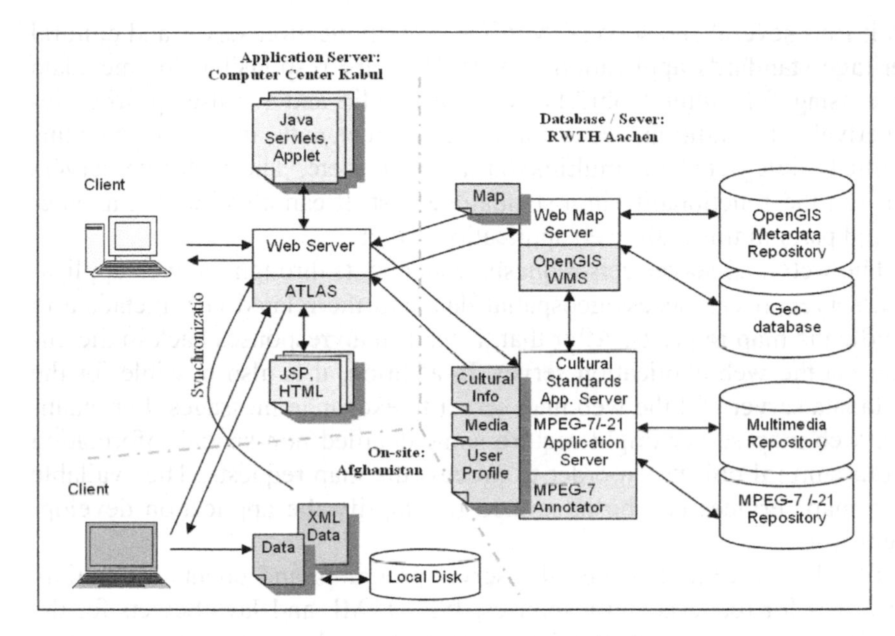

Fig. 11.4. Web Architecture of ACIS

database, while *Media* makes up the multimedia system and is stored in a multimedia database. *Object* is described with cultural heritage standards and *Person* builds up the community. The related XML files for cultural objects description and user profiles are stored in the XML database.

11.4.2 Web Architecture

With the distributed web architecture of ACIS depicted in Fig. 11.4, databases may be composed of both repositories of contents and metadata repositories to describe respective contents. They are processed with different server applications. Roughly speaking, the web server can be distributed into two sub servers: application server and metadata server. The application server can get requests from clients. Then it processes the requests, or passes the requests to the other application servers, in case that it can not fulfill the functionality. After it gets the computation results or gets the responses from the other server programs, it sends the responses back to clients. It serves as a general server between clients and the specialized servers. It is also in charge of generating user interfaces, user management and the other general server-side functionality.

The metadata server is in charge of dealing with XML files of the metadata for cultural heritage as well as metadata of MPEG-7 and MPEG-21.

So it runs several sub servers: MPEG-7/-21 application server and cultural heritage standards application server. They are responsible for metadata processing for cultural objects, for hypermedia and for user profiles respectively. In addition, some implemented metadata annotator programs might be integrated for multimedia annotation etc. The metadata servers can run the functionality in a standalone host. It can also be seen as integrated parts in this main web application server.

The web map server gets requests from clients through the web application server. It can access geospatial data and the related GIS metadata to handle the map requests. After that, it sends map responses back to the clients via the web application server. In addition, it is also possible for the metadata server and the web map server to exchange messages. For example, web map server might require some detailed non-spatial information about cultural objects, in order to process the map requests. The available web map services can be integrated to simplify the application development work.

The data streams between the server programs and clients are mainly messages for requests and responses, JSP, HTML and Java Sevlets for the user interfaces, user information, multimedia, cultural object information and various map data.

With this distributed system architecture, such a use scenarios can also be illustrated in Fig. 11.4. The databases might be maintained in Department of Information Systems at RWTH Aachen University. Meanwhile, the application server might run in the newly established modern computer centre in Kabul, the capital of Afghanistan. On-site findings at a historic site can be input into a laptop immediately and stored locally, in case that the Internet is unavailable. Later it can be synchronized with the database

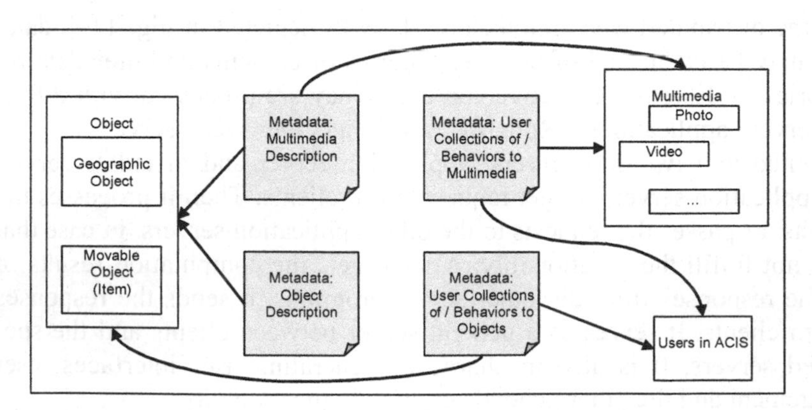

Fig. 11.5. Metadata Types in ACIS

in Aachen through the web server. Certainly, with regard to the network capacities, it might be more efficient to locate all server applications close to the database.

Summarily, this open system architecture in a three-tier model was designed based on use case analysis. The user interfaces use the packages of the business logic layer. The business logic layer accesses the database through the database interface.

11.4.3 Metadata

The suitable metadata standards must be chosen for media description, cultural objects description as well as user community. They are associated closely with both the cultural heritage objects and multimedia. In addition, users have collections both of media and of cultural objects. Hence, there are four types of metadata in ACIS (Fig. 11.5).

- Metadata to describe multimedia is associated with the multimedia itself and with the object as well. The content of multimedia refers to a cultural object usually.
- Metadata to describe objects provides a way to store information about a cultural object.
- Metadata to describe user collections and behaviors concerned with multimedia is used to trace users' access, preference and interaction of a piece of multimedia.
- Metadata to describe user collections and behaviors concerned with cultural objects is used to trace users' access, preference and interaction of a cultural object.

Various multimedia formats will be described using the MPEG-7 standard. They can include paper maps as well as DVD films. The media formats are classified into text, still image, video, audio, slides with or without audio and 3D-model (see Table 11.2). For a concise description of multimedia, different metadata template will be designed for the respective formats.

11.5 Implementation of ACIS

According to the database and system design, database was implemented with Oracle database technologies that consist of technologies for spatial database, multimedia database and XML database.

Table 11.2. Multimedia Formats in ACIS

Text	Free text documents, structured documents
Image	Maps, photographs, paintings, sketches (plan, facade, perspective), posters
Audio	Interview, radio programs, speeches, lectures
Video	Films, documentaries, news clips, lectures
(Audio)Slide	Collections of images, learning programs(lectures), tutorials
3D-model	Animations, architectural models

11.5.1 The Survey of Implementation Technologies and Tools

Correspondingly, *Oracle Spatial*, *Oracle Intermedia* and Oracle XML database utilities have been surveyed closely and applied partly.

Oracle Spatial defines various data models, indexes and queries. This overcomes the limitations of the conventional database for spatial data. In the conventional relational database the X coordinate and Y coordinate of geometries are usually stored in two columns of a table. Oracle Spatial specifies some objects in predefined scheme to store vector data. Thus, the object-oriented database technology and relational database technology are associated appropriately.

MapViewer is a Java 2 Enterprise Edition (J2EE) application for web map server integrated in Oracle Application Server 10g. It provides a Java API to render maps with spatial data stored in Oracle Spatial as well as Oracle Locator. This API hides the complexity of spatial data queries and cartographic rendering and can be integrated with GIS application in a platform-independent manner.

11.5.2 Data Survey and System Implementation

After the database was established, the original data stored in the MS Access database was migrated into the new database with SQL scripts in a semi-automated way. However, the location information of the sites and monuments is just imprecise textual information. The Afghanistan Information Management Service Website[13] (AIMS) collects useful information about Afghanistan for various purposes. The information covers a wide range of Afghan politics, culture, economics, society, and spatial data resources of the country. The spatial information is provided by AIMS to-

[13] See http://www.aims.org.af/

gether with the Central Statistics Office and the Geodesy and Cartography Office of the government of Afghanistan. Currently these institutions are using a 32 Province, 329 District Administrative Boundary model. The geographic data conforms to this model and embrace administrative divisions including the Afghan country, provincial and district boundaries, drainage including lakes, river line and river region, pass, roads, and settlements including capital, province centers, district centers and the other settlements. All the geographic data is in the format of shape files that can be browsed by the ArcExplorer[14] that is a lightweight GIS viewer developed by ESRI. In order to import the spatial data into Oracle spatial, a shape file converter tool [15] is provided by Oracle. The execute file *shp2sdo.exe* requires several parameters to give the information about the table name in Oracle Spatial, the spatial column, the spatial index, and the MBR of the geometries.

Java technologies such as Java Servlet, JSP and Java Applet have been applied for the prototype implementation. The main user interface depicted in Fig. 11.6 is made up of *Tool Panel, Map Panel* that displays maps with query results, *Spatial Search Panel* that poses queries with users' interactions, *Information Panel* that shows the textual information of the queries and *Multimedia Panel* that lists the related multimedia data.

In the basic tool panel, the map can be zoomed in and zoomed out. The map image size can also be modified according to the zooming scale. So the map panel has been implemented with a scroll bar to display the other part of the map. The center of the map can be set by giving the geographic information. The menu of the theme manager lets user select predefined themes e.g. province boundary, district boundary, main cities, main roads, main rivers and lakes.

The map with Java Applet technology is user-interactive. So the cultural heritage object's name can be shown in a label, if the users move the mouse on a site or monument in the map. If users click a site in the map, the description of the site on the interacted point can be displayed in the information panel. Meanwhile, the multimedia files related to this site can be played as thumbnail in the multimedia panel.

The spatial search panel provides different graphic tools for the users to make some spatial queries. All cultural objects that are allowed to be viewed by users as public information can be displayed in the map, when users click the button "All Monuments". The additional spatial search tools implemented in the prototype of ACIS are as follows:

[14] See http://www.esri.com/software/arcexplorer/index.html
[15] See http://otn.oracle.com/software/products/spatial/files/shp2sdo_readme.html

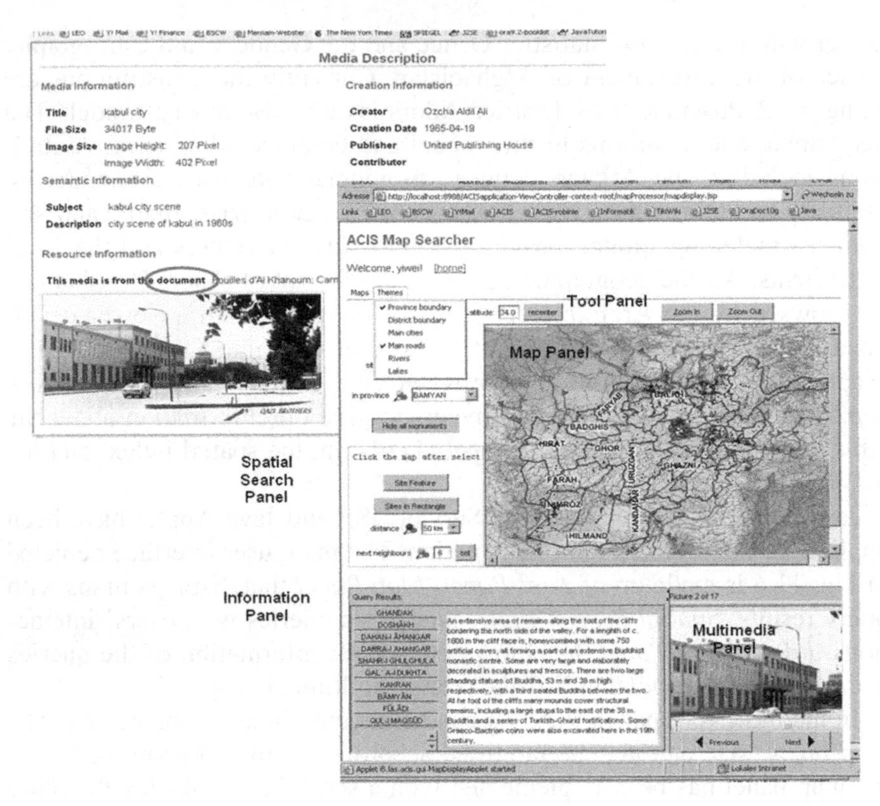

Fig. 11.6. ACIS Main User Interface

- Search a site (cultural object) with a certain site name;
- Search all sites (cultural objects) in a certain province;
- Search all sites (cultural objects) within a rectangle drawn by users;
- Search all sites (cultural objects) within a certain distance restriction, with reference to a center point clicked by users;
- Search the next *n* sites (cultural objects) in the neighborhood with reference to a center point clicked by users.

Then the spatial query results are listed in the information panel which is in form of tabs. A single site can be selected in the tab and will be marked in the map with another marker. Its related multimedia files can also be displayed in the multimedia panel. In this way, location information, site description and multimedia information are well organized and displayed.

In addition to the main user interface, multimedia files can be uploaded from users' local file systems. After the upload, some technical information about the media like file formats and sizes is extracted from ACIS automatically. They are displayed in related input field in the form.

The multimedia input interface employs the Dublin Core metadata standard. And it is generated through XSLT. The annotation information is generated as XML files using the MPEG-7 metadata standard. The XML files are stored in the Oracle XML database. In order to list the multimedia information, the XML file is transformed into normal HTML file for the browser through XSLT. Currently the technical, creation, semantic and source information are displayed by means of the style sheet file. A screenshot of the media description of an image with their meta information is also illustrated in Fig. 11.6. In the source information, the related source types including documents, fieldwork and snapshots can be displayed as well.

11.6 Evaluation of ACIS

The evaluation was carried out through polls. Several user communities have taken part in the ACIS evaluation. They are art historians at Seminar for Oriental Art History of Bonn University, architects of Department of Urban History, students at Department of Information Systems and Database Technology of RWTH Aachen University, students at Department of Traffic Engineering of Shanghai Tongji University, and Dr. R. D. Spanta, who is now a consultant of President of Afghanistan and worked for Third World Forum Aachen. In addition, both Prof. Dr. M. Jansen of Department of Urban History and Prof. PhD. S. Zheng of Institute of Architecture and Urban Space of Tongji University gave an interview. Furthermore, an on-site evaluation was also conducted by architect G. Toubekis in Kabul, Afghanistan.

Feedbacks prove that the data quality is good and the system quality of ACIS is stable. The requirements with high priorities have been fulfilled. ACIS overcomes the limitations of the MS Access-based application and can be well applied in practice for the cultural heritage management in Afghanistan by various user communities.

In the evaluation of information quality issue, the application in cultural heritage management is greatly dependant on information that it can impart. The information quality of ACIS is evaluated as interesting and instructive. However, it is not precise enough due to the limitation of the existing data. The geospatial information of sites and monuments stored in MS Access database are fairly rough. Its accuracy only reaches to the unit of minute. One minute in latitude or longitude equals approximately 1.8

kilometer on the earth. Some recent fieldwork that was conducted by Georgios Toubekis is able to provide much more precise geospatial information about monuments and sites. With those information ACIS will be able to provide more valuable information.

Moreover, the quality of multimedia is of great importance for researchers. In ACIS it keeps the quality of multimedia files as they were in the existing application.

Furthermore, the meta information of some multimedia is not comprehensive enough. Due to the mixed status of the original multimedia descriptions, most of the information can not be extracted automatically and input into the structured XML file using the MPEG-7 standard.

Finally, in order to improve the metadata quality, the annotation category using Dublin Core metadata standard needs to be organized and explained in a better way. Although ACIS presents the link to the core elements set of Dublin Core, a direct help or explanation of the categories needs to be integrated.

11.7 Future Work

ACIS proposes concepts and methods how to apply different technologies and metadata standards for multimedia, spatial information and cultural heritage in one community information system to manage Afghan monuments and sites. It proves the concept that a community application environment is capable of a sustainable support in rebuilding disaster struck communities by means of information systems. It shows the possibility to bring together networked experts from all over the world to contribute in a continuous and long-lasting sustainable relief process.

In conclusion, ACIS can be applied for cultural heritage management not only in Afghanistan, but also worldwide. In addition, its open architecture allows an easy adaptability for various other multimedia-based applications. For instance, the concepts of ACIS might also be integrated in advanced tourism information systems. The transfer of its multimedia content requires only minor adaptations in the existing architecture. Thus, tourists can search cultural information about the destinations and organize a more interesting travel via ACIS. During the travel mobile network and GPS technologies can be utilized to a greater extent. And tourists can get their location information and multimedia of the visiting place. Tourists can also use ACIS to share the newly shot pictures and videos with user communities etc. Hence, the application of ACIS in the field of e-tourism could have a good perspective in our future research work.

References

Bormans J, Hill K (2002) MPEG-21 overview. Version 5, Shanghai

CIDOC (2004) ICOM-CIDOC homepage: introduction to CIDOC, http://www.willpowerinfo.myby.co.uk/cidoc/#CIDOC

ESRI (2002) Metadata and GIS – An ESRI white paper. http://www.esri.com

Fitzke J, Greve K, Müller M, Poth A (2003) Deegree – ein Open-Source-Projekt zum Aufbau von Geodateninfrastrukturen auf der Basis aktueller OGC- und ISO-Standards. GIS, vol 9, pp 10-16

Getty (2000) A crosswalk of metadata element sets for art, architecture, and cultural heritage information and online resources. Getty Institute

HBSMR (2002) exegesis: HBSMR V2 user guide. http://www.esdm.co.uk/HBSMR.asp.

ISO (2003) MPEG-7 overview. ISO/IEC JTC1/SC29/WG11, http://www.chiariglione.org/mpeg/standards/mpeg-7/mpeg-7.htm

Jones SA, Jones DT (2004) Selecting the right GPS for a utility infrastructure GIS. GITA Conference 27, Seattle, USA

Kosch H (2003) Distributed multimedia database technologies supported by MPEG-7 and MPEG-21. CRC Press,

Kumar R, Raghavan P, Rajagopalan S, Tomkins A (2002) The web and social networks. IEEE Computer, vol 35(11), pp 32-36

MIDAS (2003) FISH: MIDAS a manual and data standard for monument inventories, http://www.english-heritage.org.uk/Filestore/nmr/standards/Midas3rd-Reprint.pdf

Momjian B (2001) PostgreSQL introduction and concepts. Addison-Wesley

NCGIA (1990) NCGIA homepage. National Center for Geographic Information & Analysis, http://www.ncgia.ucsb.edu/

OGC (2000) Open GIS abstract specification: OpenGIS metadata. Open GIS Consortium Inc., http://www.opengis.org/docs/01-111.pdf

OGC (2003) OpenGIS web map server cookbook. Version 1.0.1, Kris Kolodziej (ed), Open GIS Consortium Inc.

Oracle (2002) Oracle spatial user's guide and reference. Release 9.0, Oracle Corporation, http://www.filibeto.org/sun/lib/nonsun/oracle/9.2.0.1.0/B10501_01/appdev.920/a96630.pdf

Weibel S, Kunze J, Lagoze C, Wolf M (1998) Dublin Core metadata for resource discovery. http://www.ietf.org/rfc/rfc2413.txt (in reference to http://www.dublincore.org/documents/dces/)

Wenger E (1998) Communities of practice: learning, meaning and identity. Cambridge University Press, New York

Yasuda-Aoki (2003) Semi-automatic MPEG-7 metadata generation with a novel utilization of spatial and temporal information in content-based image retrieval. Research project at Yasuda-Aoki Laboratory, University of Tokyo, http://www.mpeg.rcast.u-tokyo.ac.jp/j/research/04-03.html.

12 Towards the Development of Open Cartographic Hypermedia Systems

Michail Vaitis, and Manolis Tzagarakis

Abstract. Despite the evolving technological developments that have influenced spatial information management and communication, digital cartography has to overcome a number of issues concerning cognition and usability. Hypermaps, integrating concepts from geographical information systems and hypermedia systems, have been proposed by many researchers as a promising solution. However, these systems are isolated, without making an attempt for information and services exchange. In this chapter, inspired by previous work on Open Hypermedia Systems, we propose a conceptual reference model for cartographic hypermedia services. As a prerequisite, a number of hypermap systems have been studied, resulting in a set of indispensable requirements.

12.1 Introduction

Analog maps have a twofold purpose: to store and communicate spatial information (Robinson *et al.* 1995). Spatial extents as well as related quantitative and qualitative characteristics should be harmoniously organized on a piece of paper or a set of transparencies. Markers and various kinds of stickers are employed by map users in order to customize map contents and express special objectives. Maps are usually accompanied by explanatory booklets, containing notes, profiles, pictures, etc. (Kübler *et al.* 1998). Digital technology has liberated cartography from the limitations imposed by the nature of the conventional underlying media. Nowadays, spatial databases are assigned the task of spatial information representation, storage and retrieval (Rigaux *et al.* 2002), while geographical information systems (GIS) offer hundreds of operations for spatial analysis and knowledge discovery (Longley 2005), accompanied by powerful visualization tools. Additionally, Internet and web technologies constitute the supporting infrastructure for spatial information exchange and dissemination. Standardization efforts are currently under way (mainly by ISO, see Kresse and Fadaie 2004; and OGC, see OGC) in order to address geographic information interchange and interoperability among geographic systems and infrastructures.

In spite of the above continually evolving developments, a number of issues regarding *cognition* and *usability* are still in early stage of maturity, especially when a cartographic system is used for educational or planning purposes (Ferschin and Schrenk 1998), rather than for spatial analysis. GIS are difficult to be used by non-expert users (Buttenfield and Weber 1993), while the demand for geographical information by many different professions or the general public has been increasing (for example, information about the environment, see Council of the European Communities 1990). As stated in (Millert-Raffort 1995), spatial cognition is based on cognitive maps, while any spatial mental model includes textual descriptions, images and other multimedia content, in addition to alphanumeric information. Various geographical applications demand on semantic associations among spatial entities and large amounts of heterogeneous multimedia information (Buttenfield and Weber 1993), with geologic (Voisard 1998) and touristic (Kraak and Driel 1997) applications being just an example. Another situation where associations among spatial entities are needed is during the map building process. Map making is an incremental process, asking for multidimensional versioning on spatial components, time and assumptions as well as annotation and metadata services. The intermediate versions of a map should be explicitly stored and organized, as they are essential means for the better understanding of the final result (Voisard 1998).

The integration of GIS and hypermedia[1] systems has been proposed by many researchers as a promising solution to the limitations and challenges stated above (Ashman and Chase 1993, Buttenfield and Weber 1993, Caporal and Viémont 1997, Catarci *et al.* 2001, Dbouk *et al.* 1996, Kraak and Driel 1997, Laurini and Thompson 1992, Millert-Raffort 1995, Perez *et al.* 1997, Voisard 1998, Wallin 1990). The term *hypermap* is commonly used to indicate this combination. Hypermedia systems and services have been inspired by the idiosyncratic associative and recall scheme within human memory (Bush 1945, Reich *et al.* 1999). In the traditional hypertext approach, the information space is composed of *nodes* interconnected with semantic *links*, in which the user is able to *navigate* in a non-linear way. The world-wide web is the most well known implementation of this conceptualization, although limited in linking functionality. In the context of digital cartography, hypermedia functionality is exploited for a number of purposes: for the incorporation of non-spatial multimedia data; for the semantic linking of geographic entities; for spatial and thematic navigation support; for the expression of users' annotations; or for the manipulation of

[1] As text is the first media type handled by hypermedia systems, the terms hypertext and hypermedia are interchangeable in the majority of the literature.

metadata and the assistance of the cartographic process. The implicit spatial relationships held by geographic entities motivate the emergence of *computations* for the instantiation of dynamic links and composite nodes (Ashman and Chase 1993), neutralizing the need for manually defined links (Boursier and Mainguenaud 1992). In other words, hypermaps attempt to support geographic knowledge building and communication in a more *convenient* and *effective* manner.

Besides the design and implementation efforts during the last decade, no attempt has been made to standardize hypermap systems' data model, functionality, or architecture. This has resulted in a number of closed monolithic systems, unable to communicate and exchange data or services. Recently, standardization efforts have been spent for both spatial data representation and linking capability over the web, but not in collaboration. On the one hand, the ISO/TC211 for "Geographic Information/Geomatics" (Kresse and Fadaie 2004) and the Open Geospatial Consortium (see OGC) have prescribed a number of standards and recommendations for the representation and exchange of spatial data. A joint outcome is the Geography Markup Language (see GML), which is an XML grammar for the modeling, transport and storage of geographic information. On the other hand, the Word Wide Web Consortium (see W3C) has developed the XLink (DeRose *et al.* 2001), XPath (Clark and DeRose 1999) and XPointer (Daniel *et al.* 2001) recommendations for the support of anchoring and linking information resources all over the web. Therefore, the combination of GML and XLink would potentially provide the ground for the standardization of hypermaps, given that a number of prerequisites would be fulfilled. Currently, GML manipulates simple links which carries the limitations of HTML uni-directional, one-ary, un-typed <a> tag (Raggett *et al.* 1999), such as the inability to manage disjoint set of links for the same document, or the dependency of link creation to write-access permissions to the data sources. XLink has been criticized as inadequate for the support of annotations or compositions (Christensen *et al.* 2003), functionalities that are useful in hypermap systems. In general, what is missing is an analysis of the hypermap domain and the formulation of a conceptual model for both the syntax and behavior of the participating constructs, before the establishment of a standardization framework using existing (such as GML and XLink) or novel infrastructures.

The purpose of this chapter is to establish a starting point for the study and standardization of a conceptual data model for *hypermap services*. The adoption of such a model will enable the common functionality of different and separately implemented hypermap systems, thus achieving interoperability among them. We have studied a number of hypermap systems appeared in the literature to infer a set of requirements for hypermap services. The research results in the field of Open Hypermedia Systems (OHS) (see OHSWG) are exploited, as they concentrate on architectural, modeling and rendering issues regarding the standardization of hypermedia services. The provision of these services through well-defined protocols and interfaces to an open set of software applications comprises the prerequisite for linking functionality in every computing environment. Furthermore, the standardization efforts of ISO/TC 211 are also taken into account. The output of our work so far is condensed in a working conceptual reference model for hypermap services, incorporating both cartographic and hypermedia concepts, while a preliminary implementation (Koukourouvli 2005) has indicated the feasibility of our approach.

The rest of the chapter is organized as follows: Section 12.2 presents the basic concepts of hypertext and open hypermedia systems; Section 12.3 briefly discusses a number of hypermap systems, basically from a modeling perspective, and tries to extract a number of needed requirements, presented in Section 12.4. Section 12.5 presents the ISO 19107 standard for spatial schema along with a well-defined (although non-standard) conceptual cartographic model. Section 12.6 proposes a reference model for open cartographic hypermedia systems and hypermap services. Finally, Section 12.7 concludes the chapter and presents our future plans.

12.2 Hypermedia Systems

Although hypermedia systems are mainly regarded as tools for interlinking information fragments and navigating into the resulting network, they are in fact an amalgam of *data*, *structure* and *behavior* (Schnase *et al.* 1993), offering *structure services* in particular *user domains* (i.e. the different ways people use to *structure* information in different situations). The provision of dedicated structure services for each domain (instead of overtyping and casting the notion of *link*) results in a more convenient and efficient utilization of its abstractions, improving in turn performance, quality and cost-effectiveness of user applications.

In the following paragraph, we are presenting the navigational domain, as it was the first one implemented in early hypertext systems, and the one that is most relevant to the hypermap field. Other domains appearing in the hypertext literature are the taxonomic and spatial (Tzagarakis *et al.* 2003). The study of their relation with the hypermap field is included in our future tasks.

12.2.1 Navigational Domain

The structuring abstraction in the navigational domain is the notion of *association* among information chunks. These associations – widely known as links – were to be made between semantically related information items. On the one hand, hypertext links condition the user to expect purposeful and important relationships between linked materials, where on the other hand, the emphasis upon this linkage stimulates and encourages habits of relational thinking. The evolution of navigational hypertext continued with an arbitrary number of systems, each one of them adopting different implementation strategies. However, as stated in (Halatz and Schwartz 1990), the basic node/link network structure is the essence of (navigational) hypertext. Every node in a hypertext network should be accompanied by enough contextual information to allow the reader to conceptualize its place in a larger discursive hierarchy. Moreover, the linking information has to be kept separate from node contents so as to allow more powerful link structures, such as bi-directional or n-ary links.

For standardization and interoperability purposes among navigational hypermedia systems, the Open Hypermedia Protocol (OHP-Nav) was proposed (Reich *et al.* 1999). It includes an abstract class (*Hypermedia Object*), providing the *node, link, endpoint, anchor*, and *context* abstractions. A node provides a wrapper for an arbitrary resource and may have several anchors associated with it, whereas an anchor can be bound to several links through different endpoints. A link is defined as an association between endpoints, with its direction clearly stated, while a context is a collection of entity references of any kind. Attribute/value pairs can be used to attach arbitrary properties to an entity. Fig. 12.1 presents the constructs and their inter-relationships in the navigational domain.

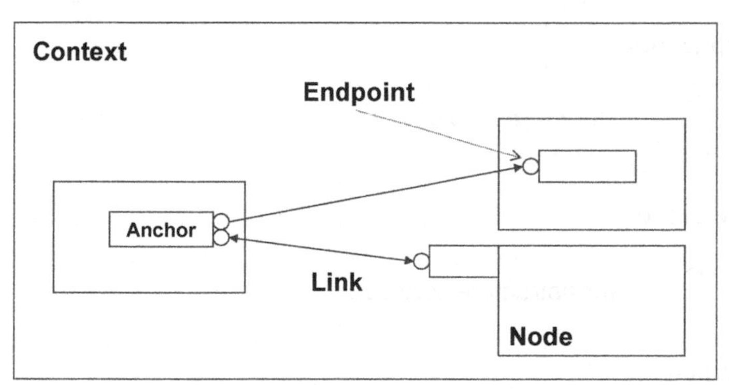

Fig. 12.1. Constructs of the Navigational Domain

12.2.2 Open Hypermedia Systems

The community of OHS has envisaged hypertext as a group of open services, offering structuring abstractions to third-party *client* applications. The term "open" refers to the ability of hypermedia systems to provide multiple, well-defined interfaces for hypertext services to all applications on the users' desktop. Component-Based OHS has emerged, comprising an open set of components, called *structure servers*, organized in a layered architecture (Fig. 12.2). Each structure server provides the foundations for supporting the structural abstractions of a single domain in a convenient and efficient way (Nürnberg and Leggett 1999). Structure servers have emerged to support well-known domains, such as navigational, taxonomic and spatial.

Infrastructure includes the fundamental services that are available to all other layers, such as persistent storage, query engine, naming and notification services, etc.

Two protocols control the communication between the layers: (a) the services offered by a particular structure server to the clients; and (b) the services offered by the infrastructure to the structural servers. To utilize structure services, clients may be either custom-build applications, or extensions to existing third-party applications. In the later case, either direct extensions are made, or wrapper programs are separately developed. In this context, our purpose is to lay the foundations for the development of a *cartographic server*, providing structural abstractions for supporting map building and navigation activities.

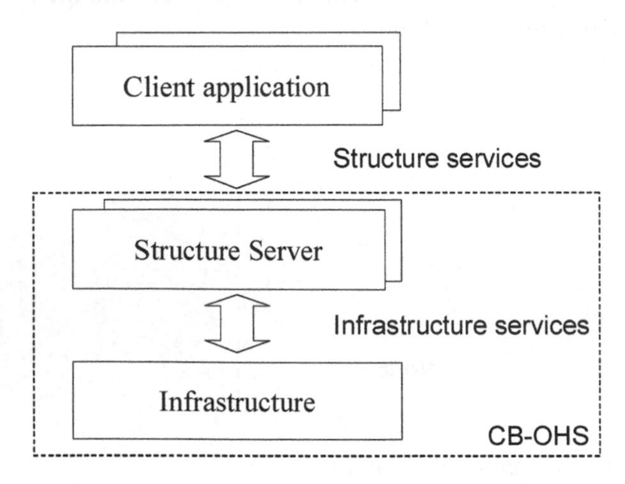

Fig. 12.2. Architecture of CB-OHS

12.3 Previous Work on Hypermaps

The work about hypermaps may be classified into two major categories: (a) Incorporation of hypermedia functionality into spatial or cartographic information systems; and (b) incorporation of spatial footprints (according to a coordinate-based referencing system) in hypermedia entities. In the following paragraphs a number of such approaches are presented, focusing mainly on their data modeling characteristics and functionality.

The conceptualization of *cartographic hyperdocuments* (*hypermaps*) (Laurini and Thompson 1992, Millert-Raffort 1995) provides a method for geographic access to multimedia information. The traditional node-anchor-link paradigm is incorporated, where nodes are augmented with one (or many) spatial reference(s), according to a coordination system. More precisely, every geographic extent related to the content of a hypermedia node is formulated as an elementary object (of type point, line, area or volume) called *zone*. A many-to-many relationship-type is maintained among nodes and zones. Membership degrees may be used to represent a partial relevance of a node to a given zone. Three types of links are envisaged: *novice*, *expert* and "*don't know*", according to user experience. Two forms of navigation are proposed: thematic navigation is based on link traversals, while spatial navigation employs spatial queries in order to retrieve hypermedia nodes related to a user-selected region or point. Three types of users are involved in hypermaps: the developer of the hypermap toolbox, the hypermap editor, and the navigator. It is pointed out that differences in spatial cognitive knowledge among them may produce misinterpretation of nodes and links structure, while the use of annotations and references is proposed as a possible solution.

The *GeoAnchor* data model (Caporal and Viémont 1997) is an extension of the Dexter hypertext reference model (Halatz and Schwartz 1990), specializing hypertext classes to encapsulate geographical data. *Map* and *Simple Geographical Object* are specializations of the Basic Node class; they are used to encompass raster maps and single features, respectively. *Localized Collection*, *Localized Structure* and *Localized Choice* are specializations of the Composite Node class. Localized Collection nodes are used to build collections of geographical nodes; localized Structure nodes permit the modeling of complex geographical objects; and Localized Choice nodes are used to store different representation alternatives of a single geographical object, according to resolution, scale or other parameters. *Deals With Link* and *Indirect Location Link* are specializations of the Link class, allowing a non-geographical node to inherit the location from a localized one. An interesting characteristic of the GeoAnchor model is the *threshold* attribute, defined for every class. Threshold holds a number, indicating the user access frequency of the object. Depending on the scale of the map view, only the objects that have a sufficient threshold value are displayed. In this way, at small scales, a high threshold permits only important features to be kept, while at larger scales, a low threshold permits the display of most localized features.

In (Kraak and Driel 1997) a simple data model is introduced, by augmenting each multimedia object with a *geo-tag*, an *attribute-tag* and a *time-tag*. These tags allow the submission of spatial, thematic and temporal searching filters, respectively. In addition, two sets of links are maintained for each multimedia object: one for the outgoing and one for the incoming links; thus, forward and backward navigation is provided.

The *HiperSIG* approach (Perez *et al.* 1997) adopts the incorporation of a hypermedia module into geographic information systems, thus enhancing them with hypermedia functionality. The traditional node-anchor-link model is supported, where nodes may hold calculations, rather than only explicit content. Links may be used to associate objects with their metadata and annotations; to connect input and output of operations; to structure or group sets of objects; to keep versions of objects; or to express spatial and temporal relationships among objects. The hypermedia module communicates with the other components of the GIS (interface, computational, and data-management modules) by intercepting messages, in order either to enrich output (forwarded from the computational module to the interface) with hypermedia links, or to calculate the destinations of activated anchors. Except navigation, HiperSIG offers a querying mechanism for determining starting points for browsing as well as *paths* and *guided tours* defining fixed traversals into the information space.

The *HyperGeo* data model (Dbouk *et al.* 1996) introduces links and multimedia representations into GIS, enhancing the spatial entities with hypermedia characteristics. HyperGeo is the kernel of the *dynamic map* – a multidisciplinary environment for the presentation and interaction of geographical data, combining object-orientation issues, hypermedia functionality and visual query languages. More specifically, the observed universe is considered as a set of heterogeneous *entities* (like regions, cities, buildings or roads). An entity is composed of a spatial part and a descriptive part. The spatial part can be represented by one or several digital spatial representations (like geometric or topological), while the descriptive part may include a number of multimedia representations (photographs, videos, satellite images, etc.).

Links are distinguished in *explicit* (static links among entities, stored into the database) and *implicit* (dynamic links, computed from the spatial relationships among the entities) ones. Explicit links are further specialized into *composition* links (used to represent composite objects) and *hypermedia* links (reflecting the hypermedia orientation of the model). Implicit links concern distance measurements, topological relationships and relative positioning between entities. All links are supplied with a *context* indicator, revealing the visibility scope of them. A dynamic map is a set of entities (along with their associated links) located at a limited geographic area, that is produced as a result of a spatial query against one or several databases. The scale, projection and presentation details of a dynamic map may be altered by the user. A number of querying and navigational functionalities accommodate the user's interaction with the system environment. Among them, the *zoom* operations are of special interest. The "intelligent" (or logical) zoom changes the level (scale), raising new entities to appear after querying the database. The "geometric" (or physical) zoom redisplays the map at a different scale, without looking for more information.

A 2-level graph model is adopted by (Voisard 1998, Kübler *et al.* 1998) to represent hypermaps in the geologic domain. Geologic objects (having a spatial component of type point, line or polygon, and an alphanumeric description), multimedia objects (photos, videos, texts, etc.), cartographic objects (legends and symbols), metadata (annotations or categorization of objects) and links are the components of a geologic hypermap. Links are used to express the various categories of relationships among objects: historic, semantic and hypothetic. Historic links keep track of the evolution of the geologic hypermap (versioning). Semantic associations offer a general linking mechanism among objects. Hypothetic associations link certain geologic objects with other geologic objects or assumptions on which have existence dependency. Queries about the hypermap contents, structure, or both, supporting recursion and negation, are proposed to assist of the end-user's navigation process.

A different approach is presented in (Wallin 1990). The author describes the underlying concepts of a *Hyper-GIS/Open System*, offering multiple levels of contexts for geographic inquiries. These levels are *systemic*, rather than thematic, and they are produced by successive mental abstractions (*causal disaggregations*) of the complete and total network of all possible events and interactions in a time-geographically defined domain. In this way, on the last level all empirical details are suppressed and the domain is presented as a featureless, infinitive and isotropic space. Hypertext functionalities assist user's navigation among the different levels of detail, while maps are the essential "*hypotexts*" serving a common ground on each level.

12.4. Requirements for Hypermaps

In the following paragraphs we are attempting to settle a set of requirements as prerequisites for the definition of a conceptual model for open cartographic services. What is deduced from the previous section is that most of the hypermap systems utilize functionality from the navigational domain (semantic associations - linking). So, the notions of the OHP-Nav interface should be customized and extended to incorporate concepts from the cartographic domain.

As another observation, a hypermap system employs three categories of users, having vague boundaries among their duties: the system developer, the map designer (or constructor) and the map reader (or navigator)[2]. Developers are responsible for constructing the tools required for the synthesis and navigation of hypermaps. Designers exploit these tools in order to produce hypermap products, while sometimes they are engaged in tool customization or extension tasks. Navigators mainly perform searching, browsing and annotating operations, but usually they are also involved in data and structure updating. The establishment of a conceptual reference model will eliminate misinterpretation and inappropriate functions produced by spatial cognitive differences among them. Additionally, such a model will support interoperability among various hypermap and geographic hypermedia systems.

According to the above observations, we have deduced a (non-exhaustive) set of requirements for open cartographic hypermedia systems.

- Demands for system developers
  - Adoption of a common conceptual reference model. The development of both client-applications and cartographic services based on such a model will enable interoperability.
  - Extension of infrastructure services to store and query spatial data. Current hypermedia systems are not able to handle geo-referenced data.
  - Establishment of a cartographic services interface. The incorporation of such an interface will provide linking functionality in GIS, spatial databases or cartographic systems.

---

[2] According to Douglas Engelbart (1998), there are three types of work that can be performed in an organization. The A-level is the work of the organization itself; the B-level is work that develops tools to improve the ability of people performing A-level work; and the C-level is work that develops tools to augment the ability of people to perform B-level work.

- Tools for map designers
  - Ability to inter-link spatial, temporal and multimedia data. Map designers should be able to inter-relate information items of any kind.
  - Typed associations among data. Types provide the means to express different semantics to links (e.g., temporal ordering, version sequence, semantic association).
  - Multiple structures of data in different contexts, resolutions and scales. Depending on the desired purpose and display capabilities, different configurations of data and links should be able to be defined.
  - Manipulation of metadata, both for data and structure elements. This is especially useful, while the cartographic process is being supported.
  - Formation of composite entities and collections. Aggregate objects express common properties or spatial configurations that are not implied (like overlapping in 2D).
  - Expression of uncertainty, both for data and structure elements. Certain applications, like geologic and historical, should be able to express doubt about relationships or boundaries, as they are based on hypothetical assumptions.
  - Versioning on multiple dimensions (e.g., time, space, or user-defined dimensions).
  - Definition of computations (for the dynamic instantiation of structure and data elements, based on queries).

- Services for map readers
  - Multiple searching capabilities (queries about spatial and non-spatial data, structure elements, and metadata).
  - Seamless navigation (link traversal) among spatial and non-spatial entities.
  - Zoom-in and zoom-out operations, with automatic adjustment of non-spatial information. The context should always be justified to include only the elements that are appropriate for the selected spatial extent and analysis.
  - Annotation. Users should be able to define personal notes and links, at a fine granularity level.

Our main concern in this work is to concentrate on modeling constructs and basic behavior (services) of a conceptual reference model for hypermaps, fulfilling many of the above requirements as possible. We have intentionally left out presentation and user-interface issues and implementation details.

12.5 Spatial and Cartographic Modeling

New standardization developments have been recently achieved in the field of *geographic information*, mainly by the ISO/TC211 and OGC. With the development of the ISO 19100 family of standards, guidelines have been established for the uniform modeling and managing of geographic information. In parallel and partly in cooperation with ISO/TC211, the OGC provides a formal framework for the development of industrial standards enabling communication and data exchange among GIS, spatial databases and the web.

The ISO 19107 (spatial schema) covers a nearly complete word of vector data. Fig. 12.3 depicts a part of the classes comprising the standard.

A *geometry object* (GM_Object) is the most general concept and depends on a Coordinate Reference System. A geometry object may be a

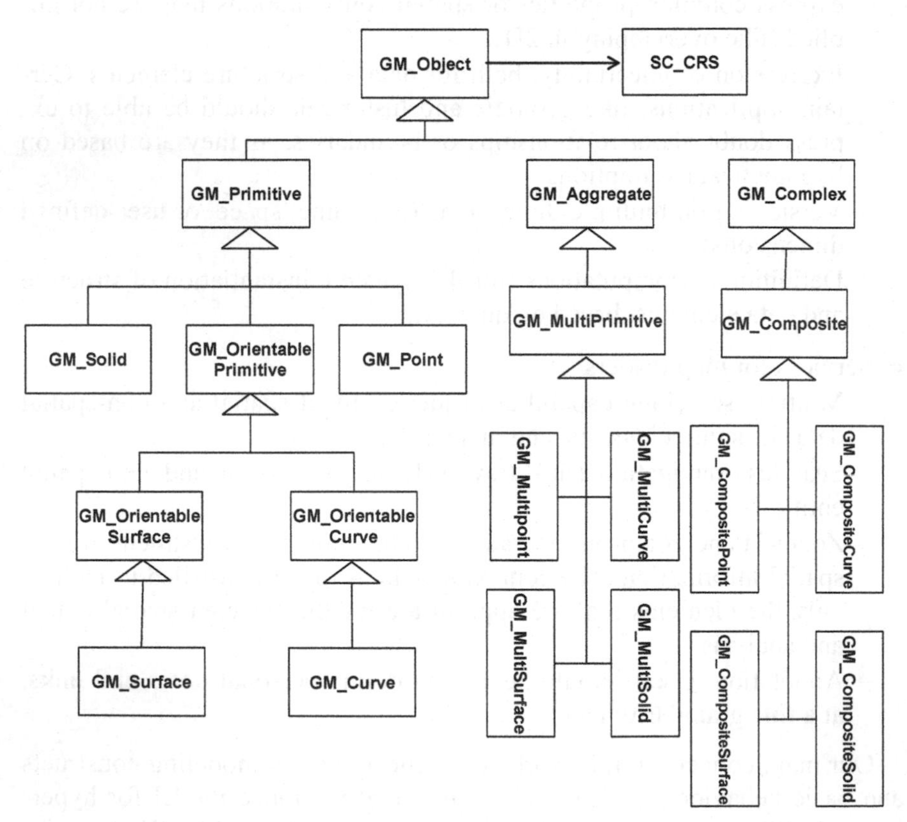

Fig. 12.3. UML class diagram of ISO 19107. This figure is a simplification of a similar one appearing in Kraak and Driel (1997).

primitive one (point, curve, surface, solid), a composition of objects (GM_Complex) or a group of objects (GM_Aggregate). Although the specialization hierarchy of GM_Curve and GM_Surface classes continues to define more specific types of curves and surfaces, for the purpose of this work we exploit only the top-level classes of Fig. 12.3.

While the ISO 19107 provides vector data with an almost complete reference model, the work of ISO/TC211on the imagery and gridded data (raster data) is still ongoing. Therefore, we utilize a reference model for digital cartography proposed by Tomlin (1990). The constructs of this model focus on the general manner in which spatial data are organized and do not dictate any specific storage structures. The model has a hierarchical structure in which the constructs of one level are composed of constructs from the successive level, till the leaves of the tree are reached.

The root of the tree is the *cartographic model* (map), which consists of a bound collection of *map layers* (themes), all pertaining to a common *study area* (extent). A map layer (or simply a *layer*) expresses the spatial variation of one characteristic of the study area. It is comprised of a number of *zones*. Every zone represents a specific variation (value) of the depicted theme. The zones of a layer should not overlap; they should rather cover the extent of the study area. The components of a zone are a set of *locations*. A location is the elementary unit of the cartographic space, for which a label and a value are recorded. The set of all locations of all zones of a given layer will encompass a grid pattern, whose overall size and shape correspond to those of the study area. A location is characterized by its *column-coordinate* and *row-coordinate*, with respect to a coordinate system of equal increments along two perpendicular axes.

12.6 Hypermap Reference Model

In this section, we describe our proposed conceptual reference model for cartographic hypermedia services, offering linking and navigation services for cartographic data. The model is based on the OHP-Nav interface for navigational hypermedia (Reich *et al.* 1999), the ISO 19107 spatial schema standard (Kresse and Fadaie 2004), and on the Tomlin's cartographic model (Tomlin 1990). The UML class diagram of the model is depicted in Fig. 12.4, Fig. 12.5 and Fig. 12.6.

In Fig. 12.4 and Fig. 12.5 we present the specialization hierarchy of the constructs of the model without the relationships among them. The *HMapObject* (hypermap object) is an abstract class representing entities that may be linked to each other. The *spObject* (spatial object) is an ab-

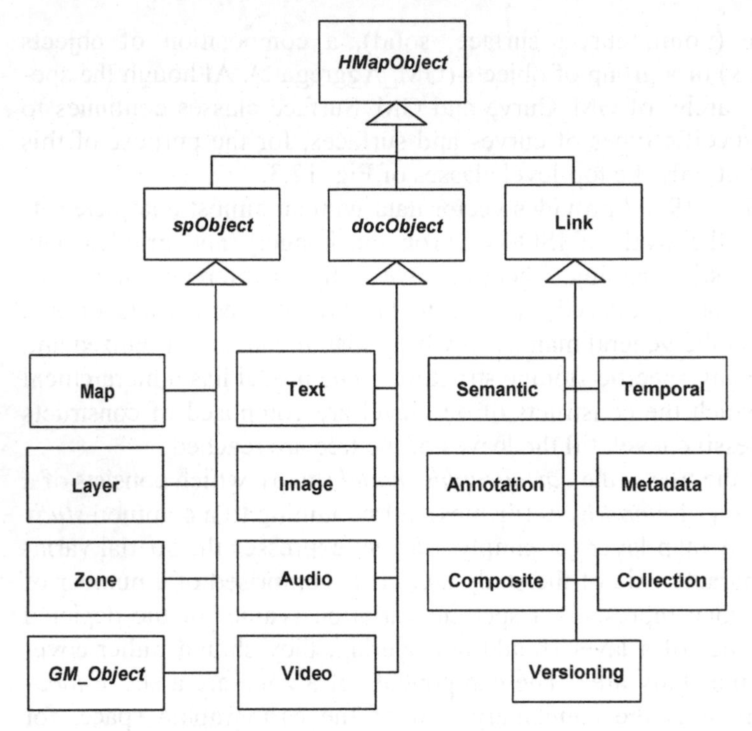

Fig. 12.4. Specialization hierarchy of the model (first part)

stract sub-class representing spatial locations and features, while the *docObject* (document object) abstract sub-class represents not geo-referenced data items. The link sub-class indicates that linking of links is permitted by the model. Geo-referencing of document objects is attained by their linking with spatial objects.

The *map* class corresponds to the cartographic model entity and represents a well-defined geographical extent (study area). The *layer* and *zone* classes correspond to the homonym entities in Tomlin's model and concern gridded data. A layer corresponds to a theme, while a zone corresponds to a specific variation of the theme. The *GM_Object* abstract class refers to the same class in ISO 19107 and concern vector data. Classes layer, zone and GM_Object enables all spatial entities to be linked (gridded and vector).

The *docObject* class represents wrappers for arbitrary non geo-referenced content. The *text, image, audio* and *video* sub-classes correspond to the most popular media types, while new media types may be included when available.

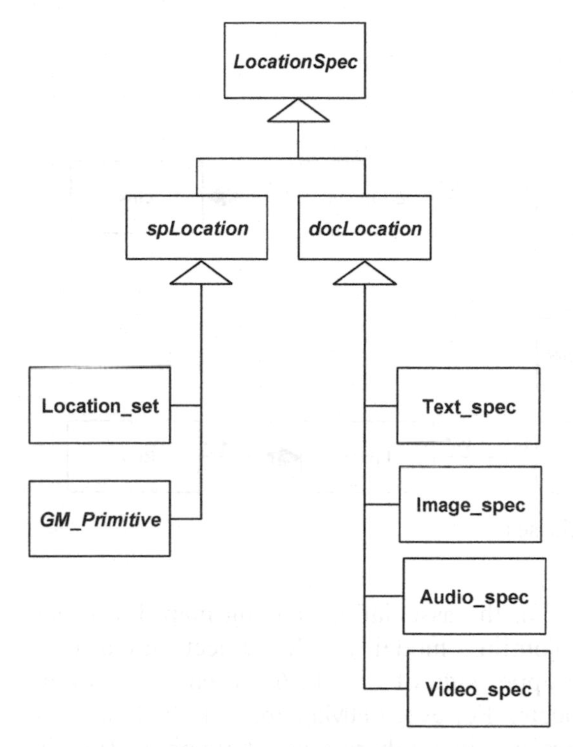

Fig. 12.5. Specialization hierarchy of the model (second part)

The *link* class implements the linking functionality in collaboration with the *anchor* and *endpoint* classes (as described below). We propose a number of link subclasses with different semantics in their behavior. *Semantic* links express various kinds of relations among HMapObjects, without being also spatially related. *Temporal* links hold temporal relations among HMapObjects, while *composite* and *collection* ones provide the means for structuring composite (complex) objects or sets of entities, respectively. *Annotation* links provide a way for navigators to express their perception about map contents. Finally, *metadata* and *versioning* links are useful tools for designers during cartographic building and maintenance. New subclasses of links may be included in the model, depending on the special needs of the application domain.

The second part of the specialization hierarchy (Fig. 12.5) is related to the anchoring abstractions of the model (class *LocationSpec* – location specification – and its sub-classes). Their description will follow that of Fig. 12.6, depicting the rest of the classes, as well as the associations among all classes of the model.

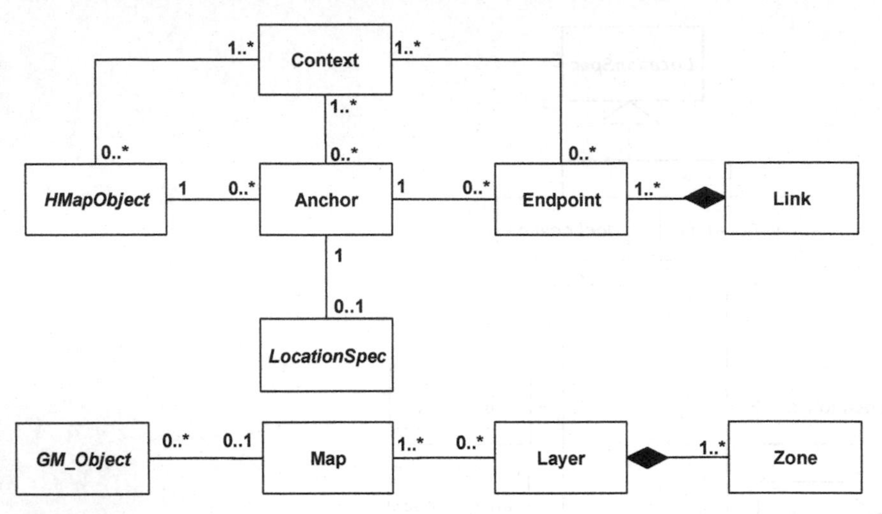

Fig. 12.6. Associations among classes

At the lower side of Fig. 12.6, the associations among map, layer, and zone classes are taken from Tomlin's model. A GM_Object may also be related to a map object. The upper side of Fig. 12.6 presents the linking mechanism among HMapObjects. For every HMapObject to be linked, an anchor should be defined, corresponding either to the whole object (i.e., its identification – *oid*), or to a certain location in the object's content. In the latter case, the anchor is associated with a *LocationSpec* (location specification), specifying the part of the HMapObject that is actually linked. A link is in turn a binding of anchors. To enable the same anchor to participate in many links with different semantics, the endpoint class is introduced, holding all the linking details (directionality, sidedness[3], role, obligation, etc.). Thus, a link is just a set of endpoints, while each endpoint corresponds to precisely one anchor.

The LocationSpec (Fig. 12.5) is an abstract class, specialized into two other abstract sub-classes – that is, *spLocation* (spatial location) and *docLocation* (document location) – for geo-referenced and non geo-referenced anchors, respectively. According to Tomlin's model, the *Location_set* class represents a set of spatial locations, referring a map, layer or zone object. For vector data, the GM_Primitive class is borrowed from ISO 19107 so as to enable the anchoring for geometry objects. What is actually permitted to be an anchor for a geometry object depends on the dimension of it. For example, a surface may be defined as an anchor for surface or solid objects, but not for points or curves. An anchor for a point is

[3] The range of the different sides of a link.

the point itself. Anchors for complex or aggregate geometry objects are defined according to their component (primitive) objects.

For the (non geo-referenced) documents, the docLocation abstract class is specialized to a set of sub-classes (*text_spec*, *image_spec*, *audio_spec*, *video_spec*), each one enabling the anchoring for each media type.

At last, the *Context* class enables the definition of specific configurations of HMapObjects, according to certain application purposes (e.g. course modules for different school classes, touristic maps and information for different travelers' profiles, etc.). From a modeling point of view, a *context* is a collection of HMapObjects, anchors and endpoints (along with their associated LocationSpec and links), belonging to a common semantic framework.

While in the previous paragraphs we have presented the static aspects of our proposed model, i.e. the classes, their specialization hierarchies and their relationships, in the following paragraph we concentrate on the dynamic aspects of it, i.e. the behavior of the model.

12.6.1 Behavior

The behavior of a cartographic hypermedia server in the context of the navigational domain (that is the subject of this chapter) naturally specializes and extends the functionality of traditional OHS in enabling the manipulation of geo-referenced gridded and vector data. In the following description of services, we emphasize their differentiating behavior influenced by the nature of spatial data.

- `CreateAnchor()`: Input parameters are the object identifier of the HMapObject with which the anchor is associated and, optionally, a LocationSpec object. In the latter case, if the anchor regards a spObject, a proper spLocation object (according to the type and dimension of the spObject) should be provided. `DeleteAnchor()` and `UpdateAnchor()` services have an obvious meaning.
- `GetAnchorList()`: Given the identifier of a HMapObject, their associated anchors are returned. If the input object is a spatial one, a special functionality of this service (controlled by proper parameters) is to return all the anchors that their spLocation objects are somehow topologically related to the spatial extent of the input object (intersection, containment, etc.).
- `CreateLink()`: As links constitute sets of endpoints, no special treatment is needed for linking spObjects. New endpoints are created for

each of the input anchors. `DeleteLink()` and `UpdateLink()` services have an obvious meaning.

- `GetLinkList()`: Given the identifier of an anchor, the associated links are returned. Proper input parameters may determine the desired link sub-class or certain properties of the attached endpoints (directionality, obligation, etc.).
- `GetDestinationList()`: Given a selected endpoint, the other endpoints of the common link are returned. Proper input parameters may determine the desired sub-class(es) of the attached HMapObject (e.g., sub-class(es) of spObject or docObject).
- `FollowLink()`: The HMapObject identifier associated with a selected destination endpoint is returned, along with the possible LocationSpec object.
- `OpenConetext()`: The context identified by the input parameter configures the execution of all successive service calls, until a new `OpenContext()` operation. Only the objects belonging to the selected context are subjects for the subsequent operations.

The development of a cartographic hypermedia server can utilize previous developments in the area of navigational OHS (especially the part concerning the manipulation of links and endpoints), as well as the specifications of the Web Feature Service interface (Vretanos 2005). Additional development is needed for the storage and retrieval of spLocation objects (and, docLocation objects, if they have not already been supported).

In addition, the client application should be able to communicate with the cartographic hypermedia server interface and to support the execution of the above services. In particular, it should control the storage of the desired HMapObject sub-classes, support their visualization, and enable the creation and visualization of anchors on them.

To support the feasibility of our proposed model, we have developed a prototype framework that implements a small part of the model (Koukourouvli 2005). In particular, using the ArcObjects and personal geodatabase products of ESRI and the Microsoft VisualBasic programming language, a client-server application has been developed, enabling anchoring and linking functionality for surface spatial objects.

12.7 Conclusions and Future Work

In this chapter, we have presented the main concepts and a first attempt for the establishment of an Open Cartographic Hypermedia System, providing hypermap services to client applications, based on the architecture of Open Hypermedia Systems. Both the data model and the behavior of a cartographic hypermedia server are described, based on previous research results in the fields of hypermap systems, open hypermedia systems, and spatial and cartographic modeling. We assert that such a model will both enhance usability and assimilation of spatial knowledge and facilitate interoperability among different hypermap systems.

Our work is at early stages. A development process is running, attempting to incarnate the proposed model using XML as the representation and communication infrastructure. Special consideration is given to the evaluation, adoption and extension of contemporary formalisms, like XLink, GML, and Web Feature Service Interface. As the web is the *de facto* hypertext framework of todays, the aforementioned standardization efforts should be adopted by contemporary implementations. In the medium term, we will concentrate on presentation and user-interface issues, not treated in this work. For this reason, the concepts and techniques incorporated in KML (see KML) will be analyzed, since it constitutes the XML grammar for the display of geographical data in the context of the widely used Google Earth system. In addition we plan to perform a more elaborate analysis of the structural abstractions in the cartographic domain to supplement navigational functionality with other forms of services (e.g. taxonomies).

References

Ashman H, Chase G (1993) Link management within a geographic hypermedia information system. In: Proceedings of the 4th Australian Conference on Information Systems, pp 327–334

Boursier P, Mainguenaud M (1992) Spatial query languages: extended SQL vs. Visual languages vs. hypermaps. In: Proceedings of the 5th Int. Conference on Spatial Data Handling, Charleston, VA, USA

Bush V (1945) As we may think. The Atlantic Monthly, vol 176 (1), pp 101–108

Buttenfield BP, Weber CR (1993) Visualization and hypermedia in geographical information systems. In: Medyckyj-Scott D, Hearnshaw HM (eds) Human factors in geographical information systems, Belhaven Press, pp 136–147

Caporal J, Viémont Y (1997) Maps as a metaphor in a geographical hypermedia system. Journal of Visual Languages and Computing, vol 8, pp 3–25

Catarci T, D'Amore F, Janecek P, Spaccapietra S (2001) Interacting with GIS: from paper cartography to virtual environments. Unesco Encyclopedia on Man-Machine Interfaces, Advanced Geographic Information Systems, Unesco Press

Christensen BG, Hansen FA, Bouvin NO (2003) Xspect: bridging open hypermedia and Xlink. In: Proceedings of 12th Int. Conference on World Wide Web (WWW '03), Budapest, Hungary, pp. 490–499

Clark J, DeRose S (eds) (1999) XML path language (XPath), version 1.0, W3C Recommendation, http://www.w3.org/TR/xpath

Council of the European Communities (1990) Directive 90/313/EEC of 7 June 1990 on the freedom of access to information on the environment, Official Journal of the European Communities, L. 158, P. 0056-0058.

Daniel R, DeRose S, Maler E (eds) (2001) XML pointer language (XPointer), W3C candidate recommendation, http://www.w3.org/TR/xptr

Dbouk M, Kvedarauskas, Boursier P (1996) Dynamic maps: an intuitive interface for naive users of spatial database systems. In: Proceedings of the 3rd Int. Workshop on Interfaces to Databases, Edinburgh

DeRose S, Maler E, Orchard D, Trafford B (eds) (2001) XML linking language (XLink), W3C Recommendation, http://www.w3.org/TR/xlink/

Engelbart D (1998) Keynote speech. In: 4th Int. Workshop on Open Hypermedia Systems (OHS4), Pittsburg, PA, USA

Ferschin P, Schrenk M (1998) Integration of GIS, hypertext and multimedia Elements for Spatial Planning Purposes. In: Proceedings of GIS'98 Conference, Brno, Czech Republic

Geography Markup Language (GML) ISO/TC 211/WG 4/PT 19136 - OGC GML RWG

Halatz F, Schwartz M (1990) The Dexter hypertext reference model. In: Proceedings of the NIST Hypertext Standardization Workshop, pp. 95–133

Keyhole Markup Language (KML), http://www.keyhole.com/kml/kml_doc.html

Koukourouvli N (2005) Infrastructure services for the development of hypermaps. Graduate thesis, University of the Aegean, Greece (in Greek)

Kraak M-J, Driel RV (1997) Principles of hypermaps. Computers & Geosciences, vol 23(4), pp 457–464

Kresse W, Fadaie K (2004) ISO standards for geographic information. Springer-Verlag

Kübler S, Skala W, Voisard A (1998) The design and development of a geologic hypermap prototype. In: Proceedings of ISPRS Commission IV Symposium on GIS, Stuttgart, Germany

Laurini R, Thompson D (1992) Fundamentals of spatial information systems. Academic Press

Longley PA (2005) Geographic information systems and science. John Wiley & Sons

Millert-Raffort F (1995) Some cognitive and technical aspects of hypermaps. In: Nyerges LT, Mark DM, Laurini R, Egenhofer MJ (eds) (1995) Cognitive aspects of human-computer interaction for geographic information systems. Kluwer Academic Publishers, pp 197–211

Nürnberg P, Leggett J (1999) A vision for open hypermedia systems. Journal on Digital Information, vol 1(2), pp. 207–248

Open Geospatial Consortium Inc. (OGC) http://www.opengeospatial.org/

Open Hypermedia Systems Working Group (OHSWG) http://www.csdl.tamu.edu/ohs/

Perez CR, Kelner J, Sadok DH, Salgado AC (1997) HyperSIG: hypermedia technology serving geographic information systems. In: Proceedings of World Multiconference on Systemics, Cybernetics and Informatics, SCI/ISAS '97, Caracas, Venezuela

Raggett D, Hors AL, Jacobs I (eds) (1999) W3C HTML 4.01 Specification, W3C Recommendation, http://www.w3.org/TR/REC-html40/

Reich S, Wiil UK, Nürnberg PJ, Davis HC, Grønbæk K, Anderson K, Millard DE, Haake JM (1999) Addressing interoperability in open hypermedia: the design of the open hypermedia protocol. The New Review of Hypermedia and Multimedia, vol 5, pp 207–248

Rigaux P, Scholl M, Voisard A (2002) Spatial databases with applications to GIS. Morgan Kaufmann Publishers

Robinson AH, Morrison JL, Muehrcke PC, Kimerling AJ, Guptill SC (1995) Elements of cartography. John Wiley & Sons

Schnase JL, Legget JJ, Hicks DL, Szabo RL (1993) Semantic data modeling for hypermedia associations. ACM Transactions on Information Systems, vol 11(1), pp. 27–50

Tomlin CD (1990) Geographic information systems and cartographic modeling. Prentice Hall

Tzagarakis M, Avramidis D, Kyriakopoulou M, Schraefel M, Vaitis M, Christodoulakis D (2003) Structuring primitives in the Callimachus component-based open hypermedia system. Journal of Network and Computer Applications, vol 26(1), pp 139–162

Voisard A (1998) Geologic hypermaps are more than Clickable Maps. In: Proceedings of the 6th ACM Int. Symposium on Advances in Geographic Information Systems, Washington, D.C., USA

Vretanos PA (ed) (2005) Web feature service implementation specification, OGC 04-094

Wallin E (1990) The map as hypertext – on knowledge support systems for the territorial concern. In: Proceedings of the European Conference on GIS (EGIS'90), vol II, pp 1125–1134

Word Wide Web Consortium (W3C), http://www.w3.org/

13 Standards for Geographic Hypermedia: MPEG, OGC and co.

Ralf Klamma, Marc Spaniol, Matthias Jarke, Yiwei Cao, Michael Jansen, and Georgios Toubekis

Abstract. Purposes of standardization in information system are interoperability, interpretability, exchangeability, and sustainability of information. Standards in geographic and hypermedia information systems are the results of complex and tedious negotiation processes on an international scale. This chapter gives an overview of the existing standards in the both fields and tries to bridge the gap between the fields. These standards include OGC OpenGIS specifications for geographic information, as well as Dublin Core, MPEG-7 and MPEG-21 for hypermedia. Meanwhile MIDAS, CIDOC and Object ID for cultural heritage will be discussed, because cultural heritage standards work with geographic hypermedia standards closely. By combining the key concepts of these technologies in an open and generic metadata framework, comprehensive geographic hypermedia systems can be deployed in various areas of applications such as cultural heritage management, e-tourism, e-government, and e-learning.

13.1 Introduction

As the outreach and scale of the information systems grow, requirements for better performance, reliability, portability, scalability of information system are more and more important and critical. The requirements also shape the main measurements to evaluate an information system. Besides the hardware enhancement data standardization is a potential method to meet these requirements. It can enhance data reliability, make data portable on different software or hardware platforms, make data available by different information systems in various scales, and enhance system performance as a result. Generally speaking, the issued standards are called metadata, data about data, which gives information at administrative, preservative, descriptive, technical and usage levels (Gilliland-Swetland 2000).

The organisations for standardization might be official organisations which are officially recognized by countries, and industry-led organisations which may or may not have collaboration with official bodies as well. Some are responsible for a certain region such as the American National Standards Institute (ANSI 2005), which is an official organisation and working on standardization in U.S. Some are for certain research fields and much of industry-led e.g. the Institute of Electric and Electronic Engineers (IEEE 2005). In addition, a great number of organizations and institutions are working on standards in information systems, such as the World Wide Web Consortium (W3C 2005), the Organization for the Advancement of Structured Information Standards (OASIS 2005), and the International Organization for Standardization (ISO 2005). The ISO is a non-government organization who is responsible for developing and deploying standards in various scientific fields. Its members come from 153 countries worldwide and make consensus on an international scale. The process of standardization gets along with requirements for standards, specifying of standards, approval and adoption of standards and deployment of standards. Moreover, standardization is often a compromise concerning political, technical and financial considerations.

Purposes of standardization in information system are interoperability, interpretability, exchangeability, and sustainability of information among others.

- *Interoperability* in the context of information describes interactions between digital information, and meanwhile leads to the question how individual metadata standards as well as domain-specific standards can be combined into coherent families of standards (Krämer *et al.* 1998).
- *Interpretability* tells how information is defined, represented and managed by means of a certain kind of metadata. At the same time, it includes the translation of information from one metadata to another one as well.
- *Exchangeability* warranties information exchange between different information systems which describes the compatibility among individual metadata standards.
- *Sustainability* is about development and reusability of information and its metadata.

Standards in information systems are both standards for information itself, namely metadata standards, and standards for information systems, also called information infrastructures in certain contexts. On the one hand, these purposes above are carried out on the base of the information level. On the other hand, it is a process between different metadata essentially. Metadata standards describe how digital items are defined, processed, and stored. And they are diverse according to the field of origin or sector the digital item is related to, such as a digital image of a library or a museum objects, a digital representation of geospatial entities or e-learning content. In the field of cultural heritage management, for instance, metadata standards clarify the relation of the virtual digital item and its association to the real physical world.

Although standards for different domains distinguish themselves greatly, most of them use the Extended Markup Language (XML) (W3C 2003) for implementation. Together with an underlying widespread standard like XML, various domain-specific standards can focus on the semantic level and make an effort to define a semantically more comprehensive standard. For the corresponding data exchange level they can take advantage of XML. As a subset of the Standard Generalized Markup Language (SGML) (Sperberg-McQueen and Burnard 1994), XML has been developed and deployed in information systems for data exchange to a wide extent shortly. XML documents are self-description data through the definable XML tags. Hence, XML documents are both machine and human readable by means of XML support tools. The Document Type Definition (DTD) and the XML Schema (W3C 2004) are used to specify XML data structures. A *schema* can formally describe the abstract structure of a set of data. An *XML Schema* developed by XML Schema Working Group is a document that describes the valid format of an XML file including what elements are and are not allowed; what the attributes for any element may be; the number of occurrences of elements etc. So XML is ensured to be written with a defined grammar and with semantic rules. XML Namespace is used in order to apply the XML tags in a certain domain without confusion. An *XML namespace* is a collection of names, identified by a URI reference, which are used in XML documents as element types and attribute names (W3C 1999). In addition, the Document Object Model (DOM) and the Simple API for XML (SAX) are used to parse XML documents, and the Extensible Stylesheet Language Transformation (XSLT) to transform XML documents. XSLT builds mapping and exchanges data among different XML data and makes XML data portable. The XPath and XQuery are used to query information in XML documents efficiently.

In short, XML serves as a basic information standard for standards on a higher level with its flexible data structure. A generalized example is the Resource Description Framework (RDF) (W3C 2003a). The metadata specification within RDF concentrates on its semantic expressiveness based on XML syntax. Meantime XML have been extended in special fields like MathML (W3C 2003d) as mathematic, GML (OGC 2003c) as geographic and ESML (2005) as earth science markup language.

Above are a brief overview on standardization organizations and a short introduction of XML that is a basis for metadata standards. We will now focus on standards for geographic information systems (GIS) and hypermedia information systems. Standards in these fields are also the results of complex and tedious negotiation processes on an international scale. As in many other areas of application of information systems, the usage of different data sources as well as different data representation like videos and audios cause heterogeneity of data that is to be managed. For that purpose, standards are used to harmonize the data on a national, regional or even global level. Responsible organizations and institutions are also structured on industry-led or official level such as the industry-led Open Geospatial Consortium (OGC) (OGC 2005b), and the official organisations, the Infrastructure for Spatial Information in Europe (INSPIRE 2005) and the Global Spatial Data Infrastructure (GSDI 2005) etc.

This chapter is organized as follows. The standards for geographic information system include system infrastructure standards and spatial data standards, which will be discussed in the next section. In Section 13.3, standards for cultural heritage management are introduced in terms of MIDAS (2003) and Object ID (Thornes *et al.* 1999) etc. In Section 13.4, efforts in the hypermedia context are discussed. Among them, the Dublin Core Metadata Initiative develops "specialized metadata vocabularies for describing resources" (Dublin 2005). The Moving Picture Experts Group (MPEG) who also specifies MPEG-7 and MPEG-21 standards deals with "the development of standards for coded representation of digital audio and video" (MPEG 2005). Finally, a comparison between these standards and mappings that allow the conversion of these standards conclude this chapter.

13.2 Standards for Geographic Information Systems

Standards for geographic information define and describe spatial data. They contain information about processing and presentation of the data as well. Basically, there are two levels of standards. One is the information-level standards for geographic data that deal with data interoperability. The other is the system-level standards for geographic infrastructures that cover the design and development of geographic information systems.

13.2.1 Standards for Geographic Data

Standards for geographic data cover aspects from spatial data structures, geographic data markup languages, spatial data search languages to spatial data rendering standards. For spatial data there are two basic data structures: *vector* data and *raster* data. It is assumed that the real world is simplified as a two dimensional space and we discuss how to map such a space to a certain data structure. In the case of vector data, information is stored in a vector that contains a set of coordinates. It represents objects of the real world by geometric elements, also called entities, such as points, lines or arcs, and polygons. Another way of storage is the raster data structure that contains spatial information in form of images. It can also be regarded as a matrix of pixels/cells. And each entry contains a value to describe the corresponding pixel of the image. So vector data represents spatial data by space boundaries, while raster data depicts the information by filling the space (Peuquet 1984). Vector data has a smaller file size and can capture each feature precisely on various scales, while raster data can be computed and displayed more quickly on screens (Rosenbaum and Tominski 2003). These different features make vector data and raster data suitable for different applications respectively. Due to the interoperability problems between the both entirely different data structures, hybrid models have been widely applied. A hybrid model refers to a physical integration of vector data and raster data in one application system on the one hand. On the other hand, it builds a hybrid data structure through some approaches to conversions between the both data structures. An approach to a hybrid data structure has been proposed in Winter (1996) and Winter (1998) to integrate vector and raster data in GIS and to translate spatial concepts in image interpretation. The use and implementation of such a hybrid data structure for geographic data integration is discussed in Isbihani (2005). The advantages of the hybrid data structure are that it can support explicit boundaries and support cells and grids. Therefore, it is useful in the field of geospatial data integration.

Another general approach to represent geographic information is the *Geographic Markup Language* (GML) specified by the Open Geospatial Consortium (OGC). Embedded in XML document, GML models, transports and stores geographic information including spatial and non-spatial properties of geographic features. Since the third version was released in 2003, GML 3.0 can represent geospatial phenomena including features with complex, non-linear, 3D geometry, features with 2D topology, features with temporal properties, dynamic features, coverages, and observations, topology, geometry, coordinate reference systems, measure units, temporal information or time, and value objects (OGC 2003c), while GML 2.0 merely represents simple 2D linear features. Dedicated GML tags represent vector data which refers to geographic features, and raster data which refers to coverages. GML documents are validated by a GML schema that is based on an XML Schema. Even more, the elements of other namespaces can be integrated into the GML document, if an additional XML schema is given. This enhances the expression power and the interoperability of GML.

For geographic information search and retrieval an extension of the *Structured Query Language* (SQL) is applied to allow spatial search. Spatial queries consist of primary and secondary filter operators (SDO_FILTER() and SDO_RELATE()), within-distance operator (SDO_WITHIN_DISTANCE()), and nearest neighbor operator (SDO_NN()) etc. In addition, the spatial query is confined with different relationships such as touch, cover, inside, covered by, any interact etc. Geographic information is stored as a geometry object in a spatial database to be processed through these SQL spatial query functions. An example of an SQL script for spatial queries is listed in Fig. 13.1. Moreover, spatial indexes are essential to accelerate spatial queries. An important type of spatial indexes is the R-tree that approximates each geometry object with the smallest single rectangle that encloses the geometry, the so-called minimum bounding rectangle (MBR) (Oracle 2003). A two-tier process is used for spatial queries to accelerate the query process with the help of a spatial index. The first is using the filter function to screen candidate rows only based on their spatial index. And the second is to check whether the candidate rows meet the query exactly.

Regarding the rendering of geospatial information the Scalable Vector Graphic (SVG) (W3C 2003c) is frequently used. SVG is a language that describes two-dimensional graphics in XML. It supports vector graphic shapes, images and text. Graphical objects can be grouped, styled, transformed and composed into previously rendered objects. Common browsers like Internet Explorer and Netscape etc. support SVG and represent SVG graphics smoothly with the MIME type is "image/svg+xml". In addition, there are special tools to read SVG such as Adobe SVG Viewer. Like XML, processing SVG uses SVG Document Object Model (DOM) to access all elements, attributes and properties. Summarily, SVG enables user interaction on the graphics completely, and is easily portable on the mobile devices with restricted resource capacities.

```
select geo_id, geo_name from geoobject g where mdsys.sdo_filter
(g.shape, mdsys.sdo_geometry(2003, 8307, null, mdsys.sdo_elem_info_array(1, 1003, 1),
mdsys.sdo_ordinate_array(30.00, 64.00, 32.00, 64.00, 32.00, 66.00, 30.00, 66.00, 30.00, 64.00)
), 'querytype=WINDOW') ='TRUE' and sdo_relate (g.shape, mdsys.sdo_geometry(2003, 8307,
null, mdsys.sdo_elem_info_array(1, 1003, 1), mdsys.sdo_ordinate_array(30.00, 64.00, 32.00,
64.00, 32.00, 66.00, 30.00, 66.00, 30.00, 64.00)), 'mask=anyinteract querytype=WINDOW')
='TRUE';
```

Fig. 13.1. SQL Script for Spatial Queries

13.2.2 Standards for Geographic Infrastructures

Standards for geographic infrastructures are applied for spatial service development and processing of spatial data. The most widespread standards for geographic infrastructures are specified by the GIS software vendor, the Environment System Research Institute (ESRI 2005), as well as the international industrial GIS consortium, the Open Geospatial Consortium (OGC 2000).

ESRI was founded as a consulting company focusing on managing and analyzing geographic information in 1969. In 1981 ESRI launched its first commercial GIS software called ArcInfo and developed a set of application tools to support geographic information systems in the following years. The main product line is the ArcGIS, which contains mainly ArcView, ArcIMS, and ArcSDE etc. Among them, ArcSDE connects its applications to databases, ArcIMS creates, supports and automates metadata complied with OGC OpenGIS specifications. ArcPad is launched especially for mobile devices. So the GIS applications can be carried out in various devices ranging from desktop to mobile and embedded devices on the client side. Besides ArcGIS, ESRI has developed ArcWeb service that provides web services for GIS users and developers. The service can be used to access users' own data directly, or can be integrated into users' web applications conveniently. In addition, ESRI also specifies a geographic data format, the shape file, which is one of the most widespread spatial data storage standard supported by many other GIS vendors and applications. The shape file related to a geometry feature consists of three types of files: a main file, an index file and a file for a dBase table. Each geometric feature must be indexed before the storage in a dBase table. ESRI passed the OGC first conformance test in 1999 and also defines standards (ESRI 2002) for GIS industry as an OGC member.

In addition, the Open Geospatial Consortium (OGC) founded in 1994 is a member-driven non-profit initiative that pursues the interoperability of GIS data among diverse commercial GIS software. OGC defines geospatial specifications and standards to design and implement GIS applications in cooperation with governments, industry and academia. OGC implements open geospatial metadata standards including dataset title, spatial representation type, geographic location of the dataset (by coordinates or by geographic identifier), dataset topic category, and spatial resolution of the dataset etc. Moreover, OGC provides the specification and various conformance tests for diverse GIS software products. Conformance testing determines whether a product implementation of a particular OpenGIS Implementation Specification fulfils all mandatory elements and whether these elements are operable (OGC 2003a).

The OGC OpenGIS specification includes four main components: Web Map Server (WMS) Interface Implementation Specification, Web Feature Server (WFS) Implementation Specification, Web Coverage Server (WCS) Implementation Specification and Geography Markup Language (GML). An OpenGIS WMS provides interfaces to simplify the web-based client and server programming for processing map requests. It defines its implementation specification as an Application Programming Interface (API). Static map images in form of GIF and JPEG etc. can be retrieved through this API. A WFS describes data manipulation operations including insertion, deletion, update, and get and query a feature / a geometric object with spatial or non-spatial constraints (OGC 2002). It enables access to individual features. A WCS provides access to potentially detailed and rich sets of geospatial information in the form of multi-themes by defining theme manipulation operations (OGC 2003b). The principle is that requests and responses are sent in the form of XML documents with the Hypertext Transfer Protocol (HTTP) between clients and web servers. They can be written in GML, as well as in any other XML schema. The request examples in OGC (2002) and OGC (2003a) indicate that WFS and WMS don't use GML elements, but the other XML Schemas to formulate the requests. That shows the query format is essentially XML-based and defined by the service (WMS or WFS). The relationships among those four main specifications and the usages are illustrated in Fig. 13.2. OGC OpenGIS has also predefined many auxiliary service specifications to aid the implementation of the main services. For instance, the Styled Layer Descriptor specifies the API for the layer rendering styles and is used by WMS. All these specifications together define essential concepts, vocabularies, and structures of geospatial services and information transfer.

Many applications have been implemented based on OGC OpenGIS specifications. For instance, the Deegree project (Fitzke *et al.* 2003) at Bonn University has launched WFS, WCS and WMS as open source to be easily integrated into other applications. Besides the GIS system specifications, the standardization of geographic information is of much importance. According to an estimation of ESRI, 80% of the costs of a GIS are associated with development and maintenance of data (FGDC 2002).

13.3 Standards for Cultural Heritage Management

Geographic standards have been widely deployed in applications for cultural heritage management (Brandon *et al.* 1999). In this research field

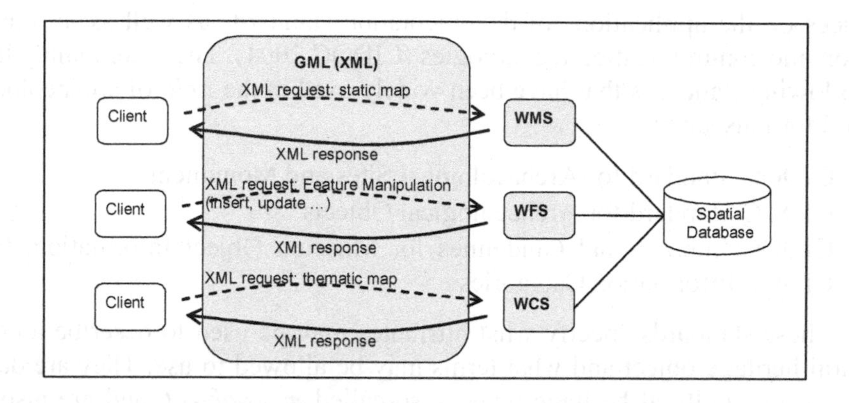

Fig. 13.2. Web Map Service Principle

great efforts have been made to specify the description of cultural heritage and to apply the existing standards (Dorninger *et al.* 2005).

Digitalization efforts with versatile categories for cultural heritage in individual country can be traced back to the late 1970's. Since that time standardization has concentrated mainly on the digitalization methods for compilation and retrieval of physical museum collections. This led to a wide variety of national standards for museum collections and their digital representation such as SPEKTRUM developed by MDA, the agency for documentation and management of museum collections for the United Kingdom. However, cultural heritage categories also include historic environment which includes built heritage, archaeology, and current and past landscapes (Lee 2005). And geographic hypermedia standards are especially crucial in this context. The political dimension of cultural diversity and the need of the information society are illustrated in the declaration of the European Union "Lund Principles" of 2001 (IST 2001). Members stated clearly to coordinate digitalization efforts and to report and publish them yearly.

Several eminent standards in the field of cultural heritage management are introduced as follows.

13.3.1 CIDOC

The International Committee for Documentation of the International Council of Museums (ICOM-CIDOC) (CIDOC 2005) is the international group for the documentation of museums and similar organizations. Aiming at gathering, management and sharing of the knowledge carried in heritage collections, CIDOC provides a discussion forum to provide advices on the applications of documentation standards as well as information and communication technologies (CIDOC 2004). There are mainly the following standards that have been widely used in the field of archaeology and for museums:

- CIDOC Standard for Archaeological Sites and Monuments
- CIDOC Standard for Archaeological Objects
- CIDOC International Guidelines for Museum Object Information: the CIDOC Information Categories

These standards specify what attributes may be used to describe a cultural heritage object and what terms may be allowed to use. They are data about real cultural heritage objects, so-called *meta-object*, and are also a kind of metadata. Hence, the CIDOC Concept Reference Model (CRM) has been developed to allow interoperability "not only on the encoding level (…) but also on the more complex level semantics level"[1]. It was on vote as Draft ISO Standard (ISO 21127) until February 2002 and was developed as core ontology model for capturing common semantics of heterogeneous data structures in order to support their semantic integration (Doer 2005). This standard is, however, text-based and can not describe multimedia data properly. So some proposals have been launched to use multimedia standards in CRM. For example, Hunter has proposed a mapping model between CIDOC CRM and MPEG-7 in Hunter (2002).

13.3.2 MIDAS

The Monument Inventory Data Standard (MIDAS 2003) is a "content standard that sets out what sort of information should be recorded, for instance, information to describe the character or location of a monument" (FISH 2005). It was developed mainly during 1996 to 1998. The late revised version was proposed in December 2003 which covered CIDOC-CRM issues. MIDAS defines an agreed list of the units of information, also called metadata elements, which should be included in an inventory or other systematic record of the built historic environment. These units of information are grouped together in *schemes* and cover areas such as monument character, events, people and organization, spatial data etc. It is developed by user communities in the field of conservation and promoted by English Heritage, the advisory body of the English Government for the

[1] According to Doer (2005) to date most attempts to bridge the gaps between incompatible information system in the cultural field have been resorted to massive simplifications, concentrating on a limited subset of core data

historic environment. As a text-based standard it generally does not support the production, presentation and dissemination of digital image or GIS technology (MIDAS 2003). This point will be considered in the next revision of the standard (Lee 2005). Nowadays MIDAS is managed by the Forum on Information Standards in Heritage (FISH 2005) established in 2000. It is a platform for software developers and standard professionals with the goal of the interoperability and exchange in the heritage sector.

13.3.3 Object ID and Core Data Standard/Index of Getty Information Institute

Object ID is an international standard for describing cultural objects initiated by Getty Information Institute in 1998. It has been developed for art and antiques through the collaboration of the museum community, police and customs agencies, the art trade, and insurance industry. The standard was defined by a combination of background research, interviews, and by major international questionnaire surveys. It is also text-based and not suitable for describing multimedia. It has a close relationship to the following two standards, because the element sets are similar among them. So the interoperability among these three standards is enhanced as they are used for different cultural heritage categories.

The *Core Data Index to Historic Building and Monuments of the Architectural Heritage* was defined by a working group with members from heritage organizations in Europe and was approved in October 1992. It specifies a minimal set of data elements to describe architectural heritage. This standard establishes consensus over 137 architectural inventories or description by 78 organizations in 26 countries (Thornes and Bold 1998). Documents of this standard comprise content such as photographs and archives, environmental files, biographical materials and so on. The information is all text-based and does not support multimedia description.

The *Core Data Standard for Archaeological Sites and Monuments* was specified by the archaeology documentation group of the Council of Europe in cooperation with the ICOM-CIDOC. It was published in 1995 and can be linked with the CIDOC standard for archaeological objects (CIDOC 1992) issued in 1992 and CIDOC's International Guidelines for Museum Object Information (CIDOC 1995) issued in 1995. This standard specifies a core data set to inventory and to document archaeological heritage. It has a close relationship to the Core Data Index to Historic Building and Monuments of the Architectural Heritage. Due to their immanent relation, the Core Data Index and Core Data Standard are similar.

13.4 Standards for Hypermedia

The aforementioned two categories of standards are used to describe and process geographic information as well as cultural heritage information. Information in both fields is in great need of multimedia for information representation. Geographic information is displayed in various digital maps, while cultural heritage has a great number of images and video files etc. Nowadays, multimedia has become an important component in a wide context that ranges from diverse multimedia information systems, enterprise information systems to geographic information systems. In the past media has been kept in the form of books, films, photos, drawings, maps and so forth. Lately, more and more multimedia in material has been digitalized and managed with modern information technology. Along with the rapid development of scanners, digital cameras and other devices, the quality of multimedia can be well kept, even enhanced.

For instance, the World Wide Web is a good platform to preserve and present multimedia. The intertwined multimedia through hyperlinks builds a new concept, the so-called hypermedia. In the same time, the management, search and retrieval of hypermedia information become key challenges. For that reason, a lot of multimedia as well as hypermedia standards have been launched to provide some solutions.

13.4.1 Dublin Core

Aiming at a better discovery for digital resources, the Dublin Core Metadata Initiative (DCMI) (Dublin 2005) began to specify Dublin Core metadata in 1995. The core set of Dublin Core, namely the Dublin Core Metadata Element Set (DCMES) (Dublin 2003), is composed of 15 elements or descriptors that try to describe resources with interdisciplinary and international consensus. Besides DCMES, Dublin Core has defined certain type vocabularies as well as refined elements set and encoding schemas, which are all brought together as the Dublin Core metadata terms. From the beginning on, DCMI has tried to specify a standard that can be simply used, can be easily understandable and can coexist with the other metadata standards (Weibel *et al.* 1998). The 15 elements of DCMES can be divided into three groups according to metadata categories:

- Descriptive metadata: *Title, Subject, Description, Type, Source, Relation, Coverage*
- Administrative and use metadata: *Creator, Publisher, Contributor, Rights*
- Technical and preservation metadata: *Date, Format, Identifier, Language*

All these elements are not defined in a hierarchical structure, but on the same level. Due to its simplicity, Dublin Core can be easily converted into more sophisticated metadata standards. DCMI specified the simplest mapping between DCMES and several other standards such as the standard of the Research Libraries Group (RLG 2002).

In conclusion, the Dublin Core metadata standard is a manageable set of core elements that can be used to describe multimedia content. However, Dublin Core is mostly suitable for print media in the fields of libraries and museums. In particular, content compliant to GIS standards can not be converted directly to and from Dublin Core.

13.4.2 MPEG-7

The Multimedia Content Description Interface (MPEG-7) defined by Motion Pictures Expert Group (MPEG) (MPEG 2005) aims at providing a rich set of audiovisual descriptions in order to enhance search and retrieval of digital multimedia content. MPEG-7 is a powerful metadata standards complex for digital media knowledge management using rich multimedia descriptions. MPEG-7 defines a multimedia library of methods and tools in a flexible and extensible framework to describe audiovisual content. MPEG-7 standardizes (MPEG 2004):

- *Descriptors (D):* A descriptor is a representation of a feature that defines the syntax and semantics of the feature representation.
- *Description Schemes (DS):* A description schema specifies the structure and semantics of the relationships between its components which may be both descriptors and description schemes.
- *Description Definition Language (DDL):* It allows for the extension and modification of existing description schemes.

Among them, *Visual Description Schemes* and *Audio Description Schemes* are particularly interesting for geographic information systems. They provide descriptors for visual and audio media to describe spatio-temporal information respectively. In addition, the MPEG-7 *Multimedia Description Schemes* (MDS) provides a standardized set of technologies to describe multimedia content. Fig. 13.3 gives an overview on the multimedia description schemes and indicates options to embed geographical information and cultural heritage data by means of MPEG-7.

The *Structural Aspects* describe audiovisual data about physical, spatial and temporal segments of the content. Physical means that an image might be analyzed and divided into several physical segments. Each segment will be described separately. The *Conceptual Aspects* describe the semantic of the content, such as events, objects, concepts and their relations. *Creation and Production* contain information about the creator, the creation date. *Media* includes information of format, coding, instances, size, and resolution etc. In *Usage* the use records, access rights and rights holder information are described. *Navigation and access* support discovery, browsing, navigation, visualization and variation adaptive to various users. *User interaction* describes the user's activities on media. It can provide worthy information of user preferences and history of audiovisual data for the purpose of data personalization. *Collection* is used for the description and organization of media collections. *Models* provide analytic models for data processing.

In practice the basic element set is usually applied within many other element sets, but might also be used standalone. MPEG-7 descriptors are specified by description schemes as well as by defining the type of each descriptor. The value of a descriptor can be either a simple type or a complex type. Simple types contain basic types such as string, integer, and date etc. A complex type itself is an MPEG-7 descriptor again. It specifies a hierarchical relationship between basic or complex descriptors as well as constraints such as min/max occurrences and sequencing etc. For example, the descriptor *CreationInformation* is a complex type that contains the descriptors *Creation*, *Classification* and *RelatedMaterial* (Fig. 13.4). Among them, the *Creation* element is a mandatory sub element, while the other two are optional. Additionally, the *RelatedMaterial* descriptor might be used more than once within any *CreationInformation* descriptor. The oc-

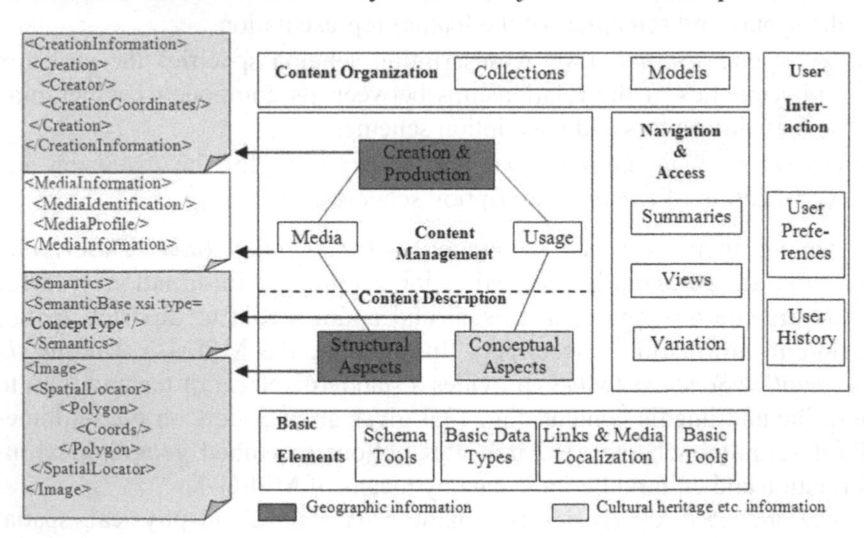

Fig. 13.3. Overview of MPEG-7 Multimedia Description Schemes (Adapted from MPEG 2004)

```
<complexType name="CreationInformationType">
  <complexContent>
    <extension base="mpeg7:DSType">
      <sequence>
    <element name="Creation" type="mpeg7:CreationType"/>
    <element name="Classification" type="mpeg7:ClassificationType" minOccurs="0"/>
    <element name="RelatedMaterial" type="mpeg7:RelatedMaterialType" minOccurs="0" maxOccurs="unbounded/>
      </sequence>
    </extension>
  </complexContent>
</complexType>
```

Fig. 13.4. Definition of CreationInformation Description Schema

currence constraints are specified with the attribute *minOccurs* and *maxOccurs*. The sub elements *Creation, Classification* and *RelatedMaterial* are complex types again and have own sub elements, too. Although each descriptor is strictly defined, it allows a flexible schema definition for versatile application scenarios.

In short, MPEG-7 is able to interoperate and exchange information among other standards easily. A mapping between MPEG-7 and Dublin Core has been proposed in Hunter (2000). A fine-grained implementation of the mapping can be found in Spaniol and Klamma (2005). In the use area of MPEG-7 Multimedia Descriptor Schema for geographic hypermedia information systems, geographic information can be described in the element sets of *Structural Aspects* as well as *Creation and Production. Conceptual Aspects* can be used to store specific information for cultural heritage management. Excerpts of the XML documents using MPEG-7 tags are listed in the left side of Fig. 13.3.

13.4.3 MPEG-21

After the release of the MPEG-7 multimedia description standard, MPEG began to recognize problems in the description of multimedia access, delivery and consumption by means of MPEG-7. For that purpose, the Multimedia Framework (MPEG-21) (MPEG 2002) has been specified since 2002, which tries to meet the requirements for multimedia processing. Some crucial outside influences have triggered the development of MPEG-21. First, many multimedia metadata standards exist simultaneously currently. Each standard provides a well-defined multimedia description possibility to consume multimedia in a certain discipline. Thus, a multimedia framework is required to enable an interdisciplinary multimedia consumption. Second, with the dramatic increase of multimedia content and its relevant access rights, it becomes increasingly difficult to only deliver the multimedia content to legitimated users. Thus, different intellectual rights have to be guaranteed and multimedia content has to be protected from illegal distribution and access. Third, the ongoing hardware development and miniaturization of end devices are rapid. An increasing number of users do have not only desktop computers, but also laptops, PDAs and cell phones. Thus, there is a need to provide and convert multimedia content for various types of hardware devices. MPEG-21 addresses this issue by variations that allow multimedia content management to be adopted for different end devices and network conditions. As mentioned in MPEG (2002), MPEG-21 aims at enabling the transparent and augmented use of multimedia resources across a wide range of networks and devices. Summarily, MPEG-21 provides a multimedia framework for multimedia access, delivery and consumption, in order to enable interoperability, content protection and content adaptation in a distributed multimedia system (Kosch 2003, MPEG 2002, Burnett *et al.* 2003).

In the following, some key concepts of MPEG-21 are introduced in detail. The *digital item* is a core feature which represents description entities of multimedia content through a container and items flexibly. A digital item can be described as a container that contains items and even containers if necessary. Each item consists of descriptors and components. Descriptors can be defined in any standard e.g. MPEG-7. An excerpt of an example file is listed in Fig. 13.5. It groups a photo and a text as digital items by using MPEG-7 descriptors. Therefore, MPEG-7 as well as the other descriptors for multimedia content can be well utilized in the open framework. At the same time, the document obtains a unified common representation through the digital item tags. Another core feature is the *user model* which represents a definition of users in a multimedia framework. A user is anyone that interacts with digital items or is involved in the MPEG-21 environment. The users include individuals, communities, organizations, corporations, consortia, government and so on. According to users' different kinds of interactions, users can be classified as consumers, creators, and providers etc.

The usage scenarios of MPEG-21 are described as follows, which correlates with the motivation to develope the MPEG-21 standard (Bormans and Grant 2002).

- The *Universal Multimedia Access* concept aims at a maximized sharing of various multimedia resources.
- *Dynamic Adaptation* for terminal devices and network resources is demanded at delivering multimedia content to various user communities with diverse needs.

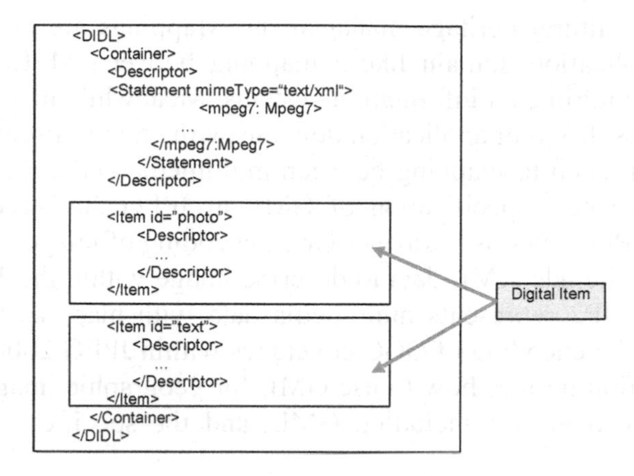

Fig. 13.5. Digital Item Declaration

- *Capacity Exchange* refers to multimedia content delivery corresponding to the capacity of the terminal devices. This could be realized through exchange the terminal profile information.
- In the aspect of *Quality of Service* (QoS) of terminal devices, MPEG-21 could be used to guarantee the multimedia stream delivery when some unexpected events during streaming occur.

Moreover, MPEG-21 can be used in many other scenarios in a combination with other multimedia standards. It allows tracing user activities in more detail by interrelating user models with digital items.

13.5 Comparison and Conclusion

Summarily, various standards have been developed in those three application areas: geography, hypermedia and cultural heritage. However, these standards are most suitable for the representation of one of the three fields: geographic data, cultural heritage management or multimedia content. Fortunately, there are a number of approaches to combining these contents through some geographic hypermedia standards or the other standards.

13.5.1 Comparison of Metadata Standards for Different Domains

Mapping is a basic and useful approach to the integration of different standards in the application area of geographic hypermedia information systems for cultural heritage management. Mappings are needed within the same application domain like a mapping between MPEG-7 and Dublin Core for multimedia information systems. Meanwhile mappings are specified across different application domains with an interdisciplinary view.

An approach to mapping between multimedia and geographic information is the new specification of GML in JPEG 2000 (GMLJP2) (OGC 2005a). JPEG 2000 is a wavelet-based encoding of images that provides an ability to include XML data to describe image within the JPEG 2000 data file. GMLJP2 represents multimedia data with maps as the content and supports the encoding of OGC coverages within JPEG 2000 data files. The specification defines how to use GML for geographic images, the packaging mechanisms for including GML, and the specific GML application schemas.

Another approach pursues a mapping between Dublin Core and Object ID (Cao 2004). This enhances the functionality of multimedia search and object search. Both the multimedia annotation and the cultural object description are saved in XML files. An appropriate mapping enables the information exchange between the both XML files (Fig. 13.6). Based on the mapping, user communities can select the related objects when they are about to annotate an image file. It works in this way as follows. Some information for the objects can be prepared in the input fields of multimedia annotation. Users just need to make some information modifications or keep it as it is. Hence, the time which is used to input some duplicated information can be saved.

13.5.2 Conclusion

Current research aims at integrating these standards including OGC OpenGIS, MPEG-7, Dublin Core, ObjectID, and Core Data Standard/Index in a comprehensive geographic hypermedia system for cultural heritage management: ACIS (Klamma *et al.* 2005a). It proves that various standards from different domains and various standards from one domain can represent data in a cooperative way through mappings among standards. Except the mapping between Dublin Core and ObjectID, a mapping between Dublin Core and MPEG-7 is employed to implement the user input interface, so that the user communities have a dedicated DCMES input mask. A further mapping to GML tags is also possible. For instance, the DCMES element *Coverage* is specified to store the location information of a hypermedia file. This information is then stored in the tag *CreationCoordinates* in an MPEG-7 XML file. An additional mapping to *gml:coordinates* can be applied. Parts of the XML documents are listed in Fig. 13.7.

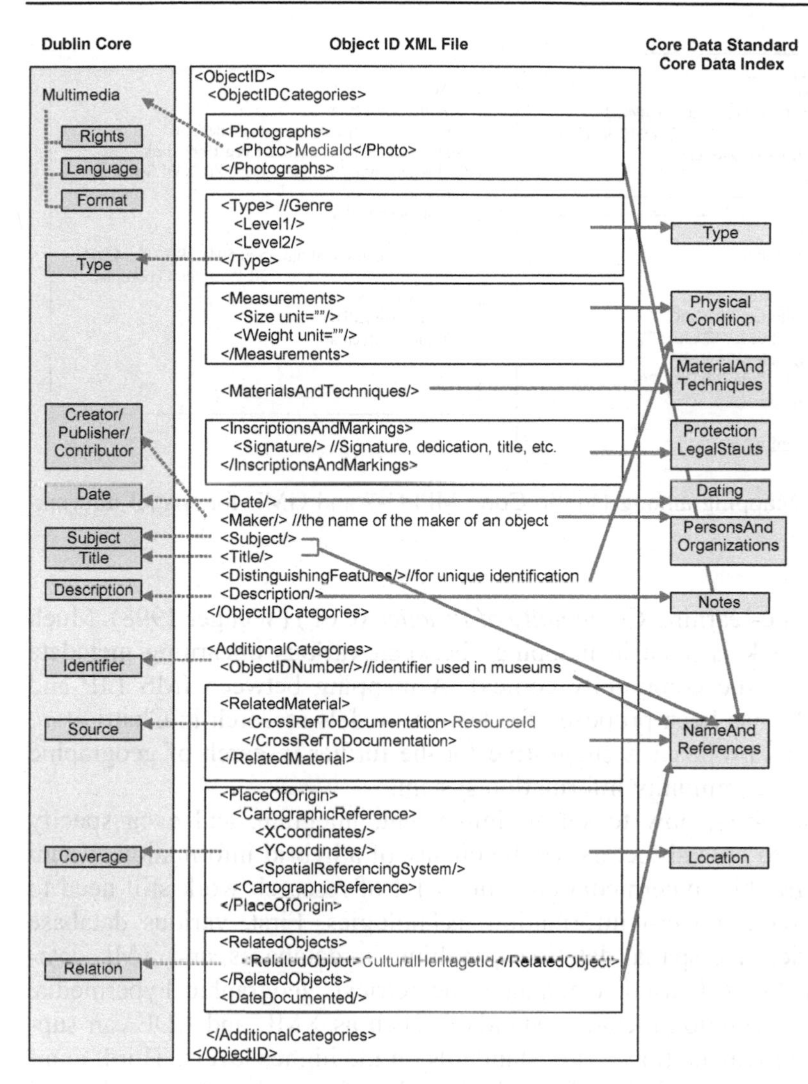

Fig. 13.6. Mapping to Core Data Index/Standard and Dublin Core

As a result, information exchange among different user communities is fostered. On the one hand, a geographic hypermedia information system is required to manage and deliver appropriate content. On the other hand, it should be able to well support the involved users, especially in a network environment (Kumar *et al.* 2002). To enhance the interaction between geographic hypermedia content and user communities, the potential usage scenarios of geographic hypermedia in e-learning are surveyed in Klamma *et al.* (2005). E-learning is another wide application area with many metadata standards for learning resources. It also involves e-learners greatly to

Dublin Core:

```
<dc:Coverage>
        <dc:Coverage.rect>60.5048,29.3615,
                        74.8941,38.4913
        </dc:Coverage.rect>
</dc:Coverage>
```

MPEG-7:

```
<CreationCoordinates>
 <Location>
<GeographicPosition>
  <Point longitude="60.5048"
          latitude="29.3615"/>
  <Point longitude="74.8941"
          latitude="38.4913"/>
</ GeographicPosition >
 </Location>
</CreationCoordinates>
```

GML:

```
<feature_collection
xmlns:gml="http://www.opengis.net/gml"
xmlns:xlink="http://www.w3.org/1999/xlink"
xmlns:xsi="http://www.w3.org/2001/XMLSchema-
instance">
- <gml:boundedBy>
- <gml:box>
        <gml:coordinates>60.5048,29.3615,74.8941,
                    38.4913</gml:coordinates>
    </gml:box>
  </gml:boundedBy>
<</feature_collection>
```

Fig. 13.7. Mapping among Dublin Core, MPEG-7 and GML for spatial information

build up an e-learning *Community of Practice* (CoP) (Wenger 1998). Much research work is done in mappings between various e-learning metadata standards in the community context. A mapping between IMS LIP and IEEE PAPI has been proposed for learner models research in Chatti *et al.* (2005). This proposal is suggestive for the further research of geographic hypermedia community information system.

In conclusion, how to select, implement, integrate, and even specify standards are crucial factors for the quality of a hybrid information system with geographic hypermedia content. A lot of research work still need to be done with modern information technologies. First, various database technologies like spatial databases, multimedia databases, and XML databases can be well used to manage and retrieve geographic hypermedia. Second, information metadata standards such as XML and RDF can support the implementation of the standards on the higher levels. Third, compared to the standards in other scientific domains such as geography and culture, multimedia standards like MPEG-7 have a more solid basis on information technology as a kind of digital information standards. Thus, they need to be investigated as a paradigm for a better interoperability with standards in domains of non-information technologies.

References

ANSI (2005) The American National Standards Institute web page. http://www.ansi.org/

Bormans J, Grant K (2002) MPEG-21 use case scenarios, ISO/MPEG N4991

Brandon RJ, Kludt T, Neteler M (1999) Archaeology and GIS, the Linux way. Linux Journal, Specialized Systems Consultants, Inc., July 1999, Seattle, USA

Burnett I, Walle Rv, Hill K, Bormans J, Pereira F (2003) MPEG-21: goals and achievements. Multimedia IEEE, vol 10(4), pp 60-70

Cao Y (2004) Open standards and architectures for community information systems in cultural heritage management. Diploma thesis, Department of Computer Science, RWTH Aachen

CIDOC (1992) CIDOC core data standard for archaeological objects. CIDOC Archaeological Sites Working Group, http://cidoc.natmus.dk/engelsk/standard_for_arch.asp

CIDOC (1995) CIDOC international guidelines for museum object information: the CIDOC information categories. Grant A, Nieuwenhuis J, Petersen T (eds), ICOMOS, USA, http://www.willpowerinfo.myby.co.uk/cidoc/guide/guide.htm.

CIDOC (2004) ICOM-CIDOC homepage: introduction to CIDOC, http://www.willpowerinfo.myby.co.uk/cidoc/#CIDOC.

CIDOC (2005) The CIDOC conceptual reference model (CRM) homepage, http://cidoc.ics.forth.gr/

Chatti MA, Klamma R, Quix C, Kensche D (2005) LM-DTM: an environment for XML-based. LIP/PAPI-compliant deployment, transformation and matching of learner models. In: Goodyear P, Sampson DG, Yang DJ-T, Kinshuk, Okamoto T, Hartley R, Chen NS (eds) Proceedings of the 5th International Conference on Advanced Learning Technologies (ICALT 2005), Kaohsiung, Taiwan, pp 567-569

Dublin (2003) Dublin Core metadata initiative usage board: Dublin Core metadata element set. Version 1.1, Reference Description, http://www.dublincore.org/documents/dces/

Dublin (2003) Dublin Core metadata initiative homepage. http://www.dublincore.org/

Dorninger P, Kippes W, Jansa J (2005) Technical push on 3D data standards for cultural heritage management. In: Schrenk M (ed) Proceeding of CORP 2005 & Geomultimedia

Doer M (2005) The CIDOC-CRM, an ontological approach to schema heterogeneity. http://drops.dagstuhl.de/opus/volltexte/2005/35/.

ESML (2005) Earth Science Markup Language homepage. http://esml.itsc.uah.edu /index.jsp

ESRI (2002) Metadata and GIS, an ESRI white paper. http://downloads.esri.com/support/whitepapers/ao_/metadata-and-gis.pdf

ESRI (2006) ESRI homepage. http://www.esri.com/

FGDC (2002) Federal Geographic Data Committee Standards: Metadata standards report. http://www.swfwmd.state.fl.us/data/gis/libraries/metadata/html/FGDC_stnds.pdf.

Fitzke J, Greve K, Müller M, Poth A (2003) Deegree – ein Open-Source-Projekt zum Aufbau von Geodateninfrastrukturen auf der Basis aktueller OGC- und ISO-Standards. GIS, vol. 9, pp 10-16

FISH (2005) The Forum of Standards in Heritage (FISH). http://www.fish-forum.info/

Gilliland-Swetland AJ (2000) Setting the stage. Getty Research Institute, http://www.getty.edu/research/conducting_research/standards/intrometadata/2_articles/index.html.

GSDI (2005) The Global Spatial Data Infrastructure (GSDI) Association. http://www.gsdi.org/

Hunter J (2000) A proposal for the integration of Dublin Core and MPEG-7. ISO/IEC JTC1/SC29/WG11 M6500, 54th MPEG Meeting, La Baule

Hunter J (2002) Combining the CIDOC CRM and the MPEG-7 to describe multimedia in museums. In: Proceedings of the International Conference about Museums and the Web Boston, Massachusetts

IEEE (2005) The Institute of Electric and Electronic Engineers (IEEE). http://www.ieee.org/portal/site

INSPIRE (2005) The Infrastructure for Spatial Information in Europe (INSPIRE). http://www.ec-gis.org/inspire/

Isbihani AE (2005) Entwicklung eines Modells für die Integration hybrider Geodaten am Beispiel der Grundwasserschutzfunktion. Diploma thesis, Department of Computer Science, RWTH Aachen

ISO (2005) The International Organization for Standardization. http://www.iso.org/iso/en/ISOOnline.frontpage.

IST (2001) The Information Society Technologies: eEurope – the Lund principles. Lund, Sweden, http://www.cordis.lu/ist/digicult/lund_p_browse.htm.

Kosch H (2003) Distributed multimedia database technologies supported by MPEG-7 and MPEG-21. CRC Press

Krämer B, Michael Papazoglou M, Schmidt HW (1998) Information systems interoperability. John Wiley & Sons Inc., New York, USA

Kumar R, Raghavan P, Rajagopalan S, Tomkins A (2002) The web and social networks. IEEE Computer, vol 35(11), pp 32-36

Klamma R, Spaniol M, Jarke M, Cao Y, Jansen M, Toubekis G (2005a) A hypermedia Afghan sites and monuments database. In: Proceedings of the First International Workshop on Geographic Hypermedia "Geographic Hypermedia: Concepts & Systems", Denver, USA, pp 59-73

Klamma R, Spaniol M, Jarke M, Cao Y, Jansen M, Toubekis G (2005) ACIS: intergenerational community learning supported by a hypermedia sites and monuments database. In: Goodyear P, Sampson DG, Yang DJ-T, Kinshuk, Okamoto T, Hartley R, Chen NS (eds) Proceedings of the 5th International Conference on Advanced Learning Technologies (ICALT 2005), Kaohsiung, Taiwan, pp 108-112

Lee E (2005) Building interoperability for United Kingdom historic environment information resource. D-Lib Magazine, vol 11(6), http://www.dlib.org/dlib/june05/lee/06lee.html

Bormans J, Hill K (eds) (2002) MPEG-21 requirement group: MPEG-21 overview. Version 5, Shanghai, http://www.chiariglione.org/mpeg/standards/mpeg-21/mpeg-21.htm.

MPEG (2004) MPEG-7 overview. ISO/IEC JTC1/SC29/WG11, http://www.chiariglione.org/mpeg/standards/mpeg-7/mpeg-7.htm

MPEG (2005) Moving Picture Experts Group homepage. http://www.chiariglione.org/mpeg/

MIDAS (2003) FISH: MIDAS a Manual and Data Standard for Monument Inventories. http://www.english-heritage.org.uk/Filestore/nmr/standards/Midas3rd-Reprint.pdf.

OASIS (2005) The Organization for the Advancement of Structured Information Standards (OASIS). http://www.oasis-open.org/home/index.php

OGC (2000) Open GIS abstract specification: OpenGIS metadata. Open GIS Consortium Inc., http://www.opengis.org/docs/01-111.pdf

OGC (2002) Web feature service implementation specification. Version 1.0.0. Vretanos PA (ed), Open GIS Consortium Inc.

OGC (2003a) OpenGIS web map server cookbook. Version 1.0.1, Kolodziej K (ed), Open GIS Consortium Inc.

OGC (2003b) Web coverage service. Version 1.0.0, Evans JD (ed), Open GIS Consortium Inc.

OGC (2003c) Geographic markup language (GML) implementation specification. Cox S, Daisey P, Lake R,, et.al. (eds) Open GIS Consortium Inc

OGC (2005a) GML in JEPG 2000 for geographic imagery (GMLJP2) implementation specification. Lake R, Burggraf D, Kyle M, Forde S (eds) Open GIS Consortium Inc.

OGC (2005b) Open GIS Consortium homepage. Open GIS Consortium Inc., http://www.opengis.org/

Oracle (2003) Oracle spatial user's guide and reference release 10g (10.1), Part No. B10826-01, Oracle Corporation

Peuquet DJ (1984) A conceptual framework and comparison of spatial data models. Cartographica, pp 66-113

RLG (2002) RLG best practice guidelines for encoded archival description. RLG EAD Advisory Group

Rosenbaum R, Tominski C (2003) Pixels vs. vectors: presentation of large images on mobile devices. IMC 2003, Rostock, Germany

Sperberg-McQueen CM, Burnard L (1994) Guidelines for electronic text encoding and interchange (TEI P3). http://www.isgmlug.org/sgmlhelp/g-index.htm.

Spaniol M, Klamma R (2005) MEDINA: a semi-automatic Dublin Core to MPEG-7 converter for collaboration and knowledge management in multimedia repositories. In: Tochtermann K, Maurer H (eds) Proceedings of I-KNOW '05, 5th International Conference on Knowledge Management, Graz, Austria, J.UCS (Journal of Universal Computer Science) Proceedings, Springer, pp 136-144

Thornes R, Dorrell P, Lie H (1999) Introduction to object ID: guidelines for making records that describe art, antiques and antiquities. Getty Information Institute

Thornes R, Bold J (1998) Documenting the cultural heritage. Getty Information Institute

W3C (1999) Namespaces in XML. W3C, http://www.w3.org/TR/REC-xml-names/
W3C (2003a) RDF Vocabulary Description Language 1.0: RDF Schema. W3C, http://www.w3.org/TR/rdf-schema/
W3C (2003b) Extensible Markup Language (XML) W3C, http://www.w3.org/XML/
W3C (2003c) Scalable Vector Graphics (SVG) 1.1 Specification. W3C, http://www.w3.org/TR/SVG/
W3C (2003d) Mathematical Markup Language (MathML) Version 2.0 (Second Edition). Carlisle D, Ion P, Miner R, Poppelier N (eds), W3C, http://www.w3.org/TR/MathML2/
W3C (2004) XML Schema Part 0: Primer Second Edition. Fallside DC, Walmsley P (eds), W3C, http://www.w3.org/TR/xmlschema-0/
W3C (2005) The World Wide Web Consortium. http://www.w3c.org.
Wenger E (1996) Communities of practice: learning, meaning and identity. Cambridge University Press, New York
Winter S (1996) Unsichere topologische Beziehungen zwischen ungenauen Flächen. DGK-C 465, München, Deutsche Geodätische Kommission, Germany
Winter S (1998) Bridging vector and raster representation in GIS. ACM GIS '98, Washington, D.C., USA
Weibel S, Kunze J, Lagoze C, Wolf M (1998) Dublin Core metadata for resource discovery. http://www.ietf.org/rfc/rfc2413.txt (in reference to http://www.dublincore.org/documents/dces/)

14 Semantically-Aware Systems: Extraction of Geosemantics, Ontology Engineering, and Ontology Integration

Marinos Kavouras, Margarita Kokla, and Eleni Tomai

Abstract. Geographic hypermedia systems include geospatial information from diverse sources. Meaningful access and utilization of such information is materialized only with semantic integration and proper documentation through ontologies. The present work presents a unified view of important research subdomains tasks related to geosemantics and ontologies, (as formal representations of geographic knowledge), such as ontology engineering, extraction of semantic information and ontology integration. Although there is a great degree of recent literature in the field, differences in (a) perspective – purpose, (b) the primary information available and (c) the methodologies and tools used, compose unrelated approaches that have not been put in the overall context. Therefore, it is extremely difficult for the wider audience to understand the difference and the applicability of available approaches in a given context. An attempt is made to draw the overall picture in order to assist users in defining their problem, selecting an appropriate approach and successfully undertaking a geosemantics or ontology-based task.

14.1 Introduction

Geographic hypermedia systems consitute complex environments that include geospatial information and services. Such systems usually contain visual, linguistic and database information. Visual information can come from maps and satellite images. Linguistic information can be met in the form of plain text providing specification of geographic concepts, or in the form of definitions giving more explicit and formal information on concepts. Finally, database information mostly presents formalized knowledge that can be easily accessed and represented.

Sophisticated services of geographic hypermedia systems consist of graphical representations of input data; therefore data visualization is a key issue of a geographic hypermedia system. Modern graphical representations are quite complex and sophisticated, providing all kinds of functionalities to the end user. To accomplish these requirements geographic data visualization should carry the essence of the input data, bear different levels of information and be conformant to the user's perception and cognition of geographic space. Therefore, usability is an important aspect of the designing process of these systems.

In order to increase usability, a geographic hypermedia system should meet two requirements. On one hand, it should contain semantically rich geographic hypermedia representations, while, on the other, it should provide semantically aware services. These requirements are further emphasized by the need for intelligent functionality. This notion reflects the ability of the system to integrate different sources of data and to provide querying and data mining services.

Such functionality cannot be served without taking into account the meaning of representations and the semantics of data during the design of a geographic hypermedia system. Semantic knowledge is formally represented by the use of ontologies. Therefore, herein, we will analyze the advantages of incorporating geospatial semantics and ontologies in such a system to meet the requirement for high functionality and usability.

The remainder of the chapter is organized as follows. Section 14.2 outlines the theoretical characteristics of geosemantics. Moreover, the notion of ontologies as a specification mechanism of geo-conceptualizations and a means to capture their semantics is presented. Section 14.3 provides an analysis of methodologies and tools relative to three important research subdomains dealing with geosemantics and ontologies: ontology engineering, extraction of semantic information and ontology integration. Section 14.4 presents conclusions and future research directions resulting from the present work.

14.2 Theoretical Issues

The semantics of concepts of a given domain are usually encapsulated, elucidated, and specified by an *ontology*. The foremost question which has to be tackled when building a geographic ontology is what constitutes the semantics of geographic concepts. Much discussion has been raised around this question due to the complexity of geographic space, the variety of the "things" it includes (Smith and Mark 1998, Varzi 2001), and their special characteristics (Egenhofer 1993). The subsequent sections address the issues of semantics and ontologies of geographic knowledge.

14.2.1 Geosemantics

Our main concern and focus is put on the semantics of geographic concepts and associated representations in order to understand the differences in conceptualizing and representing geographic reality and resolve the problems this may cause.

There is some confusion and contradiction as to what defines a geographic concept. First of all, there are concepts which among other attributes possess a spatial extent or spatial characteristics and they are called geographic or geospatial (mountains, rivers, roads, land property, boundaries, etc.). The same concepts in different contexts can be treated neglecting any of the above geospatial characteristics, appearing therefore as aspatial. This has created the position that geographic concepts are no different from other concepts and there is nothing special about geospatial. On the opposite, there is of course the generalist's view that most of the reality at least the physical (vehicles, persons, buildings, municipalities), but also the non-physical which relates to a physical one (e.g., traffic load, public views, average income, population density), being in this world, has a geographic/geospatial reference; therefore almost everything is geographic.

Both views have some right, as they are only partial conceptions. First, it is true that most concepts, being in this world, can be assigned a spatial reference. Secondly, it is clear that there are some concepts whose existence, at least in a certain context, depends vitally on their geospatial character.

As geographic concepts we do not consider everything which may be assigned a geospatial characteristic, but only those, which, in a global or local context, possess a geospatial property, which is essential to their existence. Concepts, whose geospatiality is essential in any context (i.e., context independent), are more prominent (e.g., geographic boundary, landmark, waypoint, marked land property, etc.). These are mostly related to the physical reality and present similarities to the bona-fide concepts/entities (Smith and Varzi 2000). These concepts are usually perceptually and cognitively rich (easily identified or agreed upon). Less frequently, there are also other context-depended geographic concepts (vehicles in a traffic control system, ships in electronic charts, etc.). On the other hand, there are geographic concepts that are not readily evident but are constructed by a social agreement and partial consensus. Typical such concepts are administrative units, census tracks, agglomerations, and other spatial socioeconomic units, and are akin to the fiat concepts/entities (Smith and Varzi 2000).

When referring to geographic reality, we include both prominent classical geographic concepts from the physical world, as well as constructed ones. The *principal dimensions* used in dealing with geographic concepts are the following:

- *Reference - Container*
  - *Spatial frame.* This is a datum functioning as container of space, used to represent (explicitly or implicitly) the location of geospatial entities, as well as, other spatial properties and relations.
  - *Temporal frame.* This is the temporal reference system. All temporal properties and relations are projected in this frame.
  - *Thematic frame.* This reference system is the projection space of all thematic information of geospatial semantics.
- *Semantics*
  - *Context.* This expresses a restricted conceptual milieu, also called perspective, framework, or situation, in which concepts are formed, and information is interpreted, obtains meaning, and becomes relevant.
  - *Term.* This refers to the name of a concept or entity, which by and large is considered as an essential expression of identity.
  - *Internal properties.* Geospatial concepts possess internal properties independent of the existence or relation to other concepts. These can be spatial or aspatial. Spatial properties include for example shape, form, structure, complexity, absolute location, area and volume, velocity, spatial change characteristics, etc. Aspatial properties include embodiment (material-appearance), affordance, purpose, behavior, temporality (temporal location, life, existence, periodicity), etc.
  - *External relations.* These refer to relations among two or more concepts. Spatial relations include for example relative position, distance, orientation, adjacency, proximity, containment, etc. Aspatial relations include subsumption-inheritance, dependency-association, partonomy-meronymy, role, concurrency, etc.
- *Semiotics/Pragmatics*
  - *Expression/symbolism.* Geographic concepts are associated with signs (images, words, symbols, etc.), used to capture and show their intended meaning.
- *Quality*
  - *Vagueness.* This refers to the degree of inexactness, fussiness, or indeterminable character of geographic concepts, of their properties and relations (including spatial and aspatial), as well as, of reasoning

based on vague concepts. Uncertainty, randomness, and ignorance (Sowa 2000) are important factors contributing to this parameter.

– *Approximation.* This refers to the detail or granularity of (a) the conceptualization, (b) its representation, and eventually (c) its visualization. The first refers to the way the geographic world can be perceived and mentally abstracted from different "distances". The second refers to representations of spatial and thematic content in different levels of detail. Finally, the third refers to the way detail is symbolized and shown. The last one has long concerned also the research direction of cartographic generalization.

14.2.2 Ontologies

For the last decade research on geo-information has focused on the topic of ontologies. Ontologies are considered nowadays as the most promising and efficient vehicle in Knowledge Representation. What makes ontologies appealing is their capability of grasping the semantics of the domain, as well as accounting for our conceptualizations of that domain. Ontologies are used to make explicit the concepts involved in a given domain, and the relations among them, as well as to standardize and translate between different sources of information.

Geographic ontologies belong to the category of domain ontologies as introduced in (GDDD 1994) which describe the vocabulary related to a generic domain, by specializing the terms introduced in a top-level ontology. A geographic ontology deals with the conceptualization of concepts that: interrelate, "participate" in processes, are susceptible to changes, have variety of properties and values, and have both spatial and temporal reference.

Existing "Geo-Ontologies". One would expect that the special aspects of geographic concepts as those presented in the previous section would be, by some means, addressed in current geographic ontologies. Nevertheless there is no comprehensive and agreed-upon ontology of the geographic domain so far. However, prior analyses of existent geographic ontologies – Fonseca *et al.* (2002), Kavouras *et al.* (2003), Kuhn 2002 among others - have indicated what it is been accomplished so far in ontology design.

What we have. There exist repositories of geographic information that could be considered as providing some kind of ontological information of the domain as well. The majority of these are data standards developed by certain communities. Indicative examples of these are:

1. USGS Spatial Data Transfer Standard (SDTS) (USGS 1997),
2. Geographical Data Description Directory (GDDD) (GDDD 1994).

3. CORINE LC Nomenclature of the European Environmental Agency (EEA 1995).

What we lack. According to the agreed-upon definition of what an ontology is, the previously presented repositories are far from serving the intended role of ontologies. The most important missing element from these standards is the determination of semantic relations and properties for the univocal definition of geographic concepts. This deficiency results in ill-defined hierarchies of concepts. Axioms are another missing element that should accompany concepts and relations in a well-set ontology. Axioms impose constraints, which help elucidate the meaning of concepts and disambiguate relations among them.

14.3 Methodologies and Tools

There are a number of methodologies and tools in research subdomains related to geosemantics and ontologies. Three of these domains are considered particularly important in the geographic context and are further elaborated in the subsequent sections: ontology engineering, extraction of semantic information and ontology integration.

14.3.1 Geo-ontology Engineering

Section 14.2.2, demonstrated the present situation concerning geographic ontologies revealing the deficiencies of current approaches. In what follows, we introduce some guidelines for a proper design of geographic domain ontologies.

Maedche and Staab (2001) argue that an ontology should consist of the following elements: (a) The Lexicon, (b) Concepts, (c) Relations, and (d) Axioms. Adopting this premise, we explore the role of each one of these components when building a geographic ontology.

Defining Geographic concepts

A geographic ontology should cover a variety of geographic concepts. Domain ontologies are often context-driven, that is, their concepts share the conceptualization relevant to the context tackled by the ontology. For instance, a land cover ontology would not contain geographic concepts such as *country* or *communication networks*.

The right selection of concepts in most domain ontologies is not an issue, since scientific knowledge about the given domain recognizes concepts of interest. In Fonseca *et al* (2003), it is argued that a hermeneutic analysis should guide communication between experts in the design of information system ontologies. In the geographic domain, however, the debate is still open of what experts reckon to be a geographic concept due to the diversity of the domain. Thus, the first step in building a geographic ontology is to select the concepts to be included. A discussion of proposed methodologies for determining concepts to be included in ontologies can be found in Holsapple and Joshi (2002). All these approaches result in a set of accepted for inclusion in the ontology, concepts along with their definitions.

Defining Relations

The following paragraphs deal with the notions of: (a) semantic properties, semantic relations, and their values and (b) relations among them, describing how such relations can be formally and explicitly defined in a geographic ontology.

Research in the direction of extracting semantic relations from definitions (Barrière 1997, Madsen *et al.* 2001, Vanderwende 1995) is not as wide as NLP in free text. A systematic approach to revealing semantic properties/relations specifically for the geographic domain has been introduced by Kokla and Kavouras (Kokla and Kavouras 2002, 2005).

A rigorous analysis of the findings of the case study in Kavouras *et al.* 2003 showed the deficiencies of existent geographic ontologies, in terms of semantic properties and relations (ontological and terminological deficits). More specifically, we can state the following major shortcomings:

The spatial aspect of geographic concepts is underspecified or even totally absent in existent geographic ontologies' definitions.

The hypernymic relation is present in definitions in the form of lexical information in the genus part; however the hypernymic relation is not further used to build the hierarchy of concepts.

According to the above discussion, we isolate the following sets of semantic properties and relations for the explicit definition of a geographic concept.

- *Semantic Properties (SPs)*
  - Spatiality: Covers the spatial extension of geographic concepts. It is further divided into external and internal spatiality (Kavouras 2001). The former covers relative spatial properties of the concept, such as "location" and "topology". The latter includes all the internal spatial properties of the concept, such as "size", "shape", and "form".

- Temporality: It covers the temporal aspect of geographic concepts. It is divided into the semantic property "time", and "condition/status" of the concept with reference to time.
- Nature: Accounts for the fundamental distinction of geographic concepts that denote naturally made objects and artificially made ones.
- Material/cover: Expresses the thematic cover of a geographic object, the material it is made of, the material is filled by.
- Purpose: It expresses the purpose/scope or function a geographic concept serves, or is intended for.
- Activity: This relation expresses the activities/ processes that geographic concepts undergo through space and time.
- Role: This is not a primary relation. Its existence depends rather on the existence of an activity relation. The relation role answers questions about: who/what does the activity; (agent-active role), using what; (instrument), what the activity results to; (effect).

- *Semantic Relations (SRs)*
  - Hypernymy/ hyponymy
  - Meronymy
  - Synonymy. An important issue is to determine sets of synonyms for the concepts; this is achieved by including a nominal definition, which gives synonyms for the term apart for the analytical definition, which follows the genus and differentia structure.

What provides identity to a geographic concept and makes it distinct from others however, is not only the possession or lack of a particular semantic property or relation, but also their associated values. Semantic properties and relations are materialized by their values expressed as lexical elements in Natural Language. For instance, in the following definition: "*Water bodies*: *Natural or artificial stretches of water*", the value for the extracted *hypernymic relation* is "stretches". Likewise, "natural/artificial" constitutes the value of the semantic property *nature,* and finally "water" is the value of the semantic property *material/cover*.

Relations among relations. Another element of a domain ontology is the set of relations among relations. The resulting taxonomy of relations facilitates the understanding of the semantics involved.

Defining Axioms in a geographic ontology

An exhaustive domain ontology, should contain axioms that consider: the structuring of concepts and relations, their meanings and constraints with regard to the domain itself, and the laws that make definitions of concepts and relations consistent and complete.

Examples of axioms concerning semantic relations governing geographic concepts can be found in literature. Casati *et al.* (1998) introduced axioms about the meronym relation, location, and topology (especially in respect to the theory of boundaries by Smith (1994). Eschenbach (2001) proposed an alternative of representing axioms through composition tables for mereology and mereotopology.

The Lexicon

The contents of the lexicon in a domain ontology are the descriptions of concepts in Natural Language, the documentation of semantic properties - relations, and their values for each concept and finally the list of axioms.

In addition, the lexicon should include data not related to the ontology itself, but additional information about semantic properties and relations. This refers to lexical and syntactic patterns, which help in the identification of semantic information from definitions using NLP techniques. In addition, the lexicon should include information on how the ontology should be implemented, that is, information on the ontology language, on the algebra of the axioms, etc.

Assessment

The aforementioned guidelines of geospatial ontologies engineering, exploit NLU techniques to describe the concepts included in the ontology. In what follows (Section 14.3.2) the methodology of extracting the semantics of geospatial concepts/entities is further explained.

14.3.2 Extraction of Geosemantics

Semantic information extraction is a central process in knowledge formalization and integration from available sources. Source components can be quite different, resembling somewhat to ontologies as the latter are defined loosely or rigorously. In general, source components may consist of free text, corpora, thesauri, specialized text (e.g. definitions), terms, nomenclatures, data dictionaries, hierarchical classifications, database schemata, etc. In geography, we encounter taxonomic ontologies more often than axiomatized or formal ontologies. In this wealth of source information, it is imperative to decide what constitutes semantic information so that it can be extracted with the appropriate extraction process. This is not a trivial task.

As explained before, semantic information in ontologies can be decomposed into the following semantic elements/components, such as: ontology domain/context, concepts, terms, semantic properties, semantic relations, subsumption relations, parts, functions, axioms – constraints and instances.

These elements are usually contained explicitly or implicitly in the concept definition when available, and also in the ontology hierarchical structure. It is not always the case to have all these semantic components present in the available sources. There may be just terms, i.e., class/category names, and possibly a hierarchy. In such a case, all the rest is implied from the subsumption relations and the ontology domain/context. At times, semantics are not derived from the definition but from an existing database schema (Chang and Katz 1989, Kashyap and Sheth 1996). It is not always certain however whether and to what extent such information is indeed semantic.

Given the above source components, there are empirical ad-hoc approaches attempting to formalize the concepts involved and design the associated databases. An advanced and systematic way however to extract semantics is by some semantic information extraction (SIE) approach based on natural language understanding or processing (NLU/NLP), the central notions in the area of computational linguistics and artificial intelligence (Soderland 1997, Appelt 1999, Cowie and Wilks 2000). Different levels of sophistication and automation characterize these approaches. Research on automatic acquisition of semantic information has focused more on identification of hypernyms or IS-A relations from definitions (Markowitz 1986), free text, and the WWW and less on other semantic elements.

All the previously mentioned methodologies make use of Linguistic Theories on semantics; nevertheless, the debate is still open on "where semantics comes from". Some scholars advocate that geosemantics should be examined in the light of Cognitive Science (Kuhn 2003) because much of human cognition is spatial. This opposition between cognitive and linguistic semantics presents a problematic aspect of defining what geosemantics really is, where it comes from, and on what basis it should be grounded.

Although definitions have been studied a lot in fields such as computational linguistics and lexical semantics, their potential in representing the semantics of geographic concepts only recently has been recognized. Kuhn (2001) addresses the issue of explaining the meaning of a term using the notion of conceptual integration from cognitive linguistics. Hakimpour and Timpf (2001) present an approach for schema integration based on formal definitions in Description Logic.

Kokla and Kavouras (2005) developed a method for revealing salient semantic information (semantic properties and relations) from existent geographic ontologies in order to perform concept comparison and reconciliation. A central process of this method performs semantic information extraction and formalization taking as source components the definitions of geographic concepts. Based on specific rules, definitions are analyzed in a set of semantic elements (properties and values). This information is further used in order to identify similarities and heterogeneities between geographic concepts.

14.3.3 Geo-ontology Integration

There are many issues entailed in the process of ontology integration. The term "integration" is used in literature Sowa (2000), Klein (2001), Uschold and Gruninger (2002), Kavouras (2005) as a general term to denote a number of related concepts such as association, coordination, combining, matching, mapping, translation, merging, partial compatibility, alignment, unification, fusion, mediation, true integration, and the like. Most terms refer to the way ontologies are connected, describing thus the architecture (or topology) of integration.

A fundamental objective of all approaches, no matter what methodology they subsequently employ, is more or less to compare the semantics of the given ontologies and determine the following:

- Whether the given ontologies are to some degree similar, related, or disjoint.
- How to compare concepts in overlapping or related ontologies, in order to identify similar (overlapping), related or disjoint concepts. This is a difficult but fundamental problem (Sheth *et al.* 2005) also known as *concept (or entity) identification/ disambiguation.*
- How to associate the ontologies on the basis of the findings of the previous steps and the possible architectures.

While the objectives seem clear, the perspective, and way the above issues are tackled differ a lot. Herein, some alternatives are presented which are commonly pursued in existing approaches. The objective is to clarify the principal directions and not to exhaust all variants -known or possible.

1. *Conforming to a single global ontology.* Approaches following this principle attempt to establish a single global ontology that all users employ. This severely limited principle follows the old standardization paradigm – a rather procrustean way to enforce semantic interoperability, in which the need for mappings is entirely eliminated. These approaches only suffice for very narrow applications and small community

needs, and for a limited time since they do not handle ontology evolution.

2. *Manual ad-hoc mappings.* This is a simple and commonly used approach, which lets the user/expert define mappings between concepts of the two ontologies. The vast majority of mappings are still established manually (Visser *et al.* 1999, Preece *et al.* 2000). Its major advantage is simplicity and user-controlled result. Since however it is an entirely subjective process many inconsistencies shall arise while it is not certain that semantics are preserved (Wasche *et al.* 2001). Being also laborious and error prone make it highly inefficient to deal with many, large complicated ontologies with many overlapping concepts.
3. *Intuitive mappings based on "light" lexical information.* More refined and complex) approaches exploit basic lexical information, such as terms and their synonyms, to enable a more intuitive mapping between concepts (Mena *et al.* 1998). The advantage of approaches following this direction is that they are less subjective than the first one. As a result, some parts of the process can be semi-automated in form of alternatives suggested to the user. The disadvantage is that mere term similarity is not sufficient to encapsulate the semantics of concepts.
4. *Intuitive mappings based on explication characteristics.* Some integration approaches, despite being called "semantic", attempt to solve explication problems resulting to a distortion of ontologies in order for example to make them computationally equivalent. These approaches resemble those from the field of database integration, where concepts (entities in this case) are compared/matched with respect to their syntactic similarity on explication characteristics (such as names, data types, and structures) of representation elements (attributes, relations, constraints, and instances). These approaches are very useful to integration at the explication level. Such syntactic information however is either inappropriate or insufficient to reveal semantic similarity, relation, or difference.
5. *Intuitive mappings based on structural similarity.* Some approaches originating from research on schema integration (Rahm and Bernstein 2001), are developed on the following assumption: similar ontologies also exhibit some structural (schematic) similarity. Along the same lines, if concepts in different ontologies match, then it is explored if the associated super/sub/side concepts also match.
6. *Relating (grounding) to a single shared or top-level ontology.* Another family of approaches avoids the determination of direct correspondences between concepts from different ontologies by using a single common top-level ontology. Each resource ontology only inherits superconcepts from the top-level ontology. This approach, known as *top-level ground-*

ing (Wache *et al.* 2001), has some practical advantages, the most important being that the semantics of resource ontologies remain unaltered. The fact however that only indirect correspondences are supported via more general superconcepts may create problems when exact correspondences are needed (ibid.).

7. *Direct mappings based on "deep" semantics.* Similarly to the semantic correspondences by (Wache *et al.* 2001), the objective of this direction is to avoid (a) indirect mappings via a top-level ontology, and (b) subjective direct mappings based on "shallow" semantic information. In this family of approaches, in order to support direct mappings among concepts based on "deeper" semantics (e.g., semantic relations), it is essential to derive such information from the available sources. Linguistic techniques such as NLP (Section 14.3.2) are usually applied on unstructured data, while constraint-based approaches are more suitable in the case of semi-structured data (Sheth *et al.* 2005). Formal Concept Analysis based on semantically "deep" properties, also establishes direct mappings between the concepts involved (Kokla and Kavouras 2002).
8. *Integration by view-based query processing.* This is a family of approaches in which querying mechanisms play a dominant role (Pottinger and Halevy 2001, Capezzera 2003). Integration in this context is expressed by mappings between a global and the local ontologies. Such a service is usually provided by mediators, which usually offer abstract integrated views over heterogeneous data sources.
9. *Compound similarity measures.* Concepts might be compared and matched on the basis of the available semantic components; i.e., term comparison, relation/property/attribute comparison, or instance comparison. This may conclude as to whether two concepts are equivalent, different, related, etc. Another way of dealing with semantic correspondences and concept matching is by establishing compound similarity measures among the compared concepts from different ontologies (Maedche and Staab 2002, Maedche and Zacharias 2002, Rodríguez and Egenhofer 2003, Ehrig and Sure 2004).
10. *Extensional mappings based on common reference.* Many of the above families of approaches associate concepts relying on their intensional information. There are however approaches which associate concepts on the basis of their extensional information, when of course available. The simple assumption made here is that concepts having the same instances are likely equivalent. An advantage of these approaches is that there exists an extensional base for comparison and reconciliation. There are however several disadvantages: (i) extesional information may be unavailable, unknown, incomplete or circumstantial; (ii) instances do not necessarily (or fully) describe the semantic domain of a concept; and

(iii) the degree to which extensional resemblance is directly and positively related to concept resemblance is not known or justified.

There is a case however, which requires special attention for it relates to geographic space. Space is considered to act as a reference (container) where things happen. The assumption here is that things closer in space are more associated. Furthermore, if two categories (concepts) of similar contexts/domains (ontologies) refer to the same space (common reference), then it is likely that they are similar.

14.4 Conclusions

In order to increase intelligent functionality and usability, a geographic hypermedia system should incorporate semantically rich representations of geographic concepts. The present work presents a unified view of the role of geosemantics and ontologies in designing such a system, as well as methodologies and tools for their formal representation, specifically relative to three important tasks: ontology engineering, extraction of semantic information and ontology integration. Other similar work in these areas deal with geographic information in a superficial way, taking into account syntactic (e.g., attributes) rather than semantic characteristics of geographic information. On the contrary, the present work analyzes the semantic components of geographic information (principal dimensions, semantic properties-relations, etc.) and their documentation through ontologies.

Research towards these directions is very important, and far from being complete. This, among others, includes further work on the analysis and formal representation of properties and relations, which contribute to the definition of concepts in a semantically-aware system. Furthermore, future research should focus on the development of conceptual structures suitable for representing the complexity and multidimensionality of geographic concepts.

References

Appelt DE (1999) Introduction to information extraction. AI Communications, vol 12 (3), pp 161-172

Barrière C (1997) From a children's first dictionary to a lexical knowledge base of conceptual graphs. PhD Thesis, Simon Fraser University, BC, Canada

Capezzera R (2003) Query processing by semantic reformulation. PhD Thesis, Faculty of Engineering at Modena, Universitŏ di Modena e Reggio Emilia, Italy, http://www.dbgroup.unimo.it/tesi/capezzera.pdf

Casati R, Smith B, Varzi AC (1998) Ontological tools for geographic representation. In: Guarino N (ed) Formal Ontology in Information Systems, IOS Press, Amsterdam, The Netherlands, pp 77-85

Chang EE, Katz RH (1989) Exploiting inheritance and structure semantics for effective clustering and buffering in an object-oriented DBMS, ACM Press

Cowie J, Wilks Y (2000) Information extraction. In: Dale R, Moisl H, Somers H (eds) Handbook of Natural Language Processing, Marcel Dekker Publishing

EEA (1995) European Environmental Agency: CORINE Land Cover Methodology and Nomenclature. http://reports.eea.eu.int/COR0-part1/en/land_cover Part1. pdf, http://reports.eea.eu.int/COR0-part2/en/tab_content_RLR

Egenhofer M (1993) What's special about spatial?: database requirements for vehicle navigation in geographic space. In: Proceedings of the ACM SIGMOD International Conference on Management of Data, ACM Press, New York, pp 398-402

Ehrig M, Sure Y (2004) Ontology mapping – an integrated approach. Bericht 427, Institut für Angewandte Informatik und Formale Beschreibungsverfahren, AIFB, Universität Karlsruhe (TH)

Eschenbach C (2001) Viewing composition tables as axiomatic systems. In: Proceedings of the International Conference on Formal Ontology in Information Systems FOIS'01, Ogunquit, Maine USA, ACM Press, New York, pp 93-104

Fonseca F, Davis C, Camara G (2003) Bridging ontologies and conceptual schemas in geographic information integration. Geoinformatica, vol 7(4), pp 355-378

Fonseca F, Egenhofer M, Agouris P, Camara G (2002) Using ontologies for integrated geographic information systems. Transactions in GIS, vol 6 (3), pp 231-257

GDDD (1994) Geographical Data Description Directory: The European Dataset Catalogue, http://www.eurogeographics.org/gddd/lists/features.htm.

Hakimpour F, Timpf S (2001) Using ontologies for resolution of semantic heterogeneity in GIS. In: Proceedings of the 4th AGILE Conference on Geographic Information Science, Brno, Czech Republic, http://citeseer.ist.psu.edu/ hakimpour01 using.html

Holsapple CW, Joshi KD (2002) A collaborative approach to ontology design, Communications of the ACM, vol 45, pp 42-47

Kashyap V, Sheth A (1996) Schematic and semantic similarities between database objects: A context-based approach. International Journal on Very Large Data Bases, vol 5

Kavouras M (2001) Understanding and modeling spatial change. In: Frank A, Raper J, Cheylan JP (eds) Life and Motion of Socio-Economic Units, Chapter 4, London Taylor & Francis, GISDATA Series 8

Kavouras M (2005) A unified ontological framework for semantic integration. In: Agouris P, (ed) Next Generation Geospatial Information. A.A. Balkema Publishers - Taylor & Francis, The Netherlands

Kavouras M, Kokla M, Tomai E (2003) Determination, visualization, and interpretation of semantic similarity among geographic ontologies. In: Gould M, Laurini R, Coulondre S (eds) Proceedings of 6th AGILE Conference on Geographic Information Science, Lyon France, pp 51-56

Klein M (2001) Combining and relating ontologies: an analysis of problems and solutions. In: Gomez-Perez A *et al.*, (eds): Proceedings of the IJCAI'01-Workshop on Ontologies and Information Sharing, Seattle, http://citeseer.ist.psu.edu/klein01 combining.html

Kokla M, Kavouras M (2005) Semantic information in geo-ontologies: extraction, comparison, and reconciliation, Journal on Data Semantics, accepted for publication

Kokla M, Kavouras M (2002) Extracting latent semantic relations from definitions to disambiguate geographic ontologies. In: Zavala G (ed), GIScience 2002 Abstracts, University of California Regents, pp 87-90

Kuhn W (2001) Ontologies in support for activities in geographic space. International Journal of Geographical Information Science, vol 15 (7), pp 613-631

Kuhn W (2002) Modeling the semantics of geographic categories through conceptual integration. In: Egenhofer M, Mark DM (eds) Proceedings of the Second International Conference on GIScience, Boulder, CO.

Kuhn W (2003) Why information science needs cognitive semantics and what has to offer in return, Unpublished manuscript, http://musil.uni-muenster.de/documents/WhyCogLingv1.pdf

Madsen BN, Pedersen BS, Thomsen HE (2001) Defining semantic relations for OntoQuery. In: Jensen PA, Skadhauge PR (eds) A structured overview of semantic relations and suggestions for their application in OntoQuery, pp 57-88

Maedche A, Staab S (2001) Ontology learning for the semantic web. IEEE Intelligent Systems, vol 16(2), pp 72-79

Maedche A, Staab S (2002) Measuring similarity between ontologies. In: Proceedings of the European Conference on Knowledge Acquisition and Management (EKAW-2002), LNCS/LNAI 2473, Springer-Verlag, Berlin

Maedche A, Zacharias V (2002) Clustering ontology- based metadata in the semantic web. In: Principles of Data Mining and Knowledge, Proceedings of the 6th European Conference (PKDD 2002), LNCS 2431, Springer-Verlag, Berlin

Markowitz J, Ahlswede T, Evens M (1986) Semantically significant patterns in dictionary definitions. In: Proceedings of the 24th Annual Meeting of the Association for Computational Linguistics, New York, pp 112-119

Mena E, Kashyap V, Illarramendi A, Sheth A (1998) Domain specific ontologies for semantic information brokering on the global information infrastructure. In: Guarino N (ed) Formal Ontology in Information Systems, IOS Press, Amsterdam, pp 269-283

Pottinger R, Halevy A (2001) MiniCon: a scalable algorithm for answering queries using views. VLDB Journal, vol 10 (2-3), pp 182-198

Preece AD, Hui K, Gray A, Marti P (2000) The KRAFT architecture for knowledge fusion and transformation, Knowledge Based Systems, vol 13 (2-3), pp 113-120

Rahm E, Bernstein PA (2001) On matching schemas automatically. Technical Report MSR-TR-2001-17, Microsoft Research, Microsoft Corporation, One Microsoft Way, Redmond, WA 98052-6399

Rodríguez A, Egenhofer M (2003) Determining semantic similarity among entity classes from different ontologies. IEEE Transactions on Knowledge and Data Engineering, vol 15(2), pp 442-456

Sheth A, Ramakrishnan C, Thomas C (2005) Semantics for the semantic web: the implicit, the formal and the powerful. International Journal on Semantic Web & Information Systems, vol 1(1), pp 1-18

Smith B (1994) Fiat objects. In: Guarino N, Vieu L, Pribbenow S (eds) Parts and Wholes: Conceptual Part-Whole Relations and Formal Mereology. 11th European Conference on Artificial Intelligence, European Coordinating Committee for Artificial Intelligence, Amsterdam, pp 15-23

Smith B, Mark D (1998) Ontology and geographic kinds. In: Poiker TK, Chrisman N (eds) Proceedings of the 8th International Symposium on Spatial Data Handling (SDH'98), International Geographical Union, Vancouver, pp 308-320

Smith B, Varzi AC (2000) Fiat and bona fide boundaries. Philosophy and Phenomenological Research, vol 60, pp 401-420

Soderland S (1997) Learning text analysis rules for domain-specific natural language processing. Ph.D. thesis, University of Massachusetts, Dept. of Computer Science, Amherst Massachusetts

Sowa J (2000) Knowledge representation: logical, philosophical and computational foundations. Brooks/Cole USA

Uschold M, Gruninger M (2002) Creating semantically integrated communities on the World Wide Web. Invited talk in Workshop on the Semantic Web, Co-located with WWW 2002, Honolulu HI

USGS (1997) United States Geological Survey: American National Standard for Information Systems - Spatial Data Transfer Standard (SDTS) - Part 2, Spatial Features, Annex A, Entity Types http://mcmcweb.er.usgs.gov/sdts/SDTS_standard_nov97/p2anxa.html

Vanderwende LH (1995) The analysis of noun sequences using semantic information extracted from on-line dictionaries. PhD Thesis, Georgetown University NW Washington DC

Varzi C (2001) Philosophical issues in geography – an introduction, Topoi 20, pp 119-130

Visser PRS, Jones DM, Beer MD, Bench-Capon TJM, Diaz BM, Shave MJR (1999) Resolving ontological heterogeneity in the KRAFT project. In: Proceedings of the 10th International Conference and Workshop on Database and Expert Systems Applications (DEXA '99), University of Florence, Italy

Wache H, Vögele T, Visser U, Stuckenschmidt H, Schuster G, Neumann H, Hübner S (2001) Ontology-based integration of information - a survey of existing approaches. In: Stuckenschmidt H (ed) Proceedings of IJCAI-01 Workshop: Ontologies and Information Sharing, Seattle, WA

PART IV

ANALYTICAL FUNCTIONALITY AND GEOVISUALIZATIONS

15 Towards a Typology of Interactivity Functions for Visual Map Exploration

Donata Persson, Georg Gartner, and Manfred Buchroithner

Abstract. Many interactivity functions exist in explorative map applications. This chapter provides a typology of these functions with the aim to contribute towards a standardization of the variety of interactive visualization tools. The typology contains 8 types and around 70 particular interactivity functions and is based on an evaluation of existing divisions and categorizations of interactions. The standardization implies a facilitation of both the concept of interfaces and the assessment of realized interfaces. Using the developed structure, two existing applications are evaluated showing the usefulness of the typology.

15.1 Introduction

Technological changes and new forms of data transmission lead to a separation of storage and presentation of spatial data. An increasing amount of spatial data can be visualized on-screen, mainly through interactive map use. The presentation of spatial data is combined with exploration, implying individual search for patterns, structures and trends in unknown data space. Many different functions are realized in various applications, but methodical fundamentals are still missing. The International Cartographic Association (ICA) Commission on Visualization and Virtual Environments therefore formulates the following challenge in its research agenda: "a typology of geospatial interface tasks is needed to structure both design of tools and formal testing" (Cartwright *et al.* 2001).

The aim of this chapter is to present such a typology, limited to the exploration of data of anthropogenic contents. As a consequence, this work contributes towards a standardization of the variety of interactive visualization tools. This standardization implies a facilitation of both the concept of interfaces and the assessment of realized interfaces.

The developed typology is based on an evaluation of existing divisions and categorizations of interactions, which differ substantially in their roots. Seen in the special context of explorative map use, the authors bring these heterogeneous approaches together. The development of the typology is deliberately based on theoretical issues instead of starting with the analysis of functions. In this context, the cartographic modeling processes are one of the main theoretical issues considered. Such a theoretical fundament is considered by the authors as an important base for enabling the condition for meaningful explorative map use, i.e., for knowledge generation. As a result, the authors propose in this chapter eight major interactivity types. The typology does not claim to be complete and is open to any changes. In order to demonstrate the usefulness of the typology, existing applications (both on the Internet and on discrete storage media) were evaluated using the developed typology. Two examples are shown in section 15.6.

15.2 Typecast of Interactivity Functions Concerning the Map Creation Process

Cartographic communication processes are often described as in Fig. 15.1 (e.g. Hake *et al.* 2002):

The first step of cartographic communication is the collection of topographic and thematic data of the real world. These data are represented in the primary model, which nowadays mostly is stored in databases. The major function of primary models is to structure, store and link the recorded data. To maintain the accurary, completeness and integrity of the data changes, as for instance scale-dependent operations, are not appropriate. The process for creating and deriving cartographic representation models out of primary model data includes visualization, graphic symbolization and generalization processes. The created representation model can use different media (e.g. paper map, screen map, cartographic information system), and is, in terms of the communication process, called secondary model. The user perceives this model regarding to his or her knowledge.

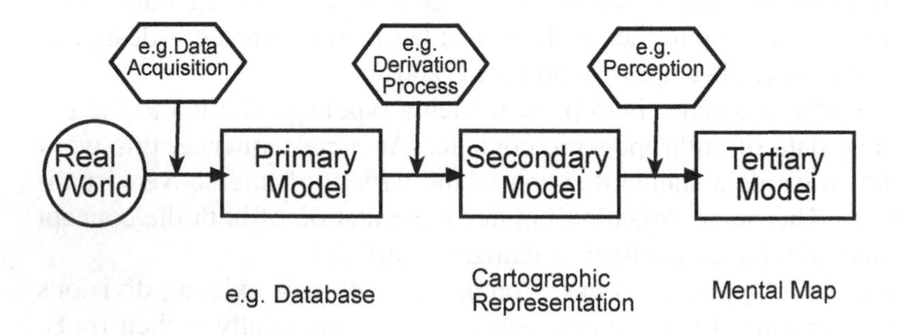

Fig. 15.1. Simplified Schema of Cartographic Communication Processes (Hake *et al.* 2002).

The characteristics of every secondary model can be described as being scale-dependent and representing the geometry and semantics of objects in a heterogeneous way. This is because of the need to enable a human (mostly visual) perception of the model. The secondary model can therefore be seen as a derivation of data from a primary (complete, accurat) model into a form, which enables human users to perceive information and to derive meaning. Finally, the model being build in the user's mental representation can be called the tertiary model of the communication process (e.g. Hake *et al.* 2002, Peterson 1995).

In existing divisions, one finds the following terms for interactivity types. In connection with the map creation process, each of these interactivity functions can be related to the secondary model of cartographic communication.

- Tool to modify the display (Fairbairn *et al.* 2002)
- Interaction with the data representation (Crampton 2002)
- Data display capabilities (Slocum *et al.* 1994)
- "graphische Interaktion" = graphic interaction (Silwester, 1998)
- Manipulation marginalia (Miller 1999)
- "Einflussnahme in das kartographische Erscheinungsbild" = Influence on cartographic appearance (Riedl 2000)

There also exists another group of interactivity types. They describe interactivity functions, which are related to early stages of the map creation process. They are named as follows:

- Tools to manipulate maps (Fairbairn *et al.* 2002)
- Interaction with the data (Crampton 2002)
- Data manipulation capabilities (Slocum *et al.* 1994)
- "Interaktion bei thematischen Darstellungen" = interaction with thematic visualizations (Silwester 1998)
- Navigation marginalia (Miller 1999)

In this second group, there is a distinction between direct access on the primary model and the access to the algorithms, which create the secondary model out of the primary model. From this analysis, the authors develop the following three types:

1. Interaction with the representation model (Type 1)
2. Interaction with the algorithms for creating a representation (Type 2)
3. Interaction with the primary model (Type 3)

These three types can be classified in the map creation process as in Fig. 15.2.

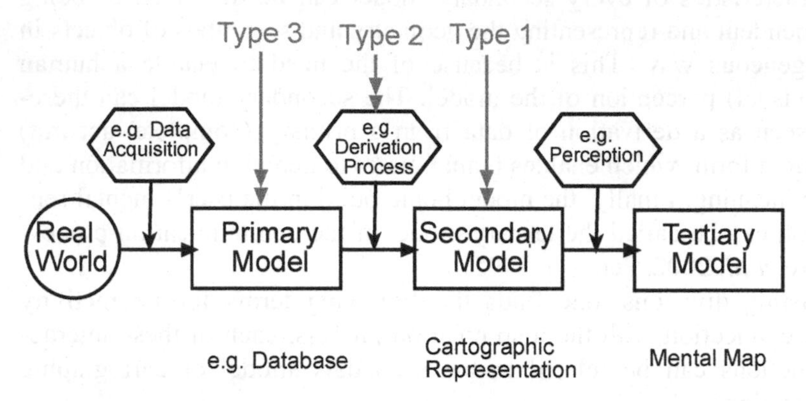

Fig. 15.2. Simplified Schema of Correlation of Interactivity Functions with Cartographic Communication Processes.

Type 1 – Interaction with the Representation Model

This interaction relates to the secondary model. Only the data representation can be changed. The developer of the system creates different selection possibilities, which concern changes of the secondary model. Interacting with the secondary model includes manipulation of the visual variables like color, shape, size and brightness as well as cartographic zooming and the integration of additional information. This makes exploration, to a limited extent, possible.

Cartographic zooming is of particular importance. Due to the consequences of cartographic generalization, several secondary models with different scales can exist. The choice of a particular scale effects the change of the used secondary model, i.e., this interaction relates to different data representations. Furthermore, in this type, the function of user digitalization is classified, i.e., the user can add signatures or mark objects in the map. Also in the latter case, only the secondary model is modified.

The following interactivity functions are classified into type 1:

- Cartographic zooming through direct manipulation
- Cartographic zooming through numerical selection
- Cartographic zooming through drawing a rectangle
- Resetting the scale
- Blending in and out of layers
- Identifying objects/ legend function
- Displaying the exact data for related objects
- Listing all values for one related area

- Listing one value for all related areas
- Displaying statistical values, e.g. maximum, minimum, median
- Modifying static visual variables for symbolizations
- Modifying the symbol scale
- Modifying the type of diagram
- Dynamic separation of qualitative values
- Emphasizing particular data through flashing
- Cartometric functions
- User digitalization

Type 2 - Interaction with the Algorithms for the Creation of a Representation

This type is related to particular algorithms that derive secondary models from primary models. Through the algorithms, the user can influence the creation of the secondary model. For thematic maps, the access to algorithms is mainly made through statistical parameters, such as the influence on class limit determination of choropleth maps. Also the creation of value ranges can be influenced through several interactivity functions and facilitates focussing on subintervals.

The following interactivity functions are classified into type 2:

- Dynamic visual comparison
- Outlier removal
- Adjusting the whole value range to a sub-value range
- Transformation of the encoding function
- Moving the boundaries of the intervals
- Modifying the number of intervals
- Modifying the classify method
- Modifying colour schemes
- Isolating an interval
- Creation of a bivariate cross-classification

Type 3 - Interaction with the Primary Model/ Database Query

Access to the primary model can usually be established. The data are structured in databases, hence the access is limited to the functions of a query language. In thematic cartography of anthropogenic contents, the databases often contain statistical data. The producer of the system has to pay attention to statistical rules and correlations of data. Complex queries allow an extensive exploration. The expressing and testing of hypotheses is possible through the linking of data sets and the generation of of new variables.

The following interactivity functions are classified into type 3:

- Data Mining – Extraction of data with special characteristics
- Logical combining of several data sets for a query
- Generation of new variables

15.3 Typecast of Interactivity Functions related to Multiple Representations

Apart from interaction with only one representation of geospatial data, there is interaction with several representations on the same screen. Functions like this are very useful for exploration. Crampton (2002) mentions the type "contextualizing interaction". Buja *et al.* (1996) distinguishes the types "linking multiple views" and "arranging many views". The authors make the same distinction as BUJA as there is an important difference between juxtaposition and dynamic linking. The following two types are developed:

1. Arranging many simultaneous views (Type 4)
2. Dynamic linking with further display types (Type 5)

Type 4 - Arranging Many Simultaneous Views

The comparison of several display types is crucial for exploration. Different display types with the same geospatial relation are juxtaposed allowing the user to compare the patterns. Several graphic versions of one data set or representations of different data sets can be juxtaposed and compared. Geospatial relation areas are identical making a comparison of pattern possible. Furthermore, data from different points in time can be juxtaposed. Monmonier (1990) calls this interactivity function "simultaneous visual comparison of time units". In distinction to type 6, the time change on the topic is not visualized with display time.

The juxtaposition can show more than two representations (multiple view). The arrangements can even be matrix-like, as in cartographic cross-classification arrays (Monmonier 1989).

The following interactivity functions are classified into type 4:

- Dynamic comparison – several graphic representations of one data set
- Dynamic comparison – several data sets at the same point in time
- Dynamic comparison – several points in time from the same data
- Cartographic cross-classification arrays

Type 5 – Dynamic Linking with further Display Types (Brushing)

These interactivity functions contain not only juxtaposition, but also a dynamic linking with further display types. This can be realized through brushing. There is a connection between the data values shown in different display types. The brushing technique was developed in statistics and is used to "relate data points, in a graph or map on a computer screen, with the corresponding entries in the spreadsheet from which the graphic was generated." (Harris 1999). If the display types are scatterplots, brushing is defined as a technique which allows the investigator to "uncover relationships among multiple variables by identifying subsets of observations on one scatterplot that depicts two of the variables. These observations are immediately highlighted on adjacent scatterplots that depict the relationships among all other pairs of variables under investigation." (MacEachren and Ganter 1990)

Different kinds of display types can be linked. If statistical display types are linked to cartographic representations, the term "geographic brush" is used (Monmonier 1989).

There is also a distinction concerning the amount of data. The values in the display types can be one-dimensional (value axis/ 1D scatterplot, histogram, choropleth map), two-dimensional (2D coordinate system, 2D scatterplot, cross classification) or three-dimensional (3D coordinate system, 3D scatterplot, smooth 3D map, stepped 3D map).

The brushing technique is similar to a query function since a selection of data sets is possible. The linking display types show the result of the query.

The following interactivity functions are classified into type 5:

- Assigning variables to several display types
- Selecting data sets by drawing a rectangle
- Linking map to scatterplot (scatterplot brushing)
- Linking map to coordinate system
- Linking map to histogram

15.4 Typecast of Interactivity Functions related to the Third and Fourth Dimension

Crampton (2002) mentions a type which he defines as functionality with "displays that change continuously". Display time can be used for both, real time as well as for motion in (pseudo) 3D space. The authors see a distinction here. The typology contains type 6 where the changing of the display is combined with a real time, and type 7 where the display changes because of motion in space:

1. Interaction with the temporal dimension (Type 6)
2. Interaction with the (pseudo) 3D visualization (Type 7)

Type 6 – Interaction with the Temporal Dimension

In this type, all interactivity functions are listed, where the display time is used for visualizing the temporal change of a topic. The level of interactivity can extend from the starting and finishing of an animation to the manipulation of the dynamic visual variables and modification of the time axis. There are continuous temporal sequences and sequences in time steps. The latter are more interesting for the visual map exploration, since the time needed for decision making depends on the individual.

The following interactivity functions are classified into type 6:

- Embedding a continuous temporal sequence of views
- Embedding a sequence of views in time steps
- Manipulation of dynamic variables
- Repeating
- Stopping
- Choosing Intervals
- Toggling of time periods
- Sorting time points
- Manipulation of the time axis (i.e. stretching and ramming)

Type 7 – Interaction with the (Pseudo-) 3D Visualization

In contrast to type 6, the display time in this type corresponds to spatial changing of the viewpoint. On one hand,motion of the observer is possible, on the other hand, the position of the visualized object can be changed. 3D interactivity functions do not necessarily need to have a time component. Also static manipulable 3D visualizations are included. Thereby, one can distinguish between three-dimensional topographic applications with thematic overlapping and statistical visualizations. In the latter one, the height of the surface is proportional to the value of the variable being represented.

The following interactivity functions are classified into type 7:

- Motion of the observer
- Motion of the object
- Changing the illumination (azimuth - wave angle of source of light)
- Changing the viewing angle direction

- Changing visual range
- Determining and dyeing of hypsometric layers
- Determining and dyeing of slope categories
- Determining and dyeing of exposition categories
- Overlapping with thematic information
- Visualizing the topic as Z-axis value – stepped 3D map
- Visualizing the topic as Z-axis value – smooth 3D map

15.5 Basic Functions

In explorative applications, there are specific cartographical interactions, but functions that result from the integration into an interactive system should also be considered.

Type 8 – System Interaction

All basic functions for panning and scrolling as well as search and help functions are classified into this type. Geometric zooming is as well listed, because generalization is lacking - in contrast to cartographic zooming. There is no modification of the secondary model at geometric zooming. All functions are classified which supply extensive deepening information in form of hypertextuality like static help instructions as well as dynamic multimedia directories. The latter includes guidance of agents.

The following interactivity functions are classified into type 8:

- Panning
- Scrolling
- Geometric zooming
- Embedding deepening information
- Static help instructions
- Dynamic help instructions
- Guidance of agents

15.6 About the Applicability of the Proposed Typology

15.6.1 Examining Applications by Using the Typology

Following the structure of the typology, seven applications in the Internet and on discrete storage media were extensively examined. Two examples are presented here:

(A) Atlas of Switzerland – interactive

This software is a commercial product developed by the Zurich ETH Institute of Cartography, the Swiss Federal Statistical Office and the Swiss Federal Office of Topography.

(B) CommonGIS Demonstration Project "Europe Statistics"

This software applying the Java applet technology has the goal to provide map-based exploratory data analysis accessible to a broad community of potential users. It is developed by the department for Spatial Decision Support (SPADE) at Fraunhofer AIS, Germany. This demo example was visited in March 2004 at http://www.CommonGIS.com.

Type 1

Type 1	A	B
Cartographic zooming through direct manipulation	-	-
Cartographic zooming through numerical selection	-	-
Cartographic zooming through drawing a rectangle	+	-
Resetting the scale	-	-
Blending in and out of layers	+	+
Identifying objects/ legend function	+	+
Displaying the exact data for related objects	+	+
Listing all values for one related area	-	+
Listing one value for all related areas	-	-
Displaying statistical values, e.g. maximum, minimum, median	-	+
Modifying static visual variables for symbolizations	+	+
Modifying the symbol scale	-	+
Modifying the type of diagram	-	+
Dynamic separation of qualitative values	-	+
Emphasizing particular data through flashing	+	-
Cartometric functions	-	+
User digitalization	-	-

A The data representation can be changed in several ways. These include basic functions as cartographic zooming, blending in and out layers and legend functions. Furthermore the static visual variable color can be manipulated. Information of statistical data is limited to displaying the data for the related object.

B The manipulation of the secondary model is centered on displaying statistical data, both in numbers and in diagrams. Furthermore, cartometric functions are provided.

Type 2

Type 2	A	B
Dynamic visual comparison	(-)	+
Outlier removal	(+)	+
Adjusting the whole value range to a sub-value range	-	+
Transformation of the encoding function	-	-
Moving the boundaries of the intervals	+	+
Modifying the number of intervals	+	+
Modifying the classify method	+	+
Modifying color schemes	+	+
Isolating an interval	-	+
Creation of a bivariate cross-classification	-	+

A By manipulating the histogram, the basic features such as the modification of interval numbers, classification method and color schemes are possible. At the same time, outlier removal is indirectly provided by the isolation of an interval. Even a dynamic visual comparison can be seen as possible, although the histogram as manipulator covers a part of the map. There are no functions concerning the value range and no bivariate visualization.

B The user has, after choosing the indicators, the possibility to decide for a cartographic presentation method, including choropleth map and bivariate cross-classification map. All main manipulation tools are provided, by buttons and by changing the statistical display type (scatterplot). Only the manipulation of the encoding function is lacking.

Type 3

Type 3	A	B
Data Mining – Extraction of data with special characteristics	+	+
Logical combining of several data sets for a query	-	+
Generation of new variables	-	+

A Data query is limited to the extraction of data with special characteristics, while the combining and the generation of of new variables is missing.

B The primary data is stored in the tables. Several kinds of data base query are possible by table calculations.

Type 4

Type 4	A	B
Dynamic comparison – several graphic representations of one data set	-	+
Dynamic comparison – several data sets at the same point in time	-	+
Dynamic comparison – several points time from the same data	-	-
Cartographic cross-classification arrays	-	-

A There is no juxtaposition of representations.

B The juxtaposition is limited to 2D and 3D representations of the same data.

Type 5

Type 5	A	B
Assigning variables to several display types	(+)	+
Selecting data sets by drawing a rectangle	-	+
Linking map to scatterplot (scatterplot brushing)	-	+
Linking map to coordinate system	-	+
Linking map to histogram	(+)	-

A The histogram is the only statistical display type used here. By selecting subsets in the histogram, these selected areas can be shown in the map. As there is no highlighting and as the function is only one dimensional, it cannot be seen as brushing.

B The user can choose among several statistical display types (scatter plot, coordinate system, 3D view), which are dynamically related among each other as well as to the 2D cartographic representation. Several variables can be assigned to the linked displays at the same time.

Type 6

Type 6	A	B
Embedding a continuous temporal sequence of views	-	-
Embedding a sequence of views in time steps	+	-
Manipulation of dynamic variables	-	-
Repeating	+	-
Stopping	+	-
Choosing Intervals	+	-
Toggling of time periods	+	-
Sorting time points	-	-
Manipulation of the time axis (i.e. stretching and ramming)	-	-

A Sequences of views in time steps are provided for some topics which supply the functions of repeating, stopping, choosing intervals and toggling of time periods. More sophisticated functions concerning the manipulation are lacking.

B The time dimension is lacking.

Type 7

Type 7	A	B
Motion of the observer	+	-
Motion of the object	|	+
Changing the illumination (azimuth and wave angle of source of light)	+	-
Changing the viewing angle direction	+	+
Changing visual range	+	-
Determining and dyeing of hypsometric layers	+	-
Determining and dyeing of slope categories	-	-
Determining and dyeing of exposition categories	-	-
Overlapping with thematic information	+	-
Visualizing the topic as Z-axis value – stepped 3D map	-	+
Visualizing the topic as Z-axis value – smooth 3D map	-	(+)

A The software provides automatically generated 3D visualization. The terrain representation can be overlapped by thematic information. Additional 3D objects (block diagram, panorama) are related to hypsometric data, but not to statistical data. There are several interactivity functions concerning the manipulation of the point of observation.

B Interaction with 3D visualization concerns the manipulation of the 3D data view, where the Z-axis shows the statistical value. There are tiered 3D maps and those with a sphere representing the value. These 3D data view can be observed from different viewing angles.

Type 8

Type 8	A	B
Panning	+	-
Scrolling	-	+
Geometric zooming	+	+
Embedding deepening information	-	-
Static help instructions	+	+
Dynamic help instructions	+	-
Guidance of agents	-	-

A The user interface design deviates from standards but provides the basic functionalities. There are animated help instructions.

B The application is built with standard user interface design. There is no deepening information because of the demonstration character.

The major strength of the "Atlas of Switzerland interactive" is the interaction with the 3D visualization (type 7), while the arranging of many simultaneous views (type 4) is missing, and the dynamic linking with further display types (type 5) is only to some extent implemented.

Dynamic linking (type 5) and data query (type 3) are the main interactivity types of the CommonGIS Demo Project, while interaction with the temporal dimension (type 6) is lacking, and the manipulation of 3D visualization (type 7) is realized only to some extent.

15.6.2 Derivation of an Application based on the Typology

In relation to the three viewpoints of user interface, which are named: conceptual, functional, and appearance level (Lindholm and Sarjakoski 1994) or conceptual, operational, and implementational level (Howard and MacEachren 1996), the main use of the typology is at the array of functions, i.e., on the functional/ operational level. On the conceptual level, while considering the result of the application, the typology is useful as well, since it provides a theoretical basis for adapting an application to a certain user group.

In the process of developing the functionality, all features implied for the investigator are defined. The array of functions can be described in a structure chart showing the connection of the user interfaces. One single user interface can provide one or several interactivity types, as it is shown in the example below (Fig. 15.3).

In the example mentioned, the brushing technique (type 5) was considered most suitable for the purpose of this application. There are three user interfaces which are based on dynamic linking of a cartographic and a statistical display type, depending on the dimension of the data space. Type 7 is included since there are 3D data displays linked. Functions of type 1 and 2 are used for the manipulation of the display types. The exploration of a temporal sequence contains type 6. For the access to the primary model (type 3), there exists an interface for the generation of new variables. Furthermore, functions of type 8 are included in all interfaces, especially in the help interface. Functions of type 4 where not considered as useful for this application.

The proposed typology can serve as a starting point to structure the operational level. It is meant to give orientation in the overwhelming and complex functionality area. The limits concerning the technical possibilities are not considered here.

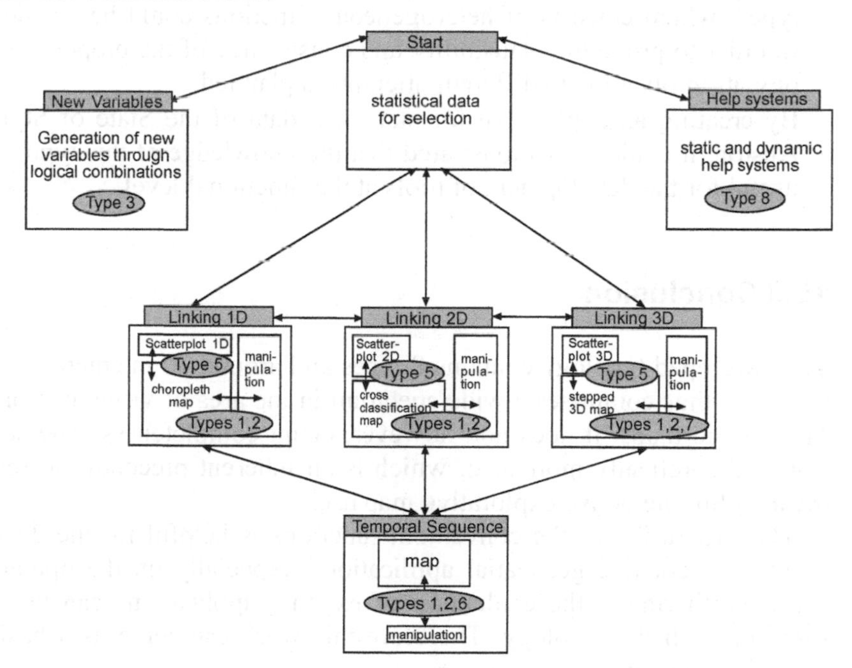

Fig. 15.3. Structure Chart showing the Functionalities of an Application for Exploration of Statistical Data of the State of Styria, Austria

15.7 Usefulness of the Typology

The aim of this chapter is to develop a typology in order to contribute to a standardization of the variety of interactive visualization tools. This standardization implies a facilitation of both the concept of interfaces and the assessment of realized interfaces. The derived typology proved its applicability and usefulness in various ways:

- The established typology simplifies an evaluation. It gives a summary of the types of interaction which dominate as crucial points for exploration. The typology helps to assess the realized interaction types and shows the amount of the particular interactivity functions which are provided.
- By proposed, the amount of interactivity functions can be shown. But the functions differ in their use for exploration. The combination of possibilities, rather than their number, is most important for knowledge generation. The exact potential for exploration should be determined empirically by user evaluation.
- The typology is ordered in a list, i.e. there is no consideration of connections between the functions. A model which contains a combination of types is left for further investigations. Furthermore, a subdivision of type 7 which consists of heterogeneous functions could be considered. In order to prove the plausibility and consistency of the proposed typology, its application to different attempts is planned.
- By creating an application for statistical data of the State of Styria in Austria, it could be demonstrated that the knowledge of this typology is useful for the development of tools at the functional level.

15.8 Conclusion

The developed typology contains 8 types and around 70 interactivity functions and thus comprises a wide spectrum in the area of visual exploration of thematic maps. It does not, however, claim completeness. The separation is theoretically motivated, which is an inherent precondition for any meaningful interactive explorative map use.

The structuring of the convenient functions is helpful for the development of interactive geospatial applications, especially on the operational level. Furthermore, the evaluation of existing applications can be structured through the typology. Therefore this work can serve as a basis for objective comparisons.

References

Andrienko GL, Andrienko NV (1999) Interactive maps for visual data exploration. International Journal of Geographic Information Science, vol 13(4), pp 355-374

Andrienko NV, Andrienko GL, Savinow A, et. al. (2001) Exploratory analysis of spatial data mining using interactive maps and data mining. Cartography and Geographic Information Science, vol 28(3), pp 151-165

Andrienko NV, Andrienko GL, Voss H, et. al. (2002) Testing the usability of interactive maps in CommonGIS. Cartography and Geographic Information Science, vol 29(4), pp 325-342

Buja, A, Cook D, Swayne DF (1996) Interactive high-dimensional data visualization. Journal of Computational and Graphical Statistics, (5), pp 78-99

Buttenfield BP, Weber CR (1994) Proactive graphics for exploratory visualization of biogeographical data. Cartographic Perspectives, (19), pp 8-19

Cartwright W, Crampton J, Gartner G, Miller S, Mitchell K, Siekierska E, Wood J (2001) Geospatial information visualization user interface issues. Cartography and Geographic Information Science, vol 28(1), pp 45-60

Chambers JM, Cleveland W, Kleiner B, Tukey P (1983) Graphical methods for data analysis. Wadsworth/Duxbury, Belmont/ Boston

Crampton JW (2002) Interactivity types on geographic visualization. Cartography and Geographic Information Science, vol 29(2), pp 85-98

Dykes JA (1997) Exploring spatial data representation with dynamic graphics. Computer & Geosciences, vol 23(4), pp 345-370

Fairbairn D, Purves R, *et. Al.* (2002) Using the Internet to deliver customisable map-based educational experience. In: Gartner G (ed.) Maps and the Internet. Geowissenschaftliche Mitteilungen 60, pp 15-20

Gartner G (1999) Interaktive Karten im Internet. In: Institut für Geographie der Universität Wien (ed) Wiener Schriften zur Geographie und Kartographie 12, pp 105-110

Hake G (1973) Kartographie und Kommunikation. In: Kartographische Nachrichten, vol 23(4), pp 137-148

Hake G, Grünreich D, Meng L (2002) Das Kartographische Kommunikationsnetz. In: Kartographie, de Gruyter, Berlin, New York, pp 22-23

Harris RL (1999) Information graphics - a comprehensive illustrated reference. Oxford University Press, New York

Howard D, MacEachren AM (1996) Interface design for geographic visualization: tools for representing reliability. In: Cartography and Geographic Information Systems, vol 23(2), pp 59-77

Huber S, Sieber R, Wipf A (2003) Multimedia in der Gebirgskartographie, 3D-Anwendungen aus dem "Atlas der Schweiz - interaktiv 2". In: Kartographische Nachrichten, vol 53(5), pp 217-224

Kelnhofer F, Lechthaler M (eds) (2000) Interaktive Karten (Atlanten) und Multimedia-Applikationen Geowissenschaftliche Mitteilungen 53

Kraak M-J, Brown A (eds) (2001) Web cartography - developments and prospects. Taylor & Francis, London, New York

Lindholm M, Sarjakoski T (1994) Designing a visual user interface. In: MacEachren AM, Taylor DRF (eds) Visualization in modern cartography. Pergamon, New York, pp 167-184

MacEachren AM, Ganter JH (1990) A pattern identification to cartographic visualization. In: Cartographica, vol 27(2), pp 64-81

MacEachren AM (1998) Design and evaluation of a computerized dynamic mapping system interface. Final Report of the National Center for Health Statistics, The Pennsylvania State University, Department of Geography

MacEachren AM, Kraak MJ (2001) Research challenges in geovisualization. In: Cartography and Geographic Information Science, vol 28(1), pp 3-12

MacEachren AM, Wachowicz M, *et al.* (1999) Constructing knowledge from multivariate spatiotemporal data: integrating geographical visualization with knowledge discovery in database methods. In: International Journal of Geographic Informations Science, vol 13(4), pp 311-334

Miller S (1999) Design of multimedia mapping products. In: Cartwright W, Peterson MP, Gartner G (eds) Multimedia cartography, Springer,Berlin, Heidelberg, pp 51-63

Monmonier M (1989) Geographical brushing: enhancing exploratory analysis of the Scatterplot Matrix. In: Cartographical Analysis, vol 21, pp 81-84

Monmonier M (1990) Strategies for the visualization of geographic time-series data. In: Cartographica, vol 27(1), pp 30-45

Peterson MP (1995) Interactive and animated cartography. Prentice Hall, New Jersey

Riedl A (2000) Virtuelle Globen in der Geovisualisierung Untersuchungen zum Einsatz von Multimediatechniken in der Geopräsentation. Dissertation, Institut für Geographie und Regionalforschung, Universität Wien

Shepherd I (1995) Putting time on the map: dynamic displays in data visualization and GIS. In: Innovations in GIS 2, (2), pp 169-187

Silwester F (1998) Kartographische Interaktionen im Internet. Diploma Thesis, Institut für Kartographie und Reproduktionstechnik, TU Wien

Slocum TR *et. Al.* (1994) Visualization software tools. In: MacEachren AM, Taylor DRF (eds) Visualization in modern cartography, Pergamon, New York, pp 91-122

Slocum TR (1999) Thematic cartography and visualization. Prentice Hall, New Jersey

Tainz P (2002) Kartographische Kommunikation. In: Bollmann J, Koch WG (eds) Lexikon der Kartographie und Geomatik in zwei Bänden. Spektrum, Heidelberg, Berlin pp 27-29

16 A Framework for Using Coordinated Displays for the Analysis of Multidimensional, Spatial, and Temporal Data

Natalia Andrienko, and Gennady Andrienko

Abstract. In geographic hypermedia, maps and other information displays are jointly used for the purpose of communicating information. We consider another role of maps: maps as instruments for data exploration and analysis. In this function, maps are also used in combination with other information displays. In order to establish links between multiple displays, various mechanisms have been developed. Some of these mechanisms might be useful in geographic hypermedia in addition to classical hyperlinks. We propose a taxonomy of generic mechanisms for linking complementary information displays and, in broader terms, complementary tools for data exploration and analysis involving maps and graphics as well as querying, data transformation, and computation-based analysis techniques. We give an example of exploration of geographically referenced data in which different mechanisms are used in cooperation.

16.1 Introduction

Geographic hypermedia is conventionally defined as integration of maps with other media (text, graphics, animation, sound, images, and video) through hyperlinks. Geographic hypermedia is used primarily as a way to organise complex information for the purposes of communication. The role of any component of a hypermedia presentation is to carry a certain portion of information or a specific message the designer of the presentation wants to convey to the audience. This refers, in particular, to maps, the most appropriate medium for communicating geographical information.

However, the functions of maps are not restricted to communication only. Maps are also used for exploration and analysis of geographical information (MacEachren and Kraak 1997). In these functions, maps are also used together with other information displays, most often with graphics such as scatterplots, parallel coordinate plots, time graphs, histograms, etc. A general requirement is that multiple displays must be somehow interlinked in order to help the user to reconstruct the whole picture from the partial views. Not only hyperlinks but also other ways of display linking are used in such systems, which are conventionally called "geographic visualisation systems " rather than "geographic hypermedia".

This chapter is about various ways of linking displays in geographic visualisation systems and, more generally, about linking tools that are jointly used for visual exploration of geographic as well as non-geographic data. While it is not strictly related to the main topic of the book, it can still be interesting to the readers by broadening their eyesight to other uses of maps and related displays and other ways of display linking than they are accustomed to. It is quite possible that some of the techniques applied in geographic visualisation can be also effectively used in geographic hypermedia for information communication.

The ideas concerning the use of multiple linked displays for data communication and exploration come from the pre-computer era. Thus, the well-known representation of Napoleon's Russian campaign of 1812 created by Minard as early as in 1861 combines a map and a time graph[1]. In order to refer locations on the map to the marks on the graph showing the temperatures at the time moments when these locations were visited, Minard connected them with lines.

It is needless to say that computer technologies provide much broader opportunities for display linking than what was possible to do on paper. This encourages the designers and developers of visualization tools to build various display combinations and look for novel ways of linking them. Display coordination (i.e. linking between interactive displays simultaneously present on a computer screen) is currently a hot topic in information visualisation, geographic visualisation, and statistical graphics. Dedicated international conferences are convened (CMV 2003-2005) and special journal issues published such as (InfoVis 2003).

The most popular method of display coordination is known as "brushing" (this term seems to originate from the paper (Newton 1978)). The basic idea is that the user selects some items in one of the displays by direct manipulation through the mouse, and this results in the corresponding items being highlighted in other displays present on the screen. Since this idea was introduced, various more sophisticated forms of brushing have been suggested, for instance, multi-colour brushing: the explorer can use distinct colours for marking different selections.

---

[1] This visualisation is frequently cited; see, for example, (Tufte 1983). It can also be found in the Web, e.g. at http://www.edwardtufte.com/tufte/posters. The site http://www.math.yorku.ca/SCS/Gallery/re-minard.html contains many references to various citations of Minard's graphics as well as suggested revisions of it with the use of modern technologies.

Besides numerous variants of brushing, there are many other ways of display linking. This prompts some researchers to systemise the existing approaches and develop general frameworks of data visualisation through multiple views. Thus, Buja *et al.* (1996) classify visualization techniques into three categories: focusing, linking, and arranging views. Focusing includes the selection of subsets and variables (projections) for viewing and various manipulations of the layout of information on the screen: choosing an aspect ratio, zooming and panning, 3-D rotations etc. Focusing results in conveying only partial information and, therefore, must be compensated by showing different aspects of data in multiple views. These multiple views need to be linked so that the information contained in individual views can be integrated into a coherent image of the data as a whole. The method of linking depends on whether the views are displayed in sequence over time or in parallel. In the first case, linking is provided by smooth animation; in the second case, brushing may be used. The purpose of arranging multiple views is to facilitate comparisons. A possible approach is to display each view in a separate window and allow the user to arbitrarily arrange the windows.

North and Shneiderman (1997) suggest a "taxonomy of multiple window coordinations", which organises the strategies for display coordination along two dimensions. First, the user can perform two types of actions in one of the displays, selecting items or navigating the view. These actions can trigger automatic selection or navigation operations in another display, with three different combinations being possible: selecting items ↔ selecting items, navigating views ↔ navigating views, and selecting items ↔ navigating views. Second, two or more displays can represent either the same collection of information items or different collections. This gives a 3×2 matrix of possible variants of display coordination.

Recently, Roberts has published a comprehensive survey of the state-of-the-art in display coordination (Roberts 2005). The author touches upon various aspects of coordination:

- what tools for interaction are available to the user, e.g. sliders and buttons or direct manipulation facilities;
- how a new view resulting from user's actions is positioned with respect to the previous view, with three possible strategies: replacement, overlay, and replication;
- what conceptual models and architectures for coordination exist[2];

[2] Roberts himself advocates a layered approach based on the dataflow model. Visualisation is considered as the flow data → selected feature set → abstract visualisation object (data features mapped onto visual features) → rendered im-

- how multiple views can be managed in order to avoid overwhelming and disorienting the user;
- what role multiple linked views play in the exploration process.

It should be noted that Roberts, like many other researchers dealing with display coordination, presents software developer's perspective on the topic and concentrates primarily on technical issues. In our book (Andrienko and Andrienko 2006), we have considered the topic in a more user-oriented way. Our goal has been to instruct the potential users on how they can consciously combine various tools for the purposes of data exploration. We have extended the scope from "pure" coordination of data displays to the combined use of various exploratory tools including, in addition to visualisation, data transformation, querying, and computation-based analysis techniques.

In the present chapter, we briefly introduce our taxonomy of tool linking modes and demonstrate an example of exploration of geographically referenced data in which different modes and mechanisms of tool linking are used complementarily.

16.2 Taxonomy of Tool Linking Modes

There is no tool for data exploration and analysis capable of everything. Each tool has different capabilities and, hence, a variety of tools have to be applied in the process of exploring a dataset. On the other hand, datasets that need to be analysed are often very large and consist of many components. It may be necessary to process data piecewise and link the fragmental information thus obtained into a coherent (mental) model of the data and the underlying phenomenon. Therefore, an explorer does not only need to apply different tools but also to apply one and the same tool several times, sequentially or concurrently, to different portions of data.

There are two basic modes of linking tools or multiple "instances" of the same tool in data analysis:

1. Sequential mode: a tool is applied to outcomes of another tool. The second tool starts its operation and produces results only after receiving the output of the first tool.
2. Concurrent mode: two or more tools or tool instances run simultaneously and independently from each other; the analyst needs to compare

---

age → window to manage the image. Coordination may occur at any level of the visualisation flow.

and/or relate their results. The tools or tool instances may be applied to the same data portion or to (partially) different data portions.

These two basic modes can be combined in various "hybrid" constructions; a detailed example is given in the next section. From the two modes, the sequential mode seems less relevant to geographic hypermedia except for cases when one or more of sequentially linked tools is *dynamic*, i.e. can change its results, for example, in response to user's actions or updates of the data. Such a dynamic tool is used in the example presented in the next section: the output of the tool is dynamically modified as the user interactively changes the tool parameters.

In general, if "Tool 1" is dynamic (i.e. may change its outputs) and "Tool 2" is applied to its results, it is necessary that any changes of the results of "Tool 1" are properly accounted for in "Tool 2" and further along the chain. This means that "Tool 2" must be re-applied to the modified results of "Tool 1", the tool following "Tool 2" in the chain of tools must be re-applied to the new results of "Tool 2", and so on. Modern software packages often do such re-application automatically. When some tool modifies its results, it notifies all other tools using these results about the change occurred. In response, these other tools automatically update their own results and, in turn, notify the following tools in the sequence, and so on. In particular, result displays are also updated.

Automatic tool re-application is not always a benefit. It may be easier to compare results of several tool runs with different input settings when the results of the previous runs remain unchanged (until the user explicitly performs certain actions for changing them) than when all the tools are very reactive, so that all the results along a chain immediately change after even a slight interaction of the user with the first tool.

In the case of concurrent tool linking, an analyst can compare and relate results of two or more tools if they are appropriately visualised and, moreover, the visual displays can be seen simultaneously. This possibility to view several displays in parallel provides an elementary level of support to the analyst's work on comparing the results and linking the information portions provided by the distinct tools into a coherent mental image.

Various mechanisms and tools can raise this level of support. We suggest the following taxonomy of the most common mechanisms for concurrent tool linking:

- Display coordination on the basis of a data *subset selection*: Several displays show information related to a selected subset of data records in such a way that the user can readily discern it from the rest. Different methods may be used to achieve this:

  - Highlighting, i.e. special marking of display items corresponding to the selected records to make them easily discernible from the remaining items, e.g. by changing their colour or increasing the size.
  - Focusing or zooming: a display is adjusted so that the information relevant to the selected subset is shown with the maximum possible expressiveness and legibility at the cost of the rest of the information being omitted or reduced in its conspicuousness.
  - Filtering, i.e. removing the display items that do not correspond to the selected records or "muting" the visual appearance of these items and restricting the user interaction with them.

  For example, classical brushing involves a query tool allowing the user to perform various selections through direct manipulation with display items. Highlighting is applied to the user-selected items in this display and to the items in the other displays corresponding to the same data records. Another example is an animation tool, which selects a particular time moment and, consequently, all data records involving this time moment. Thus, if the data are spatial time series, the animation tool selects all such pairs <*location, time*> in which *time* equals the currently chosen time moment.

- Display coordination on the basis of a data *set division*: the data set is divided into two or more non-overlapping subsets (for example, using a classification tool), and, in response, several displays show the information relevant to each subset so that it is easily recognisable and distinguishable from the information related to the other subsets. This can be achieved in following ways:
  - Multi-colour marking: each subset receives a certain unique earmark (typically a colour), which is used for marking the display elements corresponding to this subset in all coordinated views.
  - Display multiplication: a display is replaced or supplemented by several displays of the same type so that each display represents information related to one of the subsets.
  - Re-arrangement of display items: display items are positioned within the display space in such a way that the items corresponding to the same subset are situated close to each other.

- Linking on the basis of a *common visual encoding* of data in several displays such as:
  - Common scales along the display dimensions;
  - Common meanings of colours, sizes, symbols, etc. throughout the displays.

  When the encoding is changed by means of display manipulation, the changes must affect all the linked displays.

For example, data about earthquake occurrences can be represented on a map and in a space-time cube (Gatalsky *et al.* 2004) by identical circles with the sizes being proportional to the earthquake magnitudes. When the user switches from the linear to logarithmic or exponential scale of encoding of the magnitudes by circle sizes, the new scale is applied both in the map and in the space-time cube.

- Linking on the basis of a *common data transformation*: one and the same transformation technique is applied to data represented on several displays. When the user changes the transformation, this affects all the linked displays.

  For example, several maps may represent values of various attributes referring to a particular time moment. The user may apply a tool transforming the original values to differences with respect to the values for the previous time moment. The transformation takes place simultaneously on all the maps. When the user switches the tool to computing the ratios to the previous moment instead of the differences, each map is updated to reflect the change.

Besides these generic methods of tool linking, there are also methods specific for particular display types. For example, some visualisation techniques such as a map with bar charts may use colours to distinguish between attributes. If there are several displays using colours for attributes, a special coordination mechanism can maintain the consistency of the assignment of the colours to the attributes throughout these displays.

Not only sequential and concurrent modes of tool linking are often used together but also different mechanisms of sequential and concurrent linking can work simultaneously, for example, filtering together with multi-colour marking and common transformation of attribute values.

16.3 Cooperation between Sequential and Concurrent Linking: An Example

In this example, an analyst explores data concerning the health care in different counties of the state of Idaho (USA)[3] in order to determine which counties are the most in need of support for improving the availability and accessibility of health care facilities for the population. The explorer needs to evaluate the counties on the basis of multiple attributes, specifically,

[3] The example dataset was provided by Prof. Piotr Jankowski from the University of Idaho, USA. The data are described in more detail in (Jankowski *et al.* 2001).

- *N of estimated unmet visits*: the estimated number of unmet doctor visits when people coming to see a doctor cannot be attended due to doctor's overload;
- *Low-weight birth rate*: the percent of infants born with insufficient body weight averaged from a multi-year interval;
- *Burden on on-call providers*: the number of hours on call for each provider;
- *Population in >35 miles from hospital*: the number of individuals residing outside the influence zone (i.e. the 35-mile radius, according to the national standard for rural areas) of the nearest hospital.

To make the values of all attributes comparable, the analyst has transformed them into z-scores, which express the relative deviations of the values from the mean values of the respective attributes. Positive z-scores signify the original values being worse than the average values for the state of Idaho.

Now, the transformed values of the multiple attributes need to be somehow combined into integral scores characterising the situation in each county in general. For this purpose, the analyst decides to apply a special evaluation support tool capable of attribute integration. We shall not describe the specific algorithm of value combination applied in the tool since this is not relevant to our discussion. It is only important that, first, the tool produces new attributes and, second, the tool is dynamic, i.e. changes the values of these new attributes when the user modifies tool settings.

One of the new attributes produced by the tool is the integrated scores of all counties expressed as real numbers from 0 to 1; the higher the score, the more problematic is the situation in the county. Additionally to the scores, the tool produces an attribute reflecting the ranking of the counties from the most problematic (i.e. with the highest score) to the least problematic (i.e. with the lowest score). The ranks are specified as integer numbers from 1 to 44, which is the number of the counties in the state of Idaho.

The settings influencing the tool outputs are relative weights assigned to the attributes being integrated. The weights are specified as real numbers between 0 and 1. The sum of the weights of all attributes participating in the computation must equal 1. By default, all attributes receive equal weights. In our example, there are four attributes; accordingly, each of them receives the weight 0.25. The evaluation tool provides a direct manipulation interface for changing the weights. When the user modifies the weights, the tool dynamically re-computes the values of the derived attributes.

The table display in Fig. 16.1 shows the evaluation scores and ranks of the Idaho counties obtained with the attribute weights 0.3, 0.3, 0.2, and

	N of estimated unmet visits (z)	Low-weight birth rate (z)	Burden on on-call providers (z)	Population in >35 miles from hospital (z)	Evaluation score	Ranking
Washington	0.556	1.997	2.071	0.556	0.7354	1
Payette	0.650	1.099	2.071	0.570	0.6847	2
Jerome	0.678	0.873	2.071	-0.182	0.6436	3
Latah	1.172	1.617	-0.293	-0.182	0.5941	4
Madison	0.912	-0.138	-0.341	3.246	0.5926	5
Clearwater	0.540	-1.934	1.008	4.926	0.5913	6
Gooding	-1.398	2.774	1.157	-0.182	0.5849	7
Twin_Falls	1.198	1.010	-0.065	-0.182	0.5694	8
Clark	0.632	-0.356	2.071	-0.411	0.5538	9
Gem	-1.079	1.124	2.071	-0.182	0.5493	10
Power	0.358	1.997	-0.766	-0.182	0.5416	11
Blaine	0.629	1.431	-0.554	-0.411	0.5253	12
...	...	...	...	...	...	...
Caribou	-0.313	-0.567	-0.894	-0.182	0.3291	38
Bear_Lake	-0.136	-1.101	-0.453	-0.411	0.3214	39
Boise	-0.375	-0.914	-0.686	-0.182	0.3144	40
Nez_Perce	-1.340	0.185	-0.899	-0.411	0.3037	41
Adams	-2.338	0.088	-0.261	-0.411	0.2694	42
Elmore	-3.380	0.040	-0.777	-0.411	0.1729	43
Custer	-2.903	-1.214	-0.479	-0.411	0.1391	44

Fig. 16.1. The table display represents the results of evaluating the counties of Idaho on the basis of four attributes with the weights 0.3, 0.3, 0.2, and 0.2. The table columns show the values of the source four attributes (previously transformed into z-scores), the evaluation scores derived, and the ranking of the counties according to the scores. The table rows are arranged in the order of decreasing evaluation scores and, consequently, increasing ranks.

0.2, respectively. The table rows are arranged in the order of decreasing evaluation scores and, consequently, increasing ranks. Additionally to showing the attribute values as numbers, the same values are also represented visually by dark grey shading within the cells. The left edge of the shaded area indicates the relative position of the value contained in the cell between the minimum and the maximum values of the respective attribute: the higher the value, the larger the shaded area. To save the space, we did not include all the 44 rows of the table in Fig. 16.1 but only the top and bottom parts of the table. However, it may be clearly seen how the ranking is related to the evaluation scores.

Let us suppose that some limited funding for health improvement is available and can be divided between at most five counties in need. Some extra funding is expected in near future, which will allow the state administration to support additional five counties. Accordingly, the task of the analyst is to divide the 44 counties of Idaho into three classes:

1. The most needy counties, which will receive an immediate financial support.
2. The counties that will be supported later, when the additional funding comes.

3. The counties where the state of the health care is satisfactory so that they will not be funded.

A convenient way to do this division is to apply a tool for classification on the basis of a numeric attribute to the results of ranking of the counties. The counties with the ranks from 1 to 5 will be the candidates for immediate funding, the counties with the ranks from 6 to 10 will be included in the waiting list, and the remaining counties will not be funded.

This classification is shown on the map display in Fig. 16.2. The class breaks specified by the analyst are 5.01 and 10.01. The analyst uses these values rather than 5 and 10 in order to ensure that the county with the rank 5 is included in the first class and the county with the rank 10 in the second class. The counties on the map are coloured according to the classes they belong to. The black colour corresponds to the first class, the dark grey colour – to the second class, and the light grey – to the third class.

Fig. 16.3 demonstrates the effect of the classes being propagated to the table display. The display responds to the classification by appropriate colouring of its rows. Additionally, the rows of counties belonging to the same class are put together. At the top of the table, we can see 5 black rows, which correspond to the counties included in the first class. These

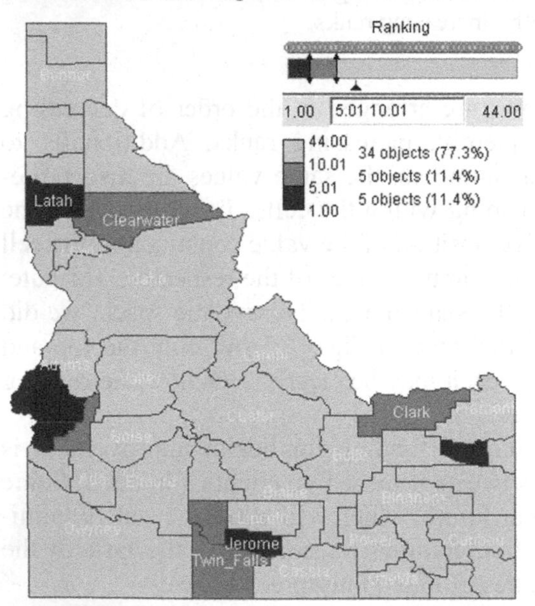

Fig. 16.2. Counties of Idaho are classified according to their ranking into the counties with the ranks from 1 to 5 (the candidates for receiving an immediate support), those with the ranks from 6 to 10 (to receive a support later), and the rest, which will not be supported.

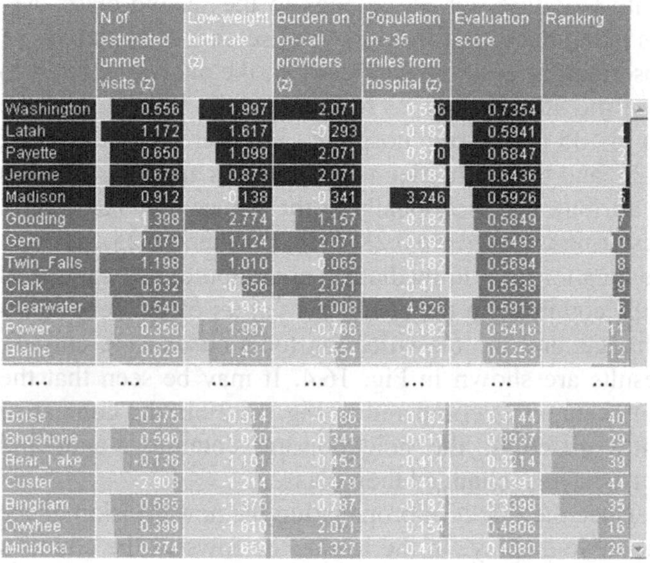

	N of estimated unmet visits (z)	Low-weight birth rate (z)	Burden on on-call providers (z)	Population in >35 miles from hospital (z)	Evaluation score	Ranking
Washington	0.556	1.997	2.071	0.556	0.7354	1
Latah	1.172	1.617	-0.293	-0.182	0.5941	4
Payette	0.650	1.099	2.071	0.570	0.6847	2
Jerome	0.678	0.873	2.071	-0.182	0.6436	3
Madison	0.912	-0.138	-0.341	3.246	0.5926	5
Gooding	-1.398	2.774	1.157	-0.182	0.5849	7
Gem	-1.079	1.124	2.071	-0.182	0.5493	10
Twin_Falls	1.198	1.010	-0.065	-0.182	0.5694	8
Clark	0.632	-0.356	2.071	-0.411	0.5538	9
Clearwater	0.540	1.934	1.008	4.926	0.5913	6
Power	0.358	1.997	-0.766	-0.182	0.5416	11
Blaine	0.629	1.431	-0.554	-0.411	0.5253	12
...	...	...	...	...	...	...
Boise	-0.375	-0.914	-0.686	-0.182	0.3144	40
Shoshone	0.596	-1.020	-0.341	-0.011	0.3937	29
Bear_Lake	-0.136	-1.101	-0.453	-0.411	0.3214	39
Custer	-2.903	-1.214	-0.479	-0.411	0.1391	44
Bingham	0.585	-1.376	-0.737	-0.182	0.3398	35
Owyhee	0.399	-1.610	2.071	0.154	0.4806	16
Minidoka	0.274	-1.659	1.327	-0.411	0.4080	28

Fig. 16.3. The classes of counties from Fig. 16.2 have been propagated to the table display. The rows of the tables are grouped according to the classes the corresponding counties belong to. The grouping is accompanied by colouring of the rows.

are followed by five dark grey rows corresponding to the second class of counties. The remaining rows of the table are coloured in light grey.

Since the effect of grouping is hardly visible when the rows are ordered according to decreasing evaluation scores or increasing ranks, another attribute, "Low-weight birth rate", has been used in Fig. 16.3 for sorting the table rows. This resulted in a different order of the rows as compared to Fig. 16.1. Grouping of table rows has a priority over ordering: the rows are first grouped according to the current division of the set of counties, and then the specified method of ordering is applied to each group individually.

In our case, the rows in each section of the table (i.e. black, dark grey, and light grey) are arranged in the order of decreasing low-weight birth rates. It can be seen that the row with the highest value of this attribute, 2.774, is put on the sixth position in the table, after a row with a much lower value, -0.138. This can be explained by the priority of grouping over ordering: the row with the value -0.138 corresponds to the county Madison, which belongs to the first class, and is put together with the rows of the other counties from the same class. The row with the value 2.774 corresponds to the county Gooding, which belongs to the second class. Accordingly, this row is put at the top of the dark grey section of the table.

Let us now suppose that, after some deliberation, the analyst decides to change the weights of the attributes so as to increase the influence of the attribute “Population in >35 miles from hospital”. She increases the weight of this attribute to 0.25, and the evaluation tool decreases automatically the weights of the other attributes proportionally to the values they had before the operation. The resulting weights are 0.28, 0.28, 0.19, and 0.25.

In response, the evaluation tool immediately re-computes the integrated scores and re-ranks the counties. In the result, the values of the derived attribute “Ranking” change, and the classification tools needs to re-classify the counties. The results are shown in Fig. 16.4. It may be seen that the class breaks have not changed as compared to Fig. 16.2 but the content of the classes is slightly different: the neighbouring counties Latah and Clearwater on the north have exchanged their classes.

In Fig. 16.5, we can see the effect of propagating the new division of the set of counties to the table display. In the result, the rows of the counties Latah and Clearwater have changed their colours and have been moved to other sections.

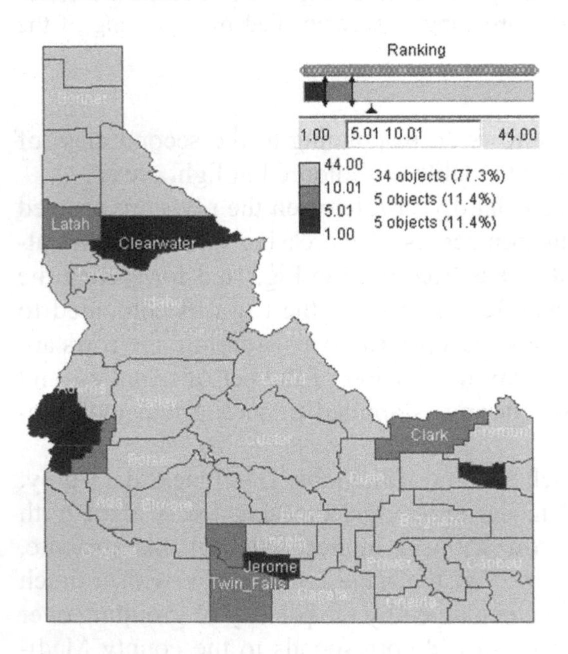

Fig. 16.4. After changing the attribute weights to 0.28, 0.28, 0.19, and 0.25, the counties have been re-evaluated and re-ranked, and the classification re-applied to the new ranks. In the result, the content of the classes has slightly changed.

	N of estimated unmet visits (z)	Low-weight birth rate (z)	Burden on on-call providers (z)	Population in >35 miles from hospital (z)	Evaluation score	Ranking
Washington	0.556	1.997	2.071	0.556	0.7017	1
Payette	0.650	1.099	2.071	0.570	0.6542	2
Jerome	0.678	0.873	2.071	-0.182	0.6071	4
Madison	0.912	-0.138	-0.341	3.246	0.5983	5
Clearwater	0.540	-1.934	1.008	4.926	0.6162	3
Gooding	-1.398	2.774	1.157	-0.182	0.5519	7
Latah	1.172	1.617	-0.293	-0.182	0.5605	6
Gem	-1.079	1.124	2.071	-0.182	0.5185	10
Twin_Falls	1.198	1.010	-0.065	-0.182	0.5373	8
Clark	0.632	-0.356	2.071	-0.411	0.5201	9
Power	0.358	1.997	-0.766	-0.182	0.5112	11
Blaine	0.629	1.431	-0.554	-0.411	0.4934	12
...	...	...	...	...	...	...
Boise	-0.375	-0.914	-0.686	-0.182	0.2978	40
Shoshone	0.090	-1.020	-0.341	-0.011	0.3743	28
Bear_Lake	-0.136	-1.101	-0.453	-0.411	0.3010	30
Custer	-2.903	-1.214	-0.479	-0.411	0.1307	44
Bingham	0.585	-1.375	-0.787	-0.182	0.3217	35
Owyhee	0.399	-1.610	2.071	0.154	0.4578	16
Minidoka	0.274	-1.659	1.327	-0.411	0.3832	26

Fig. 16.5. The altered classes of counties have been propagated to the table display, which resulted in re-colouring and re-grouping of the rows.

This example demonstrates a joint use of four different tools:

- the evaluation tool;
- the classification tool, which is applied to the output of the evaluation tool;
- the cartographic visualisation tool, which shows the output of the classification tool on a map display;
- the visualisation tool producing the table display, which shows, along with other attributes, the results of the evaluation tool. The table visualisation tool is linked to the classification tool by means of the class propagation mechanism.

The links between the tools are schematically represented in Fig. 16.6. The solid arrows represent the sequential tool linking, when results of one tool are used as an input to another tool. In this case, the evaluation tool produces new attributes, which are supplied to the table visualisation tool and to the classification tool. The latter two tools are used concurrently and coordinated by means of propagation of the division of the set of data records into subsets, or classes (in this case, there are 44 data records corresponding to the 44 counties of Idaho). This link, which is signified by the dotted arrow, allows the analyst to compare and relate the information provided in the table display (i.e. the characteristics of the counties) to the

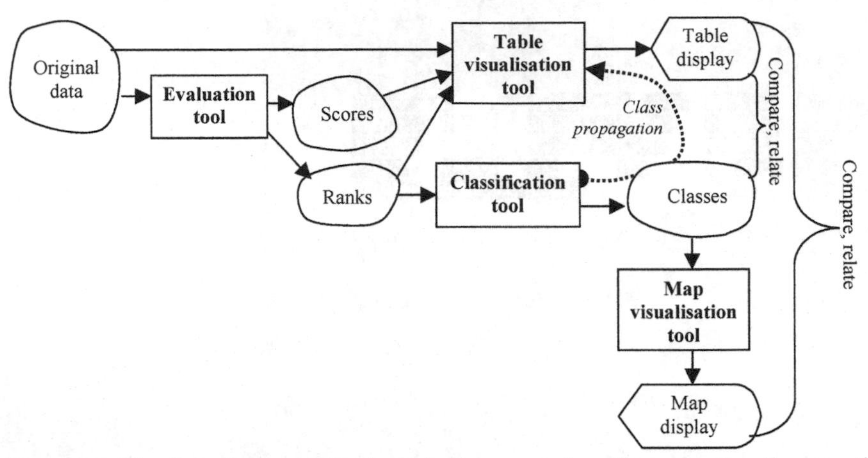

Fig. 16.6. The links between the tools in the example with evaluating the counties of Idaho. The solid arrows represent the sequential mode of tool linking, and the dotted arrow indicates the coordination of the concurrently used tools.

results of the classification. Besides, this link also allows the analyst to compare and relate the information provided by the table to the information contained in the map display, which shows the geographical positions of the counties belonging to the different classes.

16.4 Conclusion

Comprehensive data exploration and analysis essentially requires multiple tools to be used in combination. This includes diverse visual displays of various aspects of data as well as tools for data transformation, querying, computational analysis, etc. In this chapter, we have tried to define the generic mechanisms of how different tools can be used together. We expect this taxonomy to be helpful not only for users of data analysis tools and developers of such tools but also for designers of geographic hypermedia.

In the suggested general framework of tool linking, we pay a particular attention to dynamic tools, which may change their results in response to user's actions or other events. If the results are used or reflected in other tools, it is necessary that the changes were properly accounted for. Tool or hypermedia designers should care about this by devising appropriate mechanisms of change propagation.

The suggested framework does not only address spatial and temporal data but is more general and applicable to other types of data as well.

While spatial and temporal data may require specific analysis tools (e.g. maps, animated displays, time graphs, space-time cubes, etc.), the modes and mechanisms of tool linking and coordination are basically the same as for tools oriented to other data types. Moreover, the generic tool linking mechanisms allow specific tools for analysis of spatial and/or temporal data to be used together with tools that do not assume the spatial or temporal nature of the data. For example, maps or time series graphs can be used in combination with scatterplots, histograms, table displays, and so on. This provides complementary views of various data aspects and features and increases the comprehensiveness of analysis or information presentation.

References

Andrienko N, Andrienko G (2006) Exploratory analysis of spatial and temporal data, Springer, Berlin

Buja A, Cook D, Swayne DF (1996) Interactive high-dimensional data visualization. Journal of Computational and Graphical Statistics, vol 5, pp 78-99

CMV (2003) Proceedings of the 1st International Conference on Coordinated and Multiple Views In Exploratory Visualization. IEEE Computer Society, Los Alamitos, London, England

CMV (2004) Proceedings of the 2nd International Conference on Coordinated and Multiple Views In Exploratory Visualization. IEEE Computer Society, Los Alamitos, London, England

CMV (2005) Proceedings of the 3rd International Conference on Coordinated and Multiple Views In Exploratory Visualization, IEEE Computer Society, Los Alamitos, London, England

Gatalsky P, Andrienko N, Andrienko G (2004) Interactive analysis of event data using space-time cube. In: Banissi E *et al.* (eds) Proceedings of IV 2004 – 8th International Conference on Information Visualization, IEEE Computer Society, Los Alamitos, London, UK, pp 145-152

InfoVis (2003) Information Visualization. Special Issue on Coordinated and Multiple Views in Exploratory Visualization, Roberts JC (ed), vol 2(4)

Jankowski P, Andrienko N, Andrienko G (2001) Map-centered exploratory approach to multiple criteria spatial decision making. International Journal Geographical Information Science, vol 15(2), pp 101-127

MacEachren AM, Kraak MJ (1997) Exploratory cartographic visualization: advancing the agenda. Computers and Geosciences, vol 23(4), pp 335-344

Newton CM (1978) Graphics: from alpha to omega in data analysis. In: Wang PCC (ed) Graphical representation of multivariate data, Academic Press, New York, pp 59-92

North C, Shneiderman B (1997) A taxonomy of multiple-window coordinations. Technical Report CS-TR-3854, University of Maryland Computer Science Department

Roberts JC (2005) Exploratory visualization with multiple linked views. In: Dykes J, MacEachren AM, Kraak M-J (eds) Exploring geovisualization. Elsevier, Oxford, UK, pp 159-180

17 Visualization and Hypermedia for Decision Making

Peter Williams, Eva Siekierska, Costas Armenakis, Florin Savopol, Charles Siegel, and Jessica Webster

Abstract. Decision makers need access to heterogeneous, interdependent and meaningful information to obtain an understanding of the geospatial conditions for informed decision making and analysis of various options for situation assessment. Hypermedia concepts and visualization can facilitate the integration of large amounts of multi-source datasets and provide customized representations of filtered georeferenced data to enable the decision makers to explore and understand various spatial solutions without requiring advanced knowledge of geospatial technologies and systems. A specific application of decision making is the collaborative geospatial decision making, which is based on real-time data sharing, coordinated data access and synchronization between multiple geographically dispersed participants. The roles of hypermedia, visualization and geocollaboration are explored and case studies are presented to support decisions for city planning via interactive mapping, urban planning scenarios, and understanding urban sprawl over time. Two systems are discussed. The in-house developed Dynamic Visualization System (DVS), which dynamically hyperlinks to web map servers, and the GeoConference, a commercial real-time Internet-based geospatial collaborative conference system.

17.1 Introduction

Decision makers face the increasing challenge of handling and understanding large amounts of geospatial information to address complex economic, environment and social issues. Within the Natural Resources Canada, Earth Sciences Sector programs, innovative technologies and methodologies are being developed aiming at facilitating the communication of scientific information to decision makers. These technologies enhance the Natural Resources Department's capacity to implement sustainable development and help to promote the responsible use of Canada's natural resources.

Spatial decision making, as any decision making process, faces situations where the data and information may be complete, may not exist, may be incomplete or may not be precise (deterministic, stochastic, fuzzy types of information, Malczewski, 1999). Geospatial systems must support the handling of various kinds of data and information by providing decision makers with flexible problem-solving tools (Densham 1991, Feeney *et al.* 2002). These tools can assist in the analysis of the data and information, in the understanding of the situation, in the selection and evaluation of alternative action scenarios and in the communication of different views and decisions (Andrienko and Andrienko 2001). Lein (2003) breaks down the decision process into three stages: acquiring information, structuring the decision problem and evaluating the alternatives. When considering that both visual data analysis (Gahagan and O'Brien 1997) and knowledge construction (Fayyad and Uthurusamy 1999), can highly contribute to supporting spatial decision making, then we can identify methods for delivering this support.

One of the most effective tools for delivering and understanding geospatial information are the geovisualization techniques. Visualization permits the data integration and space representation through a visual and interactive reconstruction process. The organization and access of the information for knowledge elicitation can be assisted by the geo-hypermedia concept. Regarding the aspect of communication with respect to spatial decision, collaborative and participatory decision making is becoming an increasing practice using visualization, hypermedia, and the internet for interactive dialogues, understanding and communication.

17.2 Visualization, Hypermedia and Communication in Spatial Decision Making

Decision making is a systematic process of analysing sets of diverse information to develop situation assessments and reach conclusions and alternative solutions, thus leading to the selection of the optimal course of actions. Spatial decision making encompasses vast amounts of heterogeneous data, such as aerial and satellite imagery, DEM, slope, land cover/use data, transportation networks, hydrographic feature network, geophysical data, soil sample data, temporal, raster, vector, text attributes, just to mention a few. Prior to making any spatially related decision all the available spatial and non-spatial "raw" data must be transferred into meaningful information through analysis, interpretation, organization, synthesis, relationships, patterns, discovery, and presentation.

In today's information age, with the rapidly increasing volume of data, with geospatial and non spatial, there is at times an overabundance of data to be considered. Locating the right dataset may be at times problematic, or the data may need extensive cleaning, but data scarcity per se is no longer the issue. With an estimated 80% of all data collected being spatially referenced (Worrall 1991, Malczewski 1999, MacEachren *et al.* 1999), the problem is in fact quite opposite. As Hu (2005), Lévy (1999) and others have noted, the sheer volume of information available that could theoretically be integrated into any given decision is overwhelming. In addition, government and its decision making structure, was forged in the context of a much simpler information culture (Levy 1999), one that was not as fast paced and information rich. Even with the support of research scientists, media analysts and assistants, decision makers in government do not have the capability to assimilate, analyze and build knowledge on the basis of information made available to them. Add implications by numerous stakeholders in decision situations, sometimes in remote locations, arriving at a decision that everyone can accept is even more challenging (Webster *et al.* 2004).

Spatial data are among the many types of data commonly used in natural resource management. In light of sustainable development theory, economic and social dimensions are just as important. Measures of these are not always spatially referenced or even quantifiable. Different forms of information such as text documents (e.g., guidelines, reports), photographs, maps, images, videos, sounds must be taken into account when making a decision. On their own, these individual data sets or random pieces of information may at first appear unrelated. It is however the integration of information from disparate sources that offers increased analytical capabilities required for addressing the complex, ill-structured spatial decision problems, characteristic of questions of sustainable development (Smith 2002). Spatial decision making is therefore a complex process. It involves measurements, observations, analysis, interpretation, integration, modelling, understanding, information generation, knowledge discovery, synthesis, alternative actions scenarios, results presentation, impacts analysis and communication. Important stages of decision making are the understanding of spatial situations, the development and assessment of possible action-taking scenarios and the communication aspects.

The geographic space is represented as abstraction of reality in the form of symbolic graphics of data and information in both analog maps and in the spatial information systems. To understand and interpret space reality, one has mentally to reconstruct reality through the brain's synthetic ability. Relying on the cognitive pattern recognition, this ability of mental reconstruction of reality can be supported and further enhanced through the emulation of brain's synthetic process by providing virtual mental images, which can assist in the reality reconstruction process. The visualization processes handle the generation of these computer-generated but realistically looking images of space, by integrating the available heterogeneous data and information.

The process of developing actions - reaching of decision - is performed through a knowledge elicitation process by accessing, browsing and exploring multi-resource data and information, usually in a non-linear mode. This type of data discovery and exploration, which provides the ability to the user to navigate through heterogeneous data by "jumping" from one data/information source to another through a complex relationship-based indexing schema, can be accommodated by a hypermedia-type information management approach.

The communication component in decision making is twofold. First it relates to how we can integrate the different perspective views of spatial understanding into the decision process, and second, on how the decision made can be understood from the same unified perspective. This leads to participatory and collaborative environments for spatial decision making.

17.2.1 Visualization

The objective of visualization is to generate realistic-looking images enabling researchers to observe their simulations and computations, thus leveraging the scientific methods by providing new scientific insights through eye-brain interpretative patterns (McCormick *et al.* 1987). However, visualization is more that "pretty pictures", it is considered as part of data analysis and understanding process.

Visualization has a crucial role to play in the reduction of the cognitive workload. The belief that 3-dimensional visualization is one of the most natural ways to communicate (Al-Kodmany 2001a, Al-Kodmany 2001b) along with the ability of human vision to detect spatial patterns (MacEachren *et al.* 2003) would suggest that visualization is a natural representational mode for a simplified form of information analysis and scenario building towards sustainability. King *et al.* (1989) adds that visualization can be a common language between experts and non-experts. This method therefore has the potential to facilitate participation, understanding and knowledge construction by scientists, decision makers, stakeholders and general public.

Geospatial visualization has emerged as a tool for searching huge volumes of data, communicating complex patterns and providing a formal framework for data presentation and exploratory analysis of data (Gahegan *et al.* 2001). It combines the power of multimedia dynamic representation of spatial information with interactive engagement of users (experts and non-experts) to perform exploratory analysis. Data on sustainable development are characteristically heterogeneous, complex, inter-dependent, not directly comparable, and correlated in ways that may not be apparent without the use of visualization techniques. Visualization is important not only in the development of Geographic Information Systems (GIS) generally, but also as a tool to improve reliability of multiple sustainable development scenarios, and thus decision support, as well as, to improve the ability of non-experts to absorb the information presented.

17.2.2 Hypermedia

The hypermedia model has evolved from the concepts of hypertext systems, which provide the capability to create and browse through complex networks of linked documents. Hypermedia is an information management method that connects various types of media (e.g., still and animated graphics or images, text, sound, video) in a non-sequential mode (Delisle and Schwarts 1986, Hann *et al.* 1992). With the integration of additional data types into these systems the term hypermedia evolved. Hypermedia is a tool for exploration and discovery. It can be considered as a form of knowledge representation since it creates links of inheritance or other types of relationships (Delikaraoglou *et al.* 1993). Hypermedia can be used to simulate the human ability to organize and retrieve information by referential links, as well as to create a form of object oriented network that can extend knowledge representation within an intelligent system application (Bielawski and Lewand 1991). It allows the user to temporarily deviate from his main exploratory course to access various sources, and then return to the original path of the application. In cartographic applications the hypermedia approach has been applied for micro-atlases in the early 1990's (Raveneau *et al.* 1991, Armenakis and Siekierska 1991).

A hypermedia network consists of information nodes connected by the associative links. The links are created based on the logical relevance between nodes. To establish the relationship between nodes, a process is needed to identify the relevant significant information. Entity-relation models or other relational semantic models (e.g., object-oriented) can be applied for this purpose. In the case of structural links, the user can access predefined types of information via labelled buttons. In the case of interactive user-defined links, the user selects its own exploratory navigation path, for example using its own Internet browser.

The hyperlinking, and non-linear features of the Web environment (Ayersman and Reed 1995) are well suited to support knowledge sharing. Cognitive processing theorists view these features as enabling a more accurate representation of the expert's (i.e., scientist) internal structure of knowledge. These features are viewed in terms of their capacity to help users to develop unique knowledge representations (Schafer *et al.* 2005).

17.2.3 Collaborative Communication

Decision making requires communication of multiple perspectives for common knowledge formulation. Collaboration occurs when participants communicate their understanding, listen to the views of others, explore alternative perspectives, are challenged in their beliefs, and challenge others (Miller and Miller 1999). The communication and distribution of information is critical to both how groups structure a decision problem, and what type of decision will come about as a result (Artman and Watern 1999). These situations, much like a boardroom or community meetings, can be described as "same time same place" decision making (Fig. 17.1). Meetings taking place over a distributed network can be thought of as "same time different place" meetings (Jankowski and Nyerges 2001).

While paper maps can be good for group decision making in the "same time same place" context, they do not support group decision making in a "same time different place" context. Although now available on-line, thus supporting "same time different place" interaction, GIS technology is not well suited for use by non-technical experts, which decision makers often are, and therefore might prove to be an ineffective distributed decision making tool.

With the increasing availability of devices such as mobile phones, PDA's with georeferencing, and geo-conference web-based systems (http://www.marketresearch.com/map/prod/1060342.html) geospatial collaborative distributed architectures is a new field of providing a collaborative work environment among disperse users with geospatial information used for consultative decision making. Geospatial collaborative systems may be synchronous of asynchronous (MacEachren 2001). The real time demands of synchronous decision making require awareness of the loca-

	Same Time	*Different Time*
Same Place	Community planning meeting	Sustaining development planning
Different Place	Collaboration of emergency response teams	Monitoring of glaciers involving laboratory work and fieldwork

Fig. 17.1. Decision making alternative locations examples.

tion of users and real time updating of the display with users annotations and with real time data as it becomes available (http://www.geoconnections.org/projects/geoinnovations/2002/TGIS/default.TGIS.htm).

Most geo-collaborative software provides users with multiple means for dialog and interactivity. Some of these are text based chat sessions, threaded discussions, video links, and loggings of sessions for replay. User input must be synchronized during a sessions in order to pass control among users for input and feedback. The hyperlinking of geospatial features to specific annotations, textual and images and video feeds in the future will allow users to more tightly integrate multiple sources from multi-sensor geospatial data in the decision making process. This will enable decision makers to communicate in real time with scientists or community representatives and to view the problems from "on the ground" as they were there in reality.

17.3 Case Studies: Applications and Systems

In this section selected applications and systems are discussed to illustrate the applicability of visualizations and hypermedia capabilities for decision making. The systems presented have been developed and used in the Earth Sciences Sector of Natural Resources Canada or by the Canadian Geomatics Industry. The case studies described within the context of decision making cover applications from city planning, to sustainable development of natural resources and emergency responsiveness.

17.3.1 Interactive Map of City of Iqaluit

City planners, urban planners and general public need cartographic products suitable for the efficient location of city services, as well as, access to additional information. Modern web-based electronic cartography provides effective products for city planning. Interactive maps linked to databases, permit the efficient location of a particular service or group of services.

The digital map of the Canadian northern city of Iqaluit, capital of Nunavut Territory, is an example of a web-based interactive visualization, which serves to locate the existing services within the city and assists in planning of the new services. The purpose of this development was the creation of new methods of cartographic visualization and handling geographic information suitable for Internet-based use of geospatial information in remote, northern communities of Canada. The application was developed in Macromedia Flash and uses digital photography of actual buildings as well as QuickTime movies of building interiors. The application uses the "rollover" effect for the identification of buildings on the city services map. It also uses a pull down list to search categories of buildings and individual buildings by name (Fig. 17.2).

Additional information such as pictures of current city services, namely the government offices, schools, shops, restaurants, hotels, community centres, churches, etc, can be accessed though hyperlinks. An interactive interface permits selection of individual buildings, as well as groups of buildings belonging to given service category, for example hotels. The selected objects are highlighted on the map. Pictures of the individual buildings are displayed for easier identification of the buildings and (http://maps.nrcan.gc.ca/visualization).

The interactive map of Iqaluit was developed in cooperation with the municipal government of the city as well as the territorial government of Nunavut. The application was well received by users such as city planers, business corporations, government agencies and educational institutions, leading to additional applications including the narrative containing historical information about the city evolution (http://maps.nrcan.gc.ca/iqaluit). The new city of Iqaluit website will be implemented in a complete trilingual version, which will permit use of the site using the native to

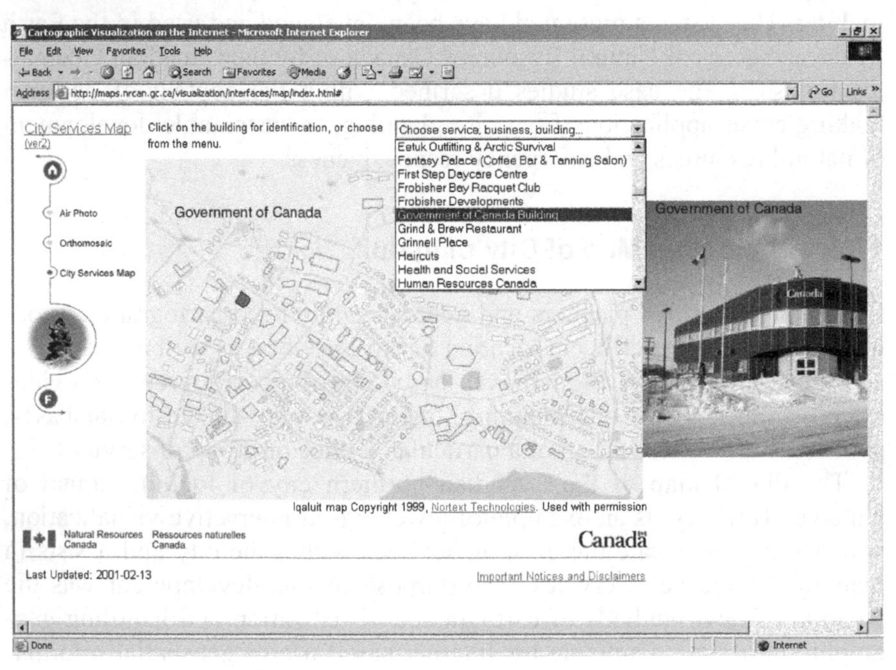

Fig. 17.2. City services map of Iqaluit.

Nunavut language – Inuktituk. The version of the website in a native language to the region will ensure the greater awareness and participation of the local population in the development of decisions concerning the city.

17.3.2 Visualization of Past, Present and Future Urban Land Use

Informed decision making for land utilization requires broad contextual information including historical land use records, present utilization of land, current development, and plans for the future. Such information is obtainable from historical maps and aerial photographs, current imagery and existing approved or proposed city plans. If this information is made available to the public via the web before decisions are finalized, feedback received from interested parties is considered and thus decisions affecting multiple stakeholders are taken into consideration.

The visualization of the Lebreton Flats area of Ottawa (Fig. 17.3) was developed based on historical aerial photographs dating back to 1923, current photographs and proposed plans of the city of Ottawa. Further, hyperlinks to websites, which provided information on current land development, namely the museum construction were added. This website was created to inform potential visitors about the upcoming exhibits planned for the museum opening – an effective promotion for tourism using mod-

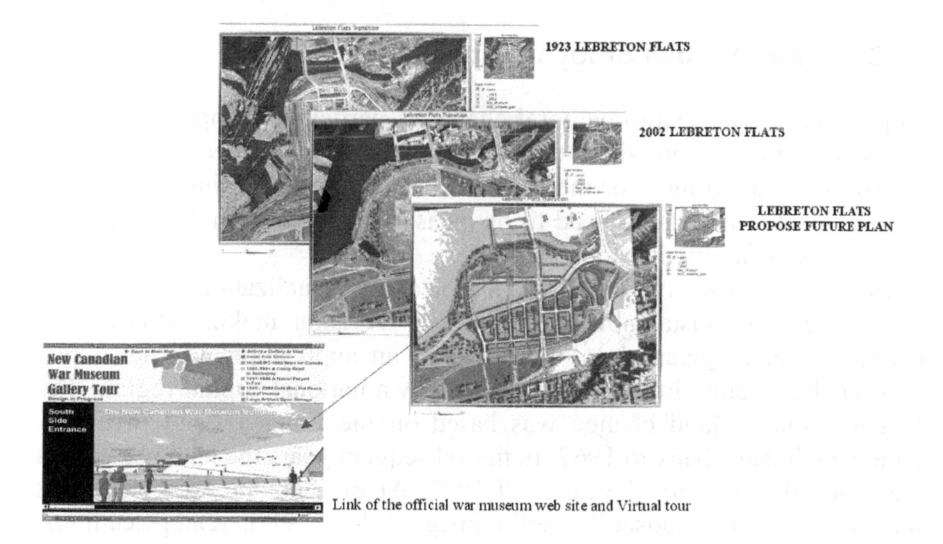

Fig. 17.3. Past present and future plans of Lebreton Flats, Ottawa.

ern technology. The potential visitors could obtain a complete tour using virtual reality techniques. This visualization uses hyperlinks to historical and current aerial photographs on which future museum and city plans for this region have been presented. Users may select or remove individual map layers and view future scenarios for the area. Rollovers are enabled on individual buildings. A particular building, under construction, may be viewed using hyperlinks to future building plans and virtual reality visualizations.

The visualization was developed based on the commercial system SVG MapMaker developed by the Canadian company, DBx Geomatics (http://www.dbxgeomatics.com) that uses SVG technology. This format is especially suitable for web-based interactive map visualization. Its strengths are: compact files size, which improves the speed of display; no degradation of line drawings, thus maintaining of cartographically correct images when enlarging or reducing the image scale; and the incorporation of multimodal formats, such as voice or sound files, to create versatile visualizations. The functionality, which is illustrated in Fig. 17.3, is a selection of individual map layers and hyper-linking to external web site.

The application was developed as a demonstration of incorporating heritage information to provide viewers a historical perspective and a broader context for current and future developments. Another aspect was to underline the strength of web-based communication to stakeholders (such as land use developers) and to potential visitors to upcoming tourist site.

17.3.3 Urban Sprawl Study in City of Ottawa

Representation of spatio-temporal change is particularly important for the decision making taken within the context of sustainable development. This approach needs to monitor the progress between specific time periods for which selected indicators need to be measured, to assess and evaluate the impact of development.

In the Natural Resource Canada's project on Visualization of Integrated Knowledge for Sustainable Development Decision making (SDKI-Vis) (http://sdki.nrcan.gc.ca/visual/index_e.php) an application was developed to visualize change in land use in the Ottawa national capital region. The visualization of land change was based on the Canada Land Inventory data, which dates back to 1967. In the subsequent years the land use census was carried out every 5 years, till 1981. At present, the land change is mapped based on Landsat 7 satellite imagery (http://sdki.nrcan.gc.ca/trans/index_e.php).

The browser-based visualization of urban sprawl of Ottawa-Gatineau (Fig. 17.4) was developed using DBxGeomatics SVG MapMaker software. The system can be used for showing land change either in a static (frame based) or dynamic (animated) way. In Fig. 17.4, the built-up areas that change between selected time periods are highlighted. Further, the SVG MapMaker has the capability of linking graphics of spatial information with attribute information using "infotip" (feature name display) and "infotool" (attribute display) functions. The "rollover" over features function displays names of features - other than permanent names. Display of statistical information using "display info tool" is possible. "Intelligent zoom" based on selective generalization of information, permitting display of variable details when changing scale of the graphics, is another capability implemented in this visualization. Using an interactive legend, the user can select and display information stored in the visualization system depending on their particular needs.

The integration of spatial and attribute data is necessary for obtaining additional insights – a drill down – capability- where more details are necessary for a particular feature. This information is not explicitly shown on display, but may be needed for informed decision making.

17.3.4 Web-based Dynamic Visualization System (DVS)

Decision makers require real time access and display of geospatial information from distributed, web-enabled data sources. Such services for web

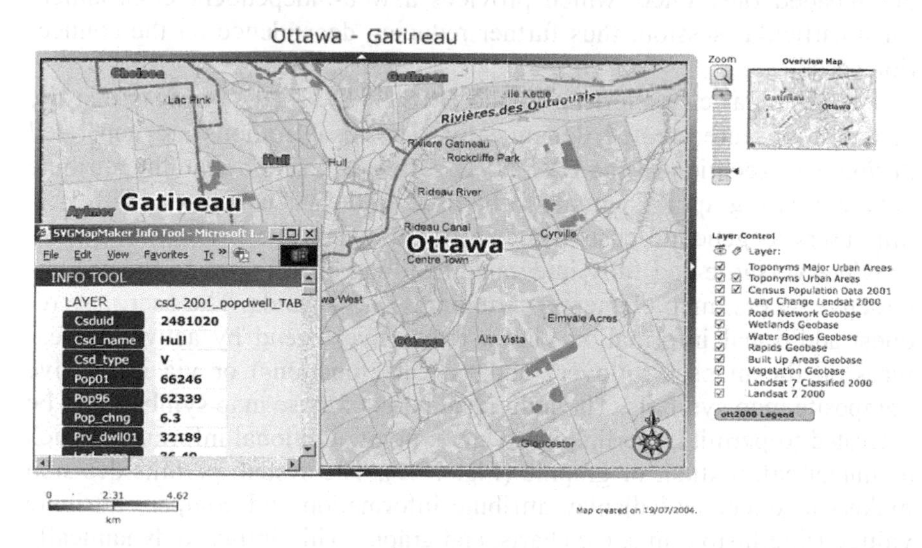

Fig. 17.4. Scalable Vector Graphics based integrated information representation.

based real time exploratory data analysis are being currently developed (http://www.galdosinc.com). Web-enabled systems provide access to geospatial information, available at remote locations, which may be downloaded by feature type or by attribute. This capability enhances the mapping, by permitting users to selectively download and incorporate information, which is most relevant to a particular application, thus avoiding excessive data processing and reducing the reliance on the current speed of Internet.

The Dynamic Visualization System (DVS) system is being developed in-house to demonstrate the applicability of the promising SVG-based technology, for real time interactive mapping and testing its potential for web based spatial decision support mapping environment. The system is based on the geoclient open source SVG mapping environment (http://www.mycgiserver.com/~amri/ geoclient.cocoon.xml). The main inspiration was the system developed by Amri Rosyada (http://www.mycgiserver.com/~amri/samples/mexico.svg). The system can display multiple user-created thematic maps, and related attribute information and images, in an integrated web-based environment in a dynamic and interactive way.

At present time, web map servers generate SVG format output directly. This data can be added without additional processing. The SVG format provides extended interactivity with geospatial data. The format permits hyperlinking between all the features on the map and their attributes and external visualizations. Geospatial information in the SVG format may be downloaded only once, which provides a web-independent environment for a particular session, thus further reducing dependence on the connection speed.

The principal characteristic of the DVS-SVG based system is that the visualizations are created dynamically. The visualizations are generated during the execution of the DVS-SVG software. Users define the attribute data and cartographic symbols to be used in the visualizations. This permits users to generate customized thematic maps. The DVS-SVG system provides versatile and fully integrated handling of information through interactive layers, map elements, cartograms and legends. The user may request additional information via an interactive legend by activating features on the maps ("rollover" and "select" functions) or via interactive composite map symbols. The individual parts of these map symbols can be activated to permit hyperlinking to any type of additional information such as numerical, textual, or graphic (Fig. 17.5). The system permits decision makers to query and display attribute information and compare attribute values using histograms, pie charts, and graphs. This ability to dynamically generate symbols and show the constituent parts of a variable, individual

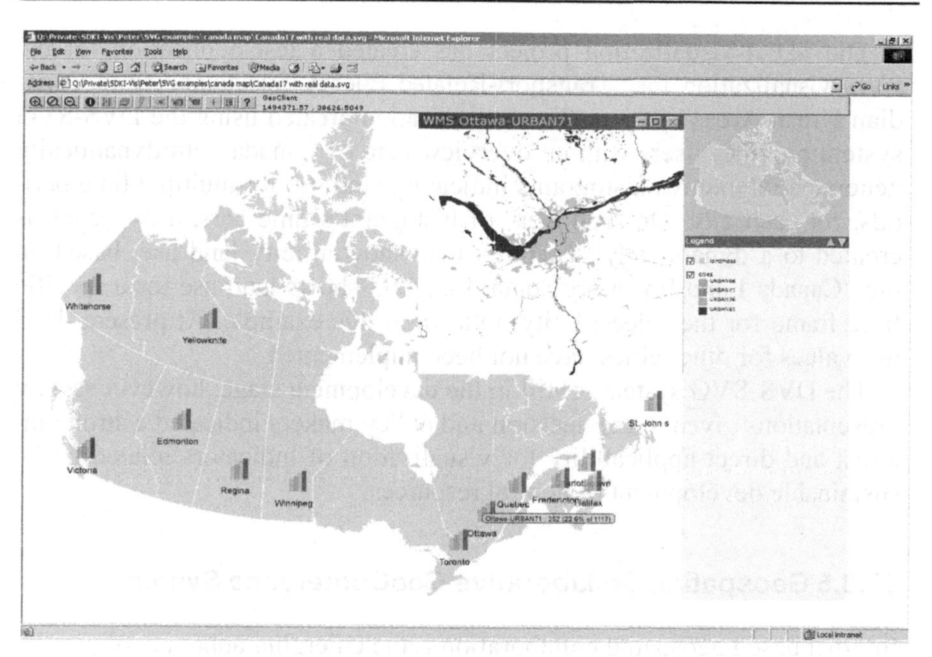

Fig. 17.5. DVS-SVG map of Canada with links to temporal land use visualizations.

segment values and percentage of the whole, are essential for data exploration. This functionality contributes to informed decision making where data analysis and comparisons of indicators are necessary for evaluating possible alternatives.

Decision makers can display maps from different periods to illustrate change from one time period to another. This can further lead to the display of multiple scenarios which have been pre-computed using modelling software and stored as map layers. In the future, animation will be used to display spatial change over time. The DVS-SVG system can display Open GIS Consortium (OGS) Web Map Service (WMS) generated web maps to show spatial temporal change for decision making purposes where change detection and the analysis of historical data is relevant to the decision making process.

The DVS system allows users to add textual or graphic annotations, notes, and comments, to displayed map elements. Further, it is possible to create hyperlinks to any map feature or to a map layer. This functionality assists users in converting tacit into explicit knowledge and to share knowledge with colleagues and policy makers. This capability offers flexibility to work in a collaborative environment whereby other users can modify maps, and complement it with additional information.

The SDKI-visualization project has created a test application for the SVG visualization for "Transport-Related Energy Sustainability in Canadian Urban Areas" project. This visualization created using the DVS-SVG system provides users with an overview map of Canada with dynamically generated interactive histograms indicating land use for multiple time periods, for each city. On "rollover" of histogram's time bars, a hyperlink is created to a dynamically generated map showing city land use, based on the "Canada Land Inventory" data. Fig. 17.5 shows land use for a specific time frame for the selected city, Ottawa in this example. At present, land use values for other cities have not been implemented.

The DVS-SVG system is still in the development stage, however several presentations given to the decision and policy makers indicated a strong interest and direct applicability for visualization of indicators related to the sustainable development of natural resources.

17.3.5 Geospatial, Collaborative GeoConference System

Internet based geospatial collaboration is the emerging approach of providing interactive collaborative work environments among geographically dispersed users (http://www.geovista.psu.edu/work/projects/geocollaboration.jsp). Geospatial collaborative software may be synchronous or asynchronous. The requirements of synchronous decision making employ multiple stakeholders and involve updating of the display to incorporate map view changes, users' annotations and real time data as it becomes available. Most geocollaborative software provides users multiple means for dialog and interactivity. Some of these are text-based chat sessions, "threaded discussions", video links, and logging of sessions for replay. User input must be mediated during a session in order to share control among users for annotation and editing purposes. The number, identification and location of users active in the session may also be available. Conferencing interaction between users, including the communication of their knowledge or expertise, has a different functionality than that called for by an analytical, asynchronous use of geospatial information.

An example of a geocollaborative conference system is GeoConference software developed by TGIS Technologies Inc. GeoConference, an Internet-based (either HTTP or binary protocols), lightweight client-server system, allows spatially distributed participation (different place, same time). The teleconference consultation is based on georeferenced material such as maps and images. Session participants share a synchronized, georeferenced cartographic view, which they control. The software also provides a text channel for instant-messaging exchanges. To accompany conversation, each participant controls a map pointer that is visible to everyone else. Multiple pointers can be active at the same time. Users can also add georeferenced annotations to the map, with changes seen in real time. Any session participant can then edit either the symbology or the georeferenced geometry of the annotations (Fig. 17.6). The bandwidth requirement of the system is light by design – conferences work even at dial-up Internet connection speeds.

GeoConference can be applied as a collaborative tool throughout the lifecycle of activities and projects that use geographic information. It provides discussion participants a map window, with zoom, pan, symbology control and layer ordering. Moreover, session participants can dynamically add OGC Web Map Service data layers, as well as include their own local georeferenced image and vector layers. Session control (control of the map view and georeferenced annotations) is shared by all participants – the GeoConference server provides the arbitration. Descriptive metadata can be provided from information stored with WMS layers or added to local sources. The georeferenced annotation facility (creation of point symbols, lines and polygons) permits experts to add explicit knowledge to the map. It also provides a way for users to incorporate field observations. The text message facility permits users to exchange opinions, register findings and discuss aspects of the visualization and map layers. GeoConference ses-

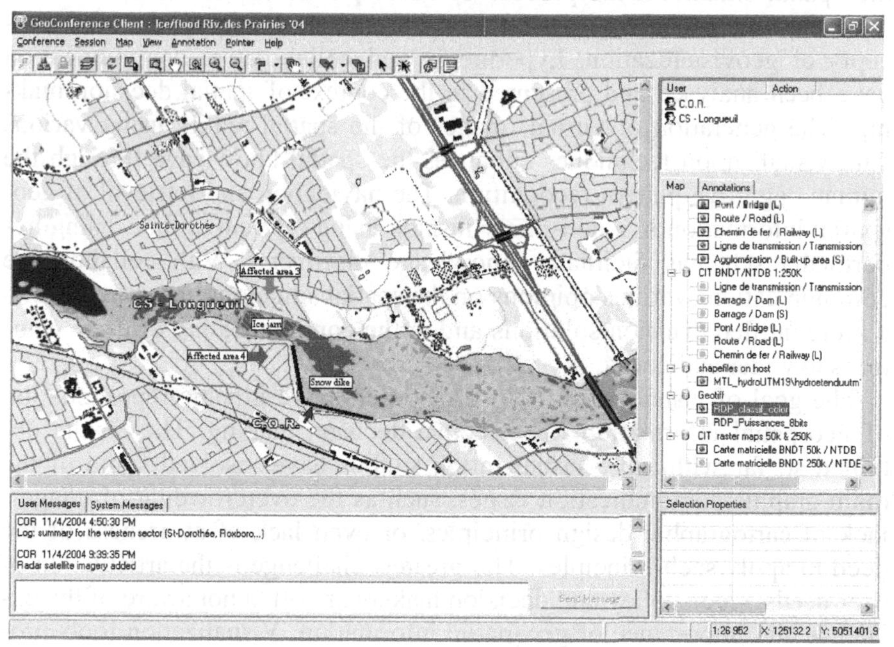

Fig. 17.6. Geoconference session for collaborative decision making.

sions are maintained in a session database and can be resumed at a later date. Presently, the text messages are logged and time stamped; it would be rather simple to add complete session logging as well.

GeoConference software has been used in various emergency management cases, both simulated and real. These include a search and rescue case, developed cooperatively with the Natural Hazards Emergency Response program in Natural Resources Canada, river ice and flood management (collaboration with Sécurité civile du Québec and the Institut national de recherche scientifique - Eau, Terre Environnement) and civil emergency management during wildfires (operational use by Sécurité civile du Québec). Fig. 17.6 illustrates the river ice and flood management application.

The unique capabilities of GeoConference software enhance geographically diverse collaborative decision making in a real time environment with multiple local and remote data sources.

17.4 Concluding Remarks

Spatial decision making encompasses the understanding and assessment of the spatial situations, the process of development of various options and their impacts, and the collaborative communication component. The concepts of geovisualization, hypermedia and collaborative communication have been analysed and presented in the context of spatial decision making. The generation of virtual images of the spatial reality using various data visualization techniques enhances the spatial knowledge through the human cognitive pattern recognition. The navigation and retrieval of geospatial information in a non-linear fashion using referential links using hypermedia schemas facilitates knowledge discovery. With collaborative communications various opinions on the spatial situations can contribute to a more unified decision solutions and to the communicating of these solutions to wider audiences.

The goal of visualization is to provide effective graphic representation of geospatial information for effective decision making. The results obtained so far within the visualization studies have resolved some of the main graphic communication issues, such as the overcrowding of images, lack of cartographic design principles, or even lack of awareness of the need to apply such principles. The greatest challenge is the articulation of user needs, since policy and decision makers are often not aware of the existence and importance of geospatial information. Visualization tools provide the means to move from static cartographic products, such as images, maps and aerial photographs, to interactive products that utilize geospatial data through dynamic and interactive graphics that employ multimedia and multi-modal technologies. By facilitating the exploration of multiple data sets, over both time and space, visualization tools can be used to develop sustainability scenarios that illustrate past or future landscape changes and assist decision makers in understanding key spatial and temporal relationships.

Hypermedia organizes the information in many purposeful structures and creates data relationships. Hyperlinking assists in the conversion of implicit knowledge to explicit knowledge. Hypermedia can therefore enhance the quality of decision making by the creation of these relationships and adding knowledge to the decision making process through the hyperlinks. The hyperlinking of geospatial features to specific annotations, textual and images and "video feeds" in the future will allow users to more broadly integrate multiple sources from multi-sensor geospatial data in the decision making process. Hyperlinking also promotes collaboration and shared understanding.

The geovisualization and hypermedia concepts for spatial understanding and decision making were implemented and evaluated in the conducted case studies. Work on DVS and Geoconferencing systems handling Internet-based interactive data exploratory analysis, mapping and collaborative decision making was discussed. The results are quite promising and is expected that this type of approaches will contribute to communicating scientific results in forms and formats that are better suitable for informed decision making.

Finally, interactive geocollaborative environments will provide users with multiple means for dialog, interactivity and simultaneous viewing of geospatial data. The system described in this chapter is based on web-based geoconferencing, which permits synchronized sessions for multiple participants "same time different place" consultation and decision making. It also includes text based chat sessions, threaded discussions, video links, and logging of a sessions for replay. In the future, geocollaborative tools may provide support for live "video camera feeds", 3-Dimensions data views, integrated attribute information, objective identification, assignment of sub tasks and goals and prioritization of tasks, risk assessments and multiple scenario with modelling capabilities.

Acknowledgements

The authors would like to acknowledge the contributions of the following persons in the development of systems and applications described in this chapter, Yves Carbonneau and Guy Trudel of TGIS Technologies Inc., Donald Fortin of the Ministry of Public Security (Québec), Yves Gauthier of the Institut national de recherché scientifique, Eau, Terre, Environnement (Québec), Benjamin Campin of DBx Geomatics, and Jean-Pierre Dostaler, Anita Muller and Ken Francis of Natural Resources Canada.

References

Al-Kodmany K (2001a) Bridging the gap between technical and local knowledge: tools for promoting community based planning and design. Journal of Architectural and Planning Research, vol 18(2), pp 110-130

Al-Kodmany K (2001b) Visualization tools and methods in community planning: from freehand sketches to virtual reality. Journal of Architectural and Planning Research, vol 17(2), pp 189-211

Andrienko G, Andrienko N (2001) Interactive visual tools to support spatial multicriteria decision making. In Proceedings of the 2nd International Workshop on User Interfaces to Data Intensive Systems, Zurich, Switzerland

Armenakis C, Siekierska E (1991) Issues on the visualization of time dependent geographic information. In: Proceedings of the Canadian Conference GIS'91, CISM, Ottawa, Canada, pp 584-595

Artman H, Watern Y (1999) Distributed cognition in an emergency. Co-Ordination Centre. Cognition, Technology and Work, vol 1, pp 237-246

Ayersman DJ, Reed WM (1995) Effects of learning styles, programming, and gender on computer anxiety. Journal of Research on Computing in Education, vol 28(2), pp 148-161

Bielawski L, Lewand R (1991) Intelligent systems design: Integrating expert systems, hypermedia, and database technologies. John Wiley & Sons, Inc., pp 302

Densham PJ (1991) Spatial decision support systems. Geographic Information Systems: Principles and Applications, John Wiley & Sons, New York, pp 403-412

Delikaraogolou D, Armenakis C, Christodoulides D (1993) Use of hypermedia tools: making global change data accessible and comprehensible for decision making and education. United Nations/Indonesia Regional Conference on Space Science and Technology for Sustainable Development, Bandung, Indonesia

Delisle N, Schwartz M (1986) A partitioning concept for hypertext. In: Proceedings of the 1986 ACM Conference on Computer-supported cooperative work, pp 403-410

Fayyad U, Uthurusamy R (1999) Data mining and knowledge discovery in databases: Introduction to the special issue. Communications of the ACM, vol 39(11)

Feenay FM-E, Williamson IP, Bishop ID (2002) The role of institutional mechanisms in spatial data infrastructure development that supports decision making. Cartography, vol 31(2), pp 33-37

Gahegan M, O'Brien D (1997) A strategy and architecture for the visualization of complex geographical datasets. International Journal of Pattern Recognition and Artificial Intelligence, vol 11(2), pp 239-261

Gahegan M, Wachowicz M, Harrower M, Rhyne T (2001) The integration of geographic visualization with knowledge discovery in databases and geocomputation. Cartography and Geographic Information Science, vol 28(1), pp 29-44

Hann BJ, Kahn P, Riley VA, Coombs JH, Meyrowitz NK (1992) IRIS hypermedia services. Communications of the ACM, vol 35(1), pp 36-51

Hu S (2005) Design issues associated with discrete and distributed hypermedia GIS. In: Proceedings 1st International Workshop on Geographic Hypermedia, Denver, CO

Jankowski P, Nyerges T (2001) Geographic information systems for group decision. Cartography and Geographic Information Systems, vol 25(2), pp 67-76

King S, Merinda C, Latimer B, Ferrari D (1989) Co-Design: a process of design participation. Van Nostrand Reinhold, New York

Lein JK (2003) Integrated environmental planning. Blackwell Publishers, Oxford, pp 228

Levy P (1999) Collective Intelligence. Robert Bononno trans. Perseus Books, Cambridge

MacEachren AM, Edsall R, Haug D, Baxter R, Otto G, Masters R, Fuhrmann S, Qian L (1999) Virtual environments for geographic visualization: potential and challenges. ACM Workshop on New Paradigms for Information Visualization and Manipulation, Kansas City, Kansas, http://www.geovista.psu.edu/publications/NPIVM99/ammNPIVM.pdf

MacEachren AM (2001) Cartography and GIS: extending collaborative tools to support virtual teams. Human Cartography, vol 25(3), pp 431-444

MacEachren AM, Brewer I, Cai G, Chen J (2003) Visually-enabled geocollaboration to support data exploration and decision making. In: Proceedings 21st International Cartographic Conference, Durban, South Africa, CD-ROM, pp 394-401

Malczewski J (1999) GIS and multicriteria decision analysis. Wiley, London.

McCormick BH, DeFanti TA, Brown MD (1987) Visualization in scientific computing. Computer Graphics, vol 21(6), pp 6-26

Miller SM, Miller KL (1999) Using instructional theory to facilitate communication in web-based courses. Educational Technology & Society, vol 2(3), pp 106-114

Raveneau J, Millar M, Brousseau Y, Dufour C (1991) Micro-Atlases and the diffusion of geographic information: an experiment with HyperCard. In: Taylor DRF (ed) Geographic information systems: the microcomputer and modern cartography, Pergamon Press, New York

Schafer WA, Ganoe GH, Coch G (2005) Designing the next generation of distributed geocolloborative tools. Cartography and Geographic Information Society, vol 32(2)

Smith J (2002) Geonomics: Bootstrap development for a sustainable Planet, UN Conference on Financing for Development, Monterrey, Mexico

Webster J, Baulch S, Gebrehana G, Müller A, Francis K, Williams P, Siekierska E (2004) Concepts in visualization in application to decision support systems for sustainable development decision making. Natural Resources Canada, Unpublished

Worrall L (1991) GIS for spatial analysis and spatial policy: developments and directions in spatial analysis and spatial policy using geographic information systems, Bellhaven Press, London, pp 1-11

18 Geovisualization of Vegetation Patterns in National Parks of the Southeastern United States

Marguerite Madden, Thomas Jordan, and John Dolezal

Abstract. Over the past ten years, the Center for Remote Sensing and Mapping Science (CRMS) at The University of Georgia has worked cooperatively with the National Park Service to create digital vegetation databases for 21 National Parks, Preserves, Home Sites and Battlefields in the southeastern United States. These databases were created from manually interpreted large-scale color infrared aerial photographs using a combination of Global Positioning System (GPS) surveys, softcopy photogrammetry and GIS modeling procedures. Throughout the process, geovisualization techniques have been used to aid in the extraction and assessment of vegetation patterns, quality control evaluation and communication of information to managers and users of park resources. In one example, three-dimensional (3D) drapes of orthorectified images and vegetation maps were used to aid in the interpretation process by providing the interpreter with multiple 3D perspective views and information on elevation range, slope and aspect. Upon completion of the databases, geovisualization techniques also were used to qualitatively identify and assess areas prone to errors in the geometric orthorectification. In this way, geovisualization contributed to improvements in both the thematic and geometric accuracy of the National Park Service vegetation data sets. In another example, animations of spatio-temporal data sets and 3D drapes were developed to assess the impacts of surrounding development and land use changes on resources within park boundaries. Geovisualizations provided to the National Park Service are being used to prioritize the acquisition of additional lands to help preserve water quality and habitats of ecological importance.

18.1 Introduction

Visualization as a tool for data exploration was highlighted in a special issue of *Computer Graphics* compiled by Bruce McCormick and colleagues (Slocum *et al.* 2005). In this issue, McCormick stated, "Visualization offers a method for seeing the unseen. It enriches the process of scientific discovery and fosters profound and unexpected insights. In many fields it is already revolutionizing the way scientists do science," (McCormick *et al.* 1987). The extension of this concept of scientific visualization to geographic data is rooted in cartography, as evidenced by the well-known 1861 paper map by Charles Minard depicting the movements of Napolean's 1812 campaign into Russia (Kraak 2002). Advancements in digital mapping and computer display created the field of geographic visualization or geovisualization defined by MacEachren and Taylor (1994) as "the use of concrete visual representations – whether on paper or through computer displays or other media – to make spatial contexts and problems visible, so as to engage the most powerful human information-processing abilities, those associated with vision." As computer technologies continue to advance, so do methods of exploratory data analysis, synthesis and presentation of geospatial data (MacEachren and Kraak 2001). An excellent example is the application of geovisualization techniques to Minard's 1861 map data to provide alternative visualizations, multiple-linked views, space-time cube depictions and animations of multiple variables and temporal information (Kraak 2002). In other applications, geovisualization has been used to recreate views of historical landscapes in National Battlefields, visualize forest cover succession following natural disturbances and display the narrow sea-land interface related to changing water levels for troop deployment (Dunbar *et al.* 2003, Fleming *et al.* 2005, Madden and Schieve 2003, Moskal 2004).

The Center for Remote Sensing and Mapping Science (CRMS) at The University of Georgia has used geovisualization techniques to assess vegetation patterns and provide information that can be used for thematic attribution, geometric quality control and assessment of human impacts on park resources. Working cooperatively with the National Park Service over the past 10 years, digital vegetation databases were created for one National Wildlife Refuge and 21 National Parks, Preserves, Home Sites and Battlefields in the southeastern United States (Madden *et al.* 1999, Welch *et al.* 1995, 1999, 2002a, Welch and Madden 2002). Manual interpretation of medium and large-scale color infrared aerial photographs was the primary method used for data development in order to maximize the amount of thematic detail that could be extracted from the remotely sensed data and increase the thematic accuracy of the resulting vegetation data sets (Remillard and Welch 1992, Welch *et al.* 1988). Geometric detail and accuracy were challenges in many parks due to their remote locations, continuous vegetation cover and, sometimes, rugged terrain. It was necessary, therefore, to use a combination of remote sensing, geographic information system (GIS), Global Positioning System (GPS), softcopy photogrammetry and rule-based modeling techniques to develop the detailed vegetation databases. For example, between 1994 and 2000, vegetation databases and hardcopy maps were created from 1:40,000-scale color infrared air photos covering over 10,000 km^2 of south Florida park units including Everglades National Park, Biscayne National Park, Big Cypress National Preserve and the Florida Panther National Wildlife Refuge (Madden *et al.* 1999, Welch *et al.* 1995, 1999, 2002a). From 1998 to 2003, the CRMS mapped both overstory and understory forest vegetation from 1:12.000 and 1:40,000 scale color infrared photos of the 2,000 km^2 Great Smoky Mountains National Park located in the southern Appalachian Mountains of Tennessee and North Carolina (Welch *et al.* 2000, 2002b). These data sets were subsequently used for visualization and analysis of vegetation patterns and fire fuel model classification to determine fuel characteristics and potential fire ignition (Dukes 2001, Madden and Jordan 2001, Madden 2003, Madden and Welch 2004).

In 2001, the CRMS and NPS began a two-phase project to map the vegetation of 13 additional National Park units within the Appalachian Highlands and Cumberland Piedmont Networks of the National Park Service. In the first two years, Phase I focused on: 1) developing detailed vegetation digital data sets in Arc/Info, ArcView and/or ArcGIS formats and associated vegetation hardcopy maps for Carl Sandburg Home National Historic Site (109 hectares), Cumberland Gap National Historic Park (8,275 hectares), Guilford Courthouse National Military Park (89 hectares), Little River Canyon National Preserve (5,546 hectares), Ninety-Six National Historic Site (405 hectares) and Stones River National Battlefield (287 hectares); 2) establishing appropriate vegetation classification systems for mapping each park unit and; 3) collaborating with vegetation experts from National Park Service and NatureServe, the research unit of The Nature Conservancy, to refine interpretation techniques and ground truthing procedures in order to maximize the information content of the vegetation databases. Phase II is currently being conducted over a three-year period using the methodologies defined in Phase I to map vegetation in Abraham Lincoln National Historic Site (138 hectares), Big South Fork National River and Recreation Area (50,607 hectares), Blue Ridge Parkway (35,910 hectares), Cowpens National Battlefield (344 hectares), Fort Donelson National Battlefield (227 hectares), Mammoth Cave National Park (21,388 hectares) and Obed Wild and Scenic River (2,024 hectares). Phase III was initiated in 2005 and includes four additional parks: Chickamauga and Chattanooga National Military Parks (3328 hectares), Kings Mountain National Military Park (1599 hectares), Shiloh National Military Park (1619 hectares) and Russell Cave National Monument (138 hectares).

Vegetation communities in all of the above mentioned National Park unit databases were mapped using manual interpretation techniques and large-scale color infrared aerial photographs. Interpreters preferred to use analog methods to digital procedures - delineating vegetation boundaries on plastic overlays registered to the film transparencies while viewing the photos in stereo with a mirror stereoscope. These traditional techniques allowed them to view relatively large areas of the terrain in stereo and in color for best determination of site conditions, (i.e., moisture, hydrology and relative soil richness) and identification of the vegetation community type. Digital manipulation of the scanned and orthorectified air photos for 3D visualization, however, can provide interpreters with additional information that can be used in the interpretation process. Drapes of the orthorectified vegetation maps and animations also can be constructed to depict changes in vegetation patterns over time. This chapter will address the use of geovisualization to: 1) assess vegetation distributions related to terrain characteristics for improving manual interpretation and thematic classification of vegetation polygons; 2) assist in quality control checks of the geometric integrity of the completed vegetation databases; and 3) convey information on development trends surrounding parks and potential impacts on park resources.

18.2 Vegetation Distributions Related to Terrain Characteristics

The spatial correlation of vegetation communities with ancillary environmental data sets can be performed in a GIS environment to determine factors influencing plant distributions (Moore *et al.* 1991, Remillard and Welch 1993, Woodcock *et al.* 2002). Overlay analysis of overstory forest vegetation polygons in Great Smoky Mountains National Park with elevation range, slope and aspect, for example, provided information on mean, range and variance in terrain variables that can be associated with individual forest and shrub classes (Madden 2004). This summary of terrain characteristics associated with southern Appalachian forest community types can provide botanists and vegetation interpreters with quantitative information defining habitat preferences for individual vegetation communities as described in the National Vegetation Classification System (Grossman *et al.* 1998, White *et al.* 2003).

Although quantitative data on vegetation distributions related to environmental factors are often desired for modeling cause-and-effect relationships, geovisualization of these same variables can glean additional information on data classification, uncertainty, thematic accuracy and geometric accuracy. To illustrate this point, a single 1:12,000-scale color infrared aerial photograph covering a small portion of Great Smoky Mountains National Park was selected (Fig. 18.1). The original positive transparency was scanned and orthorectified using the Desktop Mapping System (DMS) by R-WEL, Inc. according to procedures described in Jordan (2004) and Welch *et al.* (2002b). Summarizing these procedures, ground control points (GCPs) used in the orthorectification were largely natural features such as individual trees and forks in stream channels that could be identified on both the 1:12,000-scale color infrared transparencies and U.S. Geological Survey (USGS) Digital Orthophoto Quarter Quads (DOQQs). The horizontal Universal Transverse Mercator (UTM) coordinates of these GCPs were measured directly from the DOQQs at typical accuracies of $\pm$ 3 m, while the elevations of the GCPs were interpolated to $\pm$ 3 to 5 m from USGS Level 2 Digital Elevation Models (DEMs) with 30-m post spacing (Welch *et al.* 2002b). Analytical aerotriangulation was then performed using the AeroSys software package for blocks of up to 90 scanned photos corresponding to the area covered by one of 25 USGS 1:24,000-scale topographic quadrangles. Typical geometric accuracies for the aerotriangulated pass points was $\pm$ 7 m for X, Y and $\pm$ 10 m for Z. These pass points, in turn, were input to DMS to compute orientation parameters and differentially rectify the scanned air photos.

The orthorectification coefficients used for the scanned photos were then applied to scanned images of the plastic overlays containing delineations of manually interpreted vegetation boundaries. A raster-to-vector conversion and import to ESRI Arc/Info Workstation allowed superimposition of the interpreted vegetation polygons onto the digital orthophoto (Fig. 18.1). A visual quality control check of this 2D display was performed to confirm the orthorectification coefficients were indeed being applied adequately to geometrically correct the photo-derived vegetation boundaries. This display also can be used to ensure vegetation polygons are enclosing unique photo signatures. If separate polygons appear to be artificially or arbitrarily separating the same photo signature, the vector line work can be edited before additional processing time is spent on unnecessary attribution.

The next step in quality control assurance is performed using 3D geovisualization and visual assessment of vegetation polygons relative to the 3D terrain display. A single color infrared orthorectified air photo registered to a USGS Level-2 DEM with 30-m post spacing is draped on a 3D perspective view of the DEM to emphasize the display of relief (Fig. 18.2). In this display, it is easier for the intrepreter to visualize the photographic signature of plant communities relative to terrain characteristics. Using information previously established by GIS overlays of vegetation data sets with elevation ranges, slope and aspect, interpreters can associate the

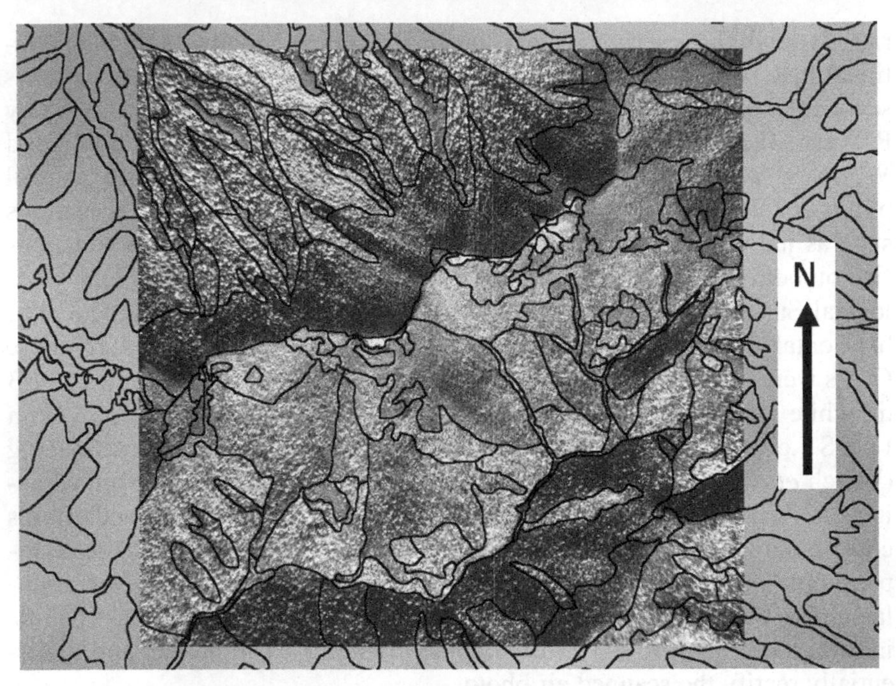

Fig. 18.1. Two-dimension display of orthorectified vegetation community boundaries superimposed on a 1998 color infrared orthophoto.

color, tone, texture, height, shape and location of vegetation signatures with descriptions of typical terrain characteristics of individual vegetation communities. For example, interpreters can see there is a central ridge of high elevation in this area and the vegetation cover is very different on one side of this ridge from the other. (Care must be taken, however, to remember the orientation of the image has changed and north is no longer at the top of the draped orthophoto display. This is very important for assessing classified and attributed vegetation polygons relative to aspect in the 3D display.) Reference to vegetation community descriptions will alert the interpreter to a list of possible communities that prefer moist and rich conditions found on concave, north-facing and moderate slopes. Other communities are listed as associated with convex, south-facing and steep slopes that reflect drier, nutrient-poor and acidic soil conditions. Geovisualization of 3D perspectives emphasizes the terrain characteristics, thereby aiding the interpreter in selecting the correct vegetation community and improving the thematic accuracy of the database.

The 3D perspective views that combine image and vegetation data set drapes on the DEM also can be used to highlight problems with georectification of the vegetation polygons. The black area on the color infrared or-

Fig. 18.2. 3D perspective view of color infrared orthophoto draped on a 30-m USGS Level-2 DEM.

thophoto marked "A" in Fig. 18.3 clearly shows the outline of an area in shadow. This area, however, does not edgematch accurately with the vegetation polygon shown in white. The arrow points to a geometric discrepancy of approximately 20 m which exceeds the expected accuracy of $\pm$ 5 to 10 m. After visual assessment of geometric accuracy, areas requiring improved ground control and subsequent re-orthorectification were identified. Although this assessment was not automated, it did provide a quality control check that was used to improve the geometric integrity of the vegetation database.

A final geovisualization technique involves systematic rotation of the 3D perspective views to create a sequence of images that, when displayed one after the other in rapid succession, can be viewed as an animation. An animated fly-through to view the 3D perspective from the south was created using DMS. Two frames from the animation sequence are shown in Fig.18.4. In this mountainous area, the fly-through was used to illustrate the changes in vegetation communities with slope and aspect as the viewpoint is rotated around the mountain top. In this way, resource managers were better able to understand the rationale by the interpreter to identify vegetation communities.

Fig. 18.3. 3D perspective view of color infrared orthophoto and attributed vegetation polygons draped on a 30-m USGS Level-2 DEM. "A" depicts an area of geometric inaccuracy.

It also is anticipated that resource managers in National Parks will use these geovisualization techniques to produce display materials for environmental education, visitor interpretation and training for researchers involved in making resource decisions to protect cultural and natural resources.

18.3 Development Trends and Potential Impacts on Park Resources

Many historic battlefields and homesites in the United States that are protected by the National Parks Service are located in towns and cities that are experiencing significant population growth and urbanization. Residential and commercial development can damage cultural viewsheds of National Parks and introduce sounds and lights that prevent visitors from experiencing the reenactment of historical events or settings (Bennett 2005). Land use changes immediately beyond park boundaries often include the re-

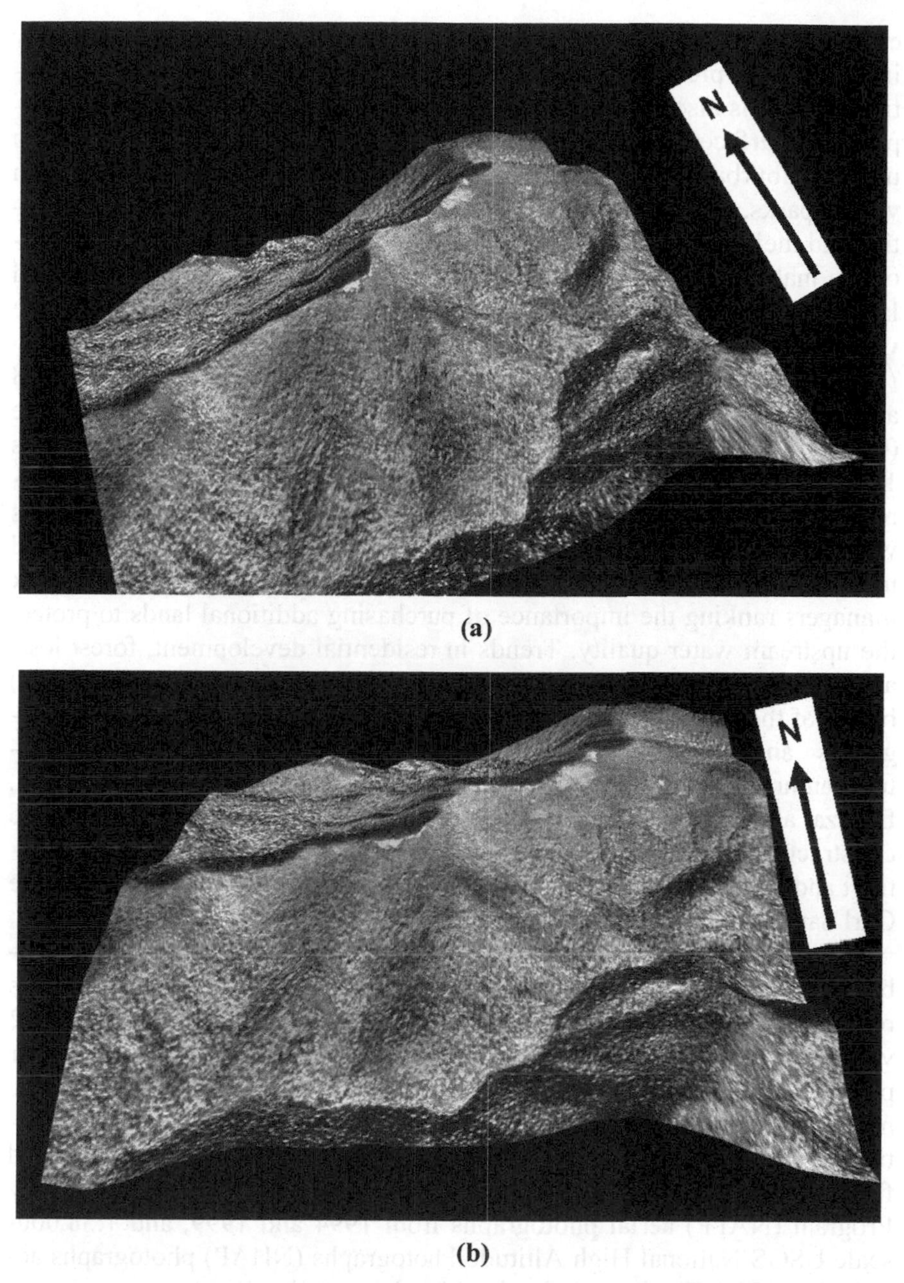

Fig. 18.4. First (a) and second (b) captured views in a sequence of 3D perspective views of color infrared orthophoto draped on a 30-m DEM.

moval of forests, construction of buildings and paving of roads and parking lots. The increased fragmentation of once natural areas and introduction of cars and people directly affects animal populations that once moved freely in and out of preserved lands. Runoff within watersheds containing National Parks is also contaminated by upstream development, increased impervious surfaces and the introduction of pesticides and fertilizers often used in suburban housing subdivisions. Baseline information on vegetation within parks, therefore, that includes both the identity of plant communities and their geographic extent is required for proper management and decision making. Spatio-temporal information on changes in vegetation and land use in areas immediately adjacent to park boundaries and within the watersheds that contain parks is equally important.

The CRMS used remote sensing, GIS and GPS to map, visualize and analyze the vegetation of Carl Sandburg Home National Historic Site (CARL) in Flat Rock, North Carolina (Madden and Jordan 2004). This 107-hectare park unit preserves the home, farm, agricultural fields and surrounding forest owned by the writer and historian Carl Sandburg and his wife, Paula. Dolezal (2005) performed an analysis of historical land use/land cover trends surrounding the park dating back to 1986 to aid park managers ranking the importance of purchasing additional lands to protect the upstream water quality. Trends in residential development, forest loss, agricultural practices and urban growth within a 1,609-m (or one-mile) buffer of the park boundary were documented from historical aerial photographs and input to an ArcGIS digital database to assess cultural/anthropogenic impacts upon water quality and natural vegetation. Dolezal also explored the use of geovisualization techniques including the construction 3D perspective views of the landscape and animations to extract and communicate trends in land use/land cover in and surrounding the Carl Sandburg Home National Historic Site.

An historical land use/land cover database that includes the Carl Sandburg Park and a 1,609-m buffer area was created from aerial photographs recorded in 1986, 1994, 1999, 2000 and 2002 (Dolezal 2005). The 2002 vegetation database was created from 1:12,000-scale color infrared air photos and contained 24 vegetation community classes with alphanumeric modifiers appended to denote evidence of disturbance, management practices or health of the vegetation. Historic information also was collected from 1:40,000-scale color infrared USGS National Aerial Photography Program (NAPP) aerial photographs from 1994 and 1999, and 1:58,000-scale USGS National High Altitude Photographs (NHAP) photographs acquired in 1986. The historic land use/land cover classification system consisted of 37 classes identifying vegetation, land cover and type and density of residential/urban development. Two change layers were created by combining land use/land cover interpretations from 1986-1994 and 1994-1999, and eight change classes were identified from the change layers to create a map showing cultural development and forest transformation (Table 18.1).

Based on the statistics provided in Table 18.1, for instance, resource managers of Carl Sandburg Home National Historic Site can be informed that there have been a 16, 8 and 23-percent loss of forest for the three time periods mapped between 1986 and 1999. They also can note that cultural encroachment on forest ranged from 13 to 82 percent while 0 to 29 percent of cultural encroachment was on non-forest land covers. Visualization of the relative magnitudes of multiple land use/land cover changes over multiple time periods, however, is difficult when based on tabular statistics alone. In a progression of spatial data visualization, a histogram graph of change data improves the communication of direction, magnitude and temporal trends in cultural encroachment on forest areas (Fig. 18.5). Maps of land use/land cover changes for the time periods 1986-1994 and 1994-1999 are better yet in depicting the extent and locations of these changes (Fig. 18.6). From these maps it is evident that the majority of cultural development since 1986 occurred in a forested area just south of the park boundary. Closer inspection of specific parcels transitioning from forest to cultural development is made somewhat difficult, however, when resource managers are forced to view first one map and then the other in a side-by-side layout. Information on cultural development of individual parcels can be enhanced by superimposing multi-temporal maps and producing a continuously cycling Flash animation (Fig. 18.7). A series of 3D perspective

Table 18.1. Change, Percent Change and Rate of Change for the Carl Sandburg Home National Historic Site Land Use/Land Cover Classifications, 1986 – 1999

	Change (ha)			**% Change**			**Change per year (ha)**		
Change Class	**86-94**	**94-99**	**86-99**	**86-94**	**94-99**	**86-99**	**86-94**	**94-99**	**86-99**
Forest	-149.4	-59.5	-208.8	-16.2	-7.7	-22.7	-18.7	-11.9	-16.1
Cultural Encroachment on Forest	117.2	41.4	158.6	60.3	13.3	81.6	14.7	8.3	12.2
Non-forest Vegetation	-22.0	2.9	-19.1	-14.1	2.2	-12.2	-2.7	0.6	-1.5
Cultural Encroachment in Non-Forest	0.0	1.3	1.3	0.0	29.0	29.0	0.0	0.3	0.1
Culturally Maintained Vegetation	26.4	6.5	32.9	8.0	1.8	10.0	3.3	1.3	2.5
Wetland	0.0	0.0	0.0	0.0	0.0	0.0	0.0	0.0	0.0
Water	0.0	0.0	0.0	0.1	0.0	0.1	0.0	0.0	0.0
Bare Ground	27.7	7.4	35.1	25.6	5.4	32.4	3.5	1.5	2.7

views produced using ArcGIS ArcScene to create a fly-through over the park provides a more advanced visualization (Fig. 18.8). This visualization can be used to assist park resource managers in the evaluation and prioritization of potential land acquisitions. Forested parcels located adjacent to and upstream of the park within the park's watershed were given highest priority due to potential impacts of soil erosion and non-point source pollution inputs to water flowing from newly developed areas into the park.

18.4 Summary

In summary, GIS analyses and geovisualization techniques are suitable for assessing patterns in National Park vegetation community distributions and conveying information on development trends and potential impacts on park resources. Visual displays in 2D and 3D of vegetation boundaries registered to a DEM provide a more clear depiction of the relationship between particular vegetation types and terrain characteristics such as slope, aspect and elevation range. The geovisualizations highlight thematic and geometric problems that can be addressed to increase database accuracy. Animation of spatio-temporal data sets and 3D drapes also improve assessment of adjacent development trends and the affect of land use changes on resources within park boundaries. Geovisualizations enhance the communication of these trends and impacts to National Park Service mangers, researchers and policy makers for land prioritization and preservation of critical habitats. Future work includes using these geovisualization techniques to further assess vegetation patterns in terms of potential

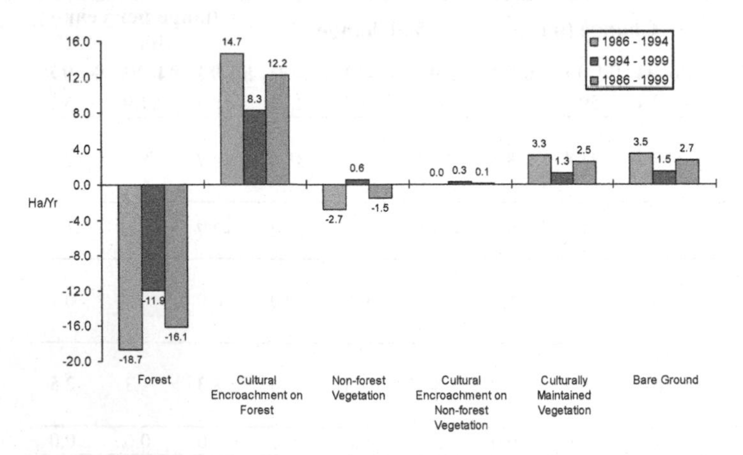

Fig. 18.5. Rates of change (hectares per year) for land use/land cover change in Carl Sandburg Home National Historic Site.

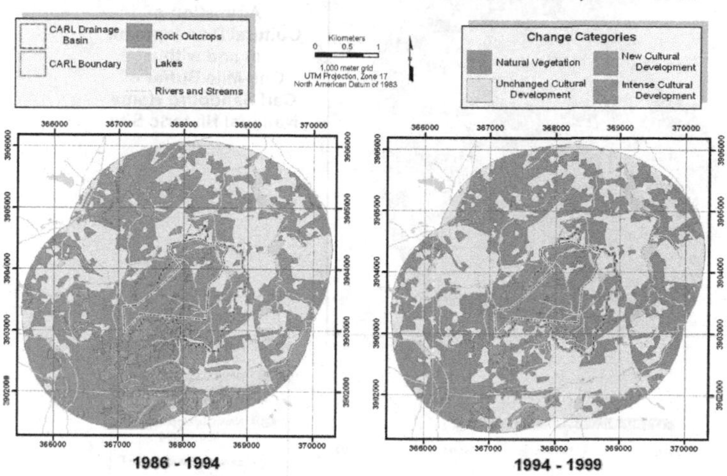

Fig. 18.6. Side by side layout of temporal maps showing cultural development at Carl Sandburg Home National Historic Site (CARL) for land use/land cover changes for 1986-1994 (left) and 1994-1999 (right).

forest vulnerability to disturbances such as wildfires, destructive exotic insects and overuse by park visitors. Overall, the combined, spatial analysis, geovisualization and GIS techniques can provide Park managers with the tools they need to conserve cultural and natural park resources for future generations.

Acknowledgements

This study was sponsored by the U.S. Department of Interior, National Park Service, (Cooperative Agreement Numbers. 1443-CA-5460-98-019 and H5028-01-0651). The authors wish to express their appreciation for the devoted efforts of the staff at the University of Georgia's Center for Remote Sensing and Mapping Science, NatureServe and National Park Service including Robert Emmott, Phyllis Jackson, Michael Jenkins, Keith Langdon, Teresa Leibfreid, Janna Masour, Nora Murdock, Jean Seavey, Rick Seavey, Chris Watson, Rickie White, Mark Whited and Nathan Vodas. We especially extend our gratitude to Dr. Roy Welch, founder and former Director of CRMS, who was the Principal Investigator of the original grants for this research.

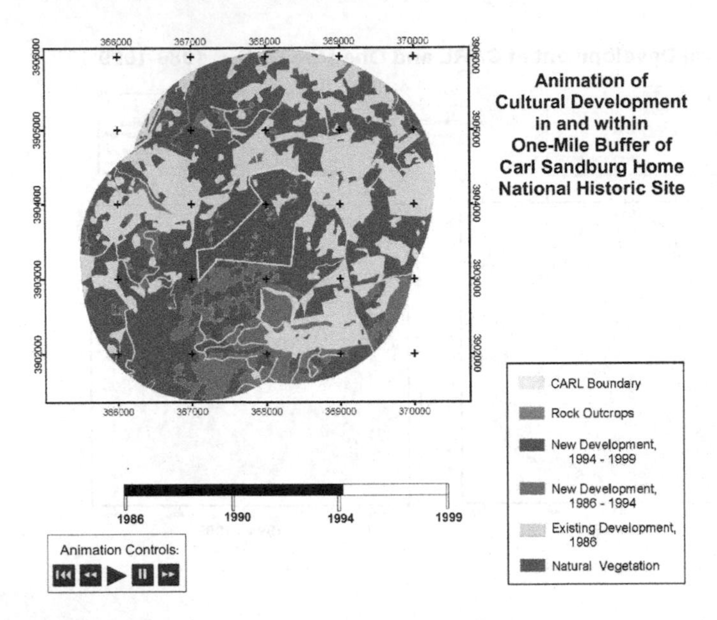

Fig. 18.7. A single frame from an animation of cultural development at Carl Sandburg Home National Historic Site depicting land use/land cover changes between 1986 and 1999. Clicking on the animation buttons gives the user full control to sequence forward, backward, stop action or continuously cycle through the multi-temporal data sets.

Fig. 18.8. A single frame from an animated fly-through over Carl Sandburg Home National Historic Site emphasizing park building locations, terrain characteristics and potential water flow into the park.

References

Bennett S (2005) Personal communication with Sue Bennett, Active Chief of Cultural Resource Management, Carl Sandburg Home National Historic Site, Flat Rock, North Carolina

Dolezal J (2005) Database development and geovisualization for managing national park resources. Master's Thesis, The University of Georgia, Athens, Georgia, pp 114

Dukes R (2001) A geographic information systems approach to fire risk assessment in Great Smoky Mountains National Park, Master's Thesis, The University of Georgia, Athens, Georgia, pp 131

Dunbar MD, Moskal LM, Jakubauskas ME, Dobson JE, Martinko EA (2003) Computer visualization of forest cover change: human impacts in northeastern Kansas and natural disturbance in Yellowstone National Park. In: Proceedings American Society for Photogrammetry and Remote Sensing Annual Conference, Anchorage, Alaska, CD-ROM

Fleming S, Jordan T, Madden M, Usery EL, Welch R (2006) GIS applications for military operations in coastal zones, ISPRS Journal for Photogrammetry and Remote Sensing, Under revision

Grossman DH, Faber-Langendoen D, Weakley AS, Anderson M, Bourgeron P, Crawford R, Goodin K, Landaal S, Metzler K, Patterson KD, Payne M, Reid M, Sneddon L (1998) International classification of ecological communities: terrestrial vegetation of the United States. Vol I. The National Vegetation Classification System: Development, Status and Applications. The Nature Conservancy, Arlington, Virginia, pp 126

Jordan TR (2004) Control extension and orthorectification procedures for compiling vegetation databases of National Parks in the southeastern United States. In: Altan MO (ed) International Archives of Photogrammetry and Remote Sensing, vol 35 (Part 4B), pp 657-662

Kraak M-J (2002) Geovisualization illustrated. ISPRS Journal of Photogrammetry and Remote Sensing, vol 57 (5-6), pp 390-399

MacEachren AM, Kraak M-J (2001) Research challenges in geovisualization, Cartography and Geographic Information Systems, vol 28(1), pp 3-12

MacEachren AM, Taylor DRF (eds) (1994) Visualization in modern cartography, Pergamon, London, pp 345

Madden M (2003) Visualization and analysis of vegetation patterns in National Parks of the Southeastern United States. In: Schiewe J, Hahn M, Madden M, Sester M (eds) Proceedings of Challenges in Geospatial Analysis, Integration and Visualization II, ISPRS Comm. IV Workshop, Stuttgart, pp 143-146, http://www.iuw.uni-vechta.de/personal/geoinf/jochen/papers/38.pdf

Madden M (2004) Vegetation modeling, analysis and visualization in U.S. National Parks. In: Altan MO (ed) International Archives of Photogrammetry and Remote Sensing, vol 35 (Part 4B), pp 1287-1293

Madden M, Jones D, Vilchek L (1999) Photointerpretation key for the Everglades vegetation classification system. Photogrammetric Engineering and Remote Sensing, vol 65(2), pp 171-177

Madden M, Jordan T (2004) Database development and analysis for decision makers in National Parks of the Southeast. In: Proceedings of the American Society for Photogrammetry and Remote Sensing Fall Conference, Kansas City, Missouri, CD-ROM, pp 10

Madden M, Jordan T (2001) Spatiotemporal analysis and visualization in environmental studies, International Archives of Photogrammetry and Remote Sensing, vol 34 (Part 4/W5), pp 71-72

Madden M, Schiewe J (eds) (2003) Editorial for theme issue: challenges in geospatial analysis and visualization. ISPRS Journal of Photogrammetry and Remote Sensing, vol 57(5-6), pp 301-303

Madden M, Welch R (2004) Fire fuel modeling in National Parks of the Southeast. In: Proceedings of the Annual American Society for Photogrammetry and Remote Sensing Conference, Denver, Colorado, CD-ROM

McCormick BH, DeFanti TA, Brown MD (1987) Visualization in scientific computing. Computer Graphics, vol 21(6)

Moore DM, Lees BG, Davey SM (1991) A new method for predicting vegetation distributions using decision tree analysis in a geographic information system. Environmental Management, vol 15, pp 59-71

Moskal LM (2004) Historical landscape visualization of the Wilson's Creek National Battlefield based on object oriented tree detection method from IKONOS imagery. In: Proceedings of the American Society for Photogrammetry and Remote Sensing Annual Conference, Denver, Colorado, CD-ROM

Remillard M, Welch R (1992) GIS technologies for aquatic macrophyte studies: I. Database development and changes in the aquatic environment. Landscape Ecology, vol 7(3), pp 151-162

Remillard M, Welch R (1993) GIS technologies for aquatic macrophyte studies: II Modeling applications. Landscape Ecology, vol 8(3), pp 163-175

Slocum TA, McMaster RB, Kessler FC, Howarc HH (2005) Thematic cartography and geographic visualization. Pearson Prentice Hall, Upper Saddle River, New Jersey, pp 518

Welch R, Jordan T, Madden M (2000) GPS surveys, DEMs and scanned aerial photographs for GIS database construction and thematic mapping of Great Smoky Mountains National Park, International Archives of Photogrammetry and Remote Sensing, vol 33 (Part B4/3), pp 1181-1183

Welch R, Madden M (2002) Vegetation mapping from remotely sensed imagery. In: Boderg A (ed) Bildteknik/Image Science Nr 2002:1, Photogrammetry Meets Geoinformatics, Invited Paper, Swedish Society for Photogrammetry and Remote Sensing, Stockholm, Sweden, pp 201-206

Welch R, Madden M, Doren R (1999) Mapping the Everglades. Photogrammetric Engineering and Remote Sensing, vol 65 (2), pp 163-170

Welch R, Madden M, Doren RF (2002a). Maps and GIS databases for environmental studies of the Everglades (Chapter 9). In: Porter J, Porter K (eds) The

Everglades, Florida Bay and Coral Reefs of the Florida Keys: an ecosystem sourcebook, CRC Press, Boca Raton, Florida, pp 259-279
Welch R, Madden M, Jordan T (2002b) Photogrammetric and GIS techniques for the development of vegetation databases of mountainous areas: Great Smoky Mountains National Park. ISPRS Journal of Photogrammetry and Remote Sensing, vol 57 (1-2), pp 53-68
Welch R, Remillard M, Doren R (1995) GIS database development for South Florida's National Parks and Preserves. Photogrammetric Engineering and Remote Sensing, vol 61 (11), pp 1371-1381
Welch R, Remillard M, Slack R (1988) Remote sensing and geographic information system techniques for aquatic resource evaluation. Photogrammetric Engineering and Remote Sensing, vol 54 (2), pp 177-185
White RD, Patterson KD, Weakley A, Ulrey CJ, Drake J (2003) Vegetation classification of Great Smoky Mountains National Park. NatureServe Final Report to BRD-NPS Vegetation Mapping Program, NatureServe, Durham, NC, pp 377
Woodcock CE, Macomber SA, Kumar L (2002) Vegetation mapping and monitoring. In: Environmental Modelling with GIS and Remote Sensing, (Skidmore, A., Ed.), Taylor & Francis, London, pp 97-120

19 Visualization of Spatial Change

Costas Armenakis, Anita Müller, Eva Siekierska, and Peter Williams

Abstract. Safety concerns and economic, environmental, scientific, and sustainable development issues require the monitoring and assessment of the various changes that occur in the geographical space. For knowledge-based decision-making and management approaches, spatial change can be better understood through representations, which enhance the cognitive eye-brain process. Work conducted in multiple programs of the Earth Sciences Sector (ESS), Natural Resources Canada is presented. This work aims to develop effective visualization of time-dependent spatial information to assist in the communication and understanding of the various geographic phenomena. Initially, elements and approaches for the effective cartographic representation of spatial changes are discussed. Subsequently, principles for the visualization of change and various modes for change representation are given along with the current web-based technological developments and trends. The visualization concepts and techniques discussed are demonstrated using several examples.

19.1 Introduction

"All is flux", Heracletos, 500BC. In ancient terms, this means that everything is variable and subject to changes with time. The environment we live in undergoes continuous change, where the earth's landscape including physical features and man-made structures are subject to variations in size, shape, position and attribute characteristics with time.

In the Earth Sciences Sector (ESS), Natural Resources Canada, multiple programs were initiated investigating the development of methods for change detection and representation regarding sustainable development, climate change, northern development and the monitoring of natural hazards. These programs focus on issues related to natural resources requirements for the use of geospatial information to support decision and policy making. Innovative technologies and methodologies are being developed that will facilitate the integration of geospatial information and the visualization of spatial changes. New technologies used by the visualization projects are being developed to enhance the Natural Resources Department capacity to implement sustainable development and help promote the responsible use of Canada's mineral, energy and forest resources.

Within the three ESS Programs: Sustainable Development through Knowledge Integration (SDKI), Geomatics for Northern Development and Reducing Canada's Vulnerability to Climate Change, methods are being developed focusing on effective representation and communication of time dependant information. The users of these methods range from policy makers to the general public as well. These methods and the prototype interfaces are used for the visualization of spatio-temporal data in order to facilitate the analysis of historical data, and for potential forecasting of future scenarios for scientists and decision makers who develop policies related to sustainable development of natural resources of Canada.

19.2 Cartographic Visualization of Spatial Changes

The representation of change can be accomplished by representing the time factor for the geographical variables. The mapping of the time of geographical elements is equivalent to the mapping of their evolution (changes, effects). Therefore, the visualization of spatial change in the cartographic process requires dynamic cartographic representations. These will enable the user to reconstruct -visualize- the temporal evolution of geographical phenomena and events by using dynamic and interactive cartographic environments.

The various approaches to visualize change depend on the process, which generates the time-dependent data. Various computer techniques can be applied to create the perception of continuous motion. Most electronic maps replicate the cartographic design principles of conventional maps, and thus portray a static abstract picture of reality using graphic forms. Although they can be used to find answers to questions dealing with location and theme, they leave the analysis procedures to the user, and they treat time as a static variable. The effective portrayal of the spatial changes on maps requires special attention to overcome its unique properties. The hypermedia concept and the visualization of change can create an interactive active cartographic environment (Armenakis 1996). The current methodologies for change visualization need to utilize the opportunities provided by modern computer technologies but at the same time should incorporate the legacy of conventional cartography (Andrienko *et al.* 2003).

The dissemination of the geographical information of change should not only be limited to graphic representation. The available web browser technology should be seen as a means of enhancing the communication capabilities of cartographic systems. The involvement of the users in the visualization process must also be considered. Interactive systems based on predefined processes provide quick response to user queries. Proactive approaches allow users to decide their own exploratory paths, and to access, retrieve, and process various sources of information during the analysis. One approach to building an active cartographic environment is to integrate motion -animated maps- and various multiple types of data (images, text, sound, numerical data) -hypermedia- with conventional and innovative cartographic representations. The concept of hypermedia can provide the tool for exploration, while the visualization of spatio-temporal data will facilitate dynamic representations (Siekierska and Armenakis 1999).

The human brain perceives change either as movement or as change in shape, texture, colour, sound or feel. Thus, historically maps were very limited to show change through time because of their static nature and their two dimensional space representations. They showed the temporal phenomenon as single slice of time map. Also challenging was to show interactions or flows between places (Tufte 1990). Most commonly representation of change was demonstrated by several states at once through two of more maps, displayed side by side. For example, change in urban growth would have been represented by a map showing the spatial extend of a city in specific time period and a second map showing current boundaries of a city. In this type of change visualization users are to make visual comparison and interpretation of presented juxtaposed data. Another type of change visualization in static maps is to indicate amount or rate of change using a single choropleth map, where change is represented as a ratio and thus assigned colour or texture. For example, census subdivisions with the darkest colour would symbolize the greatest change in population numbers, while the light areas would signify the least change.

With advances in computer technology, multimedia is used to represent a multitude of changes in a dynamic environment. Cartographic animation has become a very effective visualization technique to intuitively represent dynamic geographical phenomena (Buziek 1999, Harrower 2002). Through animation one can show interrelations amongst geospatial data components, location, attribute and time and 2,3 and 4 dimensional representations. Spatio-temporal changes are best visualized through dynamic maps, enabled by computer technology (Peterson 1995). In dynamic maps, real time is compressed or scaled into a changing display. Change may be represented by non-moving occurrences (events added and deleted at places through time) or by moving objects (movement is animated on the screen).

Cartographically, change can be visualized through various graphical variables in the form of animation. These graphical variables, as first studied in the 1960s by Jacques Bertin are: location, value, hue, size, shape, spacing, and orientation (Bertin 1967, Slocum *et al.* 2005). Location emphasizes where the symbol is and thus is determined primarily by geography. It is also a primary means of showing spatial relations. The human brain computes relations such as "is within", "crosses" instantly from the eye's perceived image of the map. Hue, value and saturation are three dimensions of colour. Hue, frequently referred to as colour, is important aesthetically and usually represents qualitative differences. When using hue in change visualization, it is important to be aware of colour association and use appropriate hues, for example green for positive change and red for negative change. Value is associated with the lightness or darkness of the colour. In visualization of change, value is very important as the eye tends to be led by patterns of light and dark. Value is usually used to represent quantitative differences and traditionally darker symbols are associated with "more". The third dimension of colour, saturation, is a measure of the vividness of a colour. There are also other terms describing saturation with slightly varied definitions: chroma, colourfulness, purity, and intensity (Brewer 1999).

Size is another visual variable and the size of the symbol conveys quantitative differences. The human brain has difficulty inferring quantity accurately from the size of a symbol, for example, if proportional circles are used to portray city population. Doubling the radius of a circle (quadrupling its area) is perceived as indicating more than twice the population, but not four times. Thus, the brain infers population from the combination of the radius and the area of the symbol. Shape is a geometric form of the symbol used to differentiate between object classes. Shape is also used to convey nature of the attribute, for example population indicated by images of people, or urban density by house symbols. Spacing, arrangement and density of symbols in a pattern are used to show quantitative differences. A classic example is a dot density to show population, while the orientation of a pattern is used to show qualitative differences.

When visualizing change through maps one must take into consideration graphical limits (Vasiliev 1996). In terms of spatial limits, symbols of phenomena that are changing must be in close proximity for a human eye to notice. If it is shape, the change must be observable. If for some reason it is not, colour or change in scale may enhance the ability to perceive change, however, displayed pixels have a set size and thus a finite number of spatial locations. Aliasing line (or point) mapped may also play an important role in distinguishing change.

Visualizing change through colour requires particular attention and careful planning. Because computer monitors are able to display millions of possible colours, it is important that the selection of colour is limited in order to be comprehensive. Also, the changes in colour need to be sufficiently significant for a human eye to distinguish (DiBiase *et al.* 1991). Depending on the purpose of visualization, certain changes may need to be highlighted or emphasized through expanding or limiting the range of luminance and contrast using brighter, more vibrant colours. Important to consider is colour association in order to prevent faulty analysis, especially in the case of complex and long animations. Particular attention needs to be paid to non-interactive animation where viewers do not have the tools to control sequence or control the speed of animation (Peterson 1995).

Change is directly linked to time. Time is typically treated as an attribute to be mapped. However, according to some researchers (MacEachren 1994), treating time as an attribute limits the potential of dynamic maps displays. MacEachren (1994) advocated treating time as a cartographic variable to be manipulated, the same as size, colour hue and space itself. DiBiase *et al.* (1992), Szego (1987) and MacEachren (1994) identified four fundamental dynamic variables: duration, rate of change, order and phase. The duration is controlled by user interaction (viewing time) of a given image (short or long). The rate of change is depicted as duration of different animation frames (e.g., slow constant, fast constant or steadily increasing). Matching animation frame order with temporal order of the depicted phenomena is the most natural way of ordering dynamic variables. However, with dynamic maps one can use time order to represent in a symbolic way any order of interest based on selected attributes. The phase has been defined as a rhythmic repetition of certain events (Szego 1987). According to MacEachren (1994) the addition of new variables to the classical set of cartographic variables as defined by Bertin (1967) will likely make the most substantial impact on maps as a visualization tool.

19.3 Concepts for Change Representation

The visualization of change should provide answers to basic questions, such as: "what is the current situation", "what was the situation at a past time t", "what are the differences between the geographic state at t_1 and the geographic state at t_2", "what caused the changes", "how have the changes occurred", and "what will be the situation in the future time t_f under certain conditions". With respect to the change itself we should be able to represent and address queries on the magnitude, direction, rate, and duration of change. The appropriate visualization of the spatio-temporal changes enhances the inference process and allows the users to generate, display, view and manipulate data relying on the cognitive pattern recognition. It also offers the ability to intuitively explore and navigate through multi-resource databases, which enhances data exploration and the understanding process. Thus, the visualization of spatial change systems should include the following functionalities.

1. the most recent geographical state must be readily available for visualization. This requires the visualization of the desirable current geographical data based on all past states and on all additional current data, which are available at the instant of query. This can be accomplished by having in place a *filtering* mechanism, which searches and interrogates all the available data sources and finally estimates and represents the current geographical situation.
2. the ability to reconstruct past geographical states and to recreate the evolution of past processes and events, when the time-queries fall within the span of the up-to-date available spatio-temporal data. The estimation and visualization of a state at previous times requires the presence of an *interpolation* mechanism. In cases where the estimation of previous states needs to be improved based on new data, then a *smoothing* process must be available.
3. the ability to forecast, simulate and visualize future states and developments in the geographical domain, based on present information, state dynamics and future scenarios. A *prediction* mechanism is therefore required to estimate and visualize geographical states or events that occur after the last available data.
4. the visualization of the *comparison* between absolute states of the geographical domain, which provides the changes that occur over a period of time between two or more time snapshots.
5. the visualization of the *transitional* procedures, which provide the reasons for the changes, that is the manner in which an absolute state of the geographical domain is transferred to another one over a period of time.
6. the visualization of the *elements* of spatial changes, such as magnitude, direction, rate and duration.

The following definitions are related to the change and its visualization:

- *State:* the condition (status) of the geographical space at a given point or interval in time. It is composed of the geographical objects, which are concurrent at this instant or period in time.
- *Event:* the instant in time at which an occurrence causes changes to the state.

- *Version:* the form (configuration) of a spatial object during a time period.
- *Mutation:* the instant in time at which a object undergoes change.
- *Duration:* the time interval during which a version or a state lasts.

19.3.1 Change Representation Modes

The determination of information about absolute states, changes and transitional procedures utilizing the above mentioned estimation mechanisms could be done according to the following estimation modes. Certain parts of these representation modes have been developed based on the ideas of Langran and Chrisman (1988).

Static Frame Mode

In the static mode, the absolute individual states are visualized in consecutive time-ordered sequence. The most recent state is readily available. The data is organized, stored and displayed in time-snapshots:

$$S_0, S_1, S_2, \ldots, S_t \quad (19.1)$$

where, S_i is the individual time-snapshot; and $i = 0, 1, 2, \ldots, t$, with 0 indicating the initial state and t the current state.

It is therefore a series of concatenated time snapshots of the geographical domain. All elements comprising each state carry the same time-stamp. The data structure of the static mode allows the direct visualization of a geographical state at discrete and specific instants time. Unless the sampling time points are taken at frequent time intervals, it is difficult to represent accurately and completely the spatio-temporal paths of the geographical data. This affects the visualization of sequentially displayed static states (succession of frames) as they will not be perceived by the viewers as a continuous form of presentation without creating interpolated states during the display.

The comparison between states to visualize changes, occurring within a given time interval is determined as:

$$\Delta S_{ij} = S_j - S_i \quad (19.2)$$

When comparing two datasets to determine changes, the domain of comparison has to be defined. For example, vector data provide a classified abstract representation of the landscape, while imagery is an unclassified continuous but resolution-dependent generalized representation of the landscape. If comparing data of the same nature it may be possible that the change is represented at the *data level domain* (comparison of pre-processed data), while when comparing heterogeneous datasets the change representation is at the *information level domain* (comparison of analyzed/processed data). The change representation modes may be different between data storage and data display.

The storage of all temporal time snapshots is also of a concern. No matter the degree of changes, a great portion of unchanged data will have to be stored every time an updated state is produced. This creates an unnecessary storage redundancy.

Differential Frame Mode

In the differential mode, the initial time snapshot frame (state) and the subsequent differential files representing the changes from one state to the next are used for the visualization of changes. The stored and displayed data will be:

$$S_0, \Delta S_1, \Delta S_2, \ldots, \Delta S_n \tag{19.3}$$

where, S_0 is the initial state; ΔS_i are the recorded changes over time; i=1, 2,..., n indicates that changes occur at time 1,..., n between states i and i-1.

The current state is not readily available for visualization and it has to be determined indirectly in 'real-time' as:

$$S_n = S_0 \cup \Sigma_{[i=1\ldots n]} \cup \Delta S_i \tag{19.4}$$

The visualization of changes between t_i and t_j, for example the changes between t_1 and t_3 can be determined as:

$$\Delta S_{1\text{-}3} = \Delta S_{2\text{-}1} \cup \Delta S_{3\text{-}2} \Rightarrow \Delta S_{1\text{-}3} = [S_2 - (S_1 \cap S_2)] \cup [S_3 - (S_2 \cap S_3)] \tag{19.5}$$

For the determination of the changes, the required processing is less than that required in the case of absolute states (static mode). In the latter, the entire state files need to be merged and compared to produce the differences, while in the other only the differential files need to be retrieved.

The involved differential files should have independent topological structure, that is, they must not depend on the initial state. This is necessary for the visualization of the changes within the 'delta' files and for the existence of error-checking mechanisms. The storage requirements for the differential mode are minimal.

Accumulated Differential Frame Mode

Besides the step-by-step approach of the differential method, there is also the so-called accumulated differential approach (Langran 1988). In this alternative, the summation of the individual differences between the initial (or current) state and the states under consideration constitute the total differential file. Thus, instead of moving step by step from one state to the next (or previous) by adding the difference, we leap from the initial (or current) state to any state. This approach is expressed as:

$$S_i = S_c \cup \Delta S_{c\text{-}i\,,\,tc > ti} \tag{19.6}$$

with t_c referring to current state and t_i referring to any past state.

While the accumulated differential mode may duplicate elements stored in the differential files, it also allows the direct determination and visualization of the state of interest without state-to-state hopping. However, the estimation of the changes between any two or more states still requires extra processing for their visualization.

Sequential Updating Mode

The sequential updating mode attempts to reduce some of the disadvantages of the static and differential approaches. Its objective is to maintain the current state readily available, while providing mechanisms for spatio-temporal navigation using minimum storage and processing cost. Depending on the organization of the data (which depends on the requirements for the manipulation of the time-varying data), two types of sequential updating processes can be distinguished.

Sequential Differential Frame Updating Mode

This approach is based on a continually updated state. That is, whenever a change occurs it is immediately incorporated into the current state, and it stored as a time-indexed differential file. The process is expressed as:

$$S_t = S_{t\text{-}1} \cup \Delta S_{t\text{-}1,t} \tag{19.7}$$

At each update cycle of the state of the geographical domain, the corresponding difference file is stored with the appropriate reference pointer (index). Thus, the stored elements are: (i) the current state file: $S_{t;}$ and (ii) the difference files: $\Delta S_{t\text{-}1,t}$, $\Delta S_{t\text{-}2,t\text{-}1}$,..., $\Delta S_{1,2}$, $\Delta S_{0,1}$

The approach is quite attractive as the present state is there when needed and there is no redundant storage of data. In addition, it is easy to determine the changes, which occurred between time periods, and it is possible to indirectly reconstruct any previous geographical situation. It is possible to avoid storing the delta files if the space-time composite approach (Langran and Chrisman 1988) is followed. In this context, the time factor characterizing the changes is attached to each geographical unit as a temporal attribute. A mechanism for retro-version is then required to allow traveling through the history of the attributes and to reconstruct any time state by selecting and building the units with the same time reference.

Sequential Objects Versioning Updating Mode

This is similar to the previous approach. Once again the current state is continually updated as changes occur. The difference between these two approaches is that instead of keeping the changes (delta files), which may turn out to be impractical, the object-units of the geographical domain that existed in the previous time period are stored. This means that each unaltered geographical domain within a time interval consists of a set of geographical objects. When an event affects one object, its until now current status is superseded by its new version, while the immediate previous object is stored using link mechanisms and temporal indexing. The term object is defined here as any geographic unit that possesses specific characteristics defined by the user or the data structure. It is dependent on the organization of the data and the structure of the database. For example, it may be a geographic entity (or simple object), a cartographic element, or a thematic layer.

This approach offers immediate access and visualization of the current state, reconstructs for visualization past states by substituting current objects for their previous versions, determines the changes and their visualization by comparing only the components that have been changed (comparison operations at parts of the database), and does not duplicate unchanged cartographic objects.

19.4 Technology for Change Visualization

The proliferation of the World Wide Web has increased the demand for highly interactive maps for visualization of geospatial temporal change. The evolution from static historical maps to real time dynamically generated cartographic visualizations of spatio-temporal data has been driven by new methods of delivery based on Extensible Markup Language (XML) technologies, open computing standards, higher degrees of interoperability, global positioning system, mobile computing, decrease in the cost of remote sensing imagery, increased data sharing, data streaming for efficient data transmission, and increased data bandwidth availability. In addition, spatio-temporal databases developments, geo-collaboration tools, 3D visualization tools, virtual and augmented reality modelling, and web based spatial decision support systems have lead to an increased demand for temporal spatial analysis tools.

19.4.1 Open GIS Consortium

The main goals of the Open GIS Consortium (OGC) are to maximize accessibility, provide consistent framework data, and develop partnerships to promote OGC and ISO standards. The OGC Web Map Service (WMS) allows users to transparently access data stored on any OGC compliant Web Map Server in a distributed environment (Open GIS Consortium 2000). The OGC WMS specification permits data providers and users to specify colouring, line styles, labelling, scale level dependant entity depiction for both raster and vector geospatial data. The portrayal of web maps is based on Styled Layer Descriptors (SLD), which define the graphic representation of geographic features as generated by the Web Map Server and displayed on a users browser. A Styled Layer Descriptor is represented by an XML encoding of rendering rules used as input by a WMS. The Styled Layer Descriptor is identified to a WMS by using a Uniform Resource Locator (URL) address. A Styled Layer Descriptor may be composed of a combination of named styles associated with a named layer. A palette of styles may be created for the common depiction of like features regardless of the data provider. A named layer is a collection of features that can be accessed from a WMS using a name and retrieved by a "get" capabilities request. Users can also specify the minimum and maximum scales for the depiction of features and as well filter features by attribute values. There is no provision for automatic placement of labels and de-confliction. These features will be addressed in a future version of the specification.

The OGC Web Feature Service (WFS) provides data manipulation based on individual features. A WFS supports information queries based on spatial and non-spatial attributes, additions and deletions of features and browsing of data content (discovery operations) (Open GIS Consortium 2002). WFS transactions are returned to the user in the Geography Markup Language (GML), another OGC standard developed for the definition of point, line and area data. The on-line access to distributed databases is essential for the live-monitoring information relevant to decision making regarding sustainable development.

The SDKI's-Visualization project is implementing the use of Styled Layer Descriptors for the depiction of spatio-temporal data using temporal attributes of geospatial data for the depiction and rendering of spatial tem-

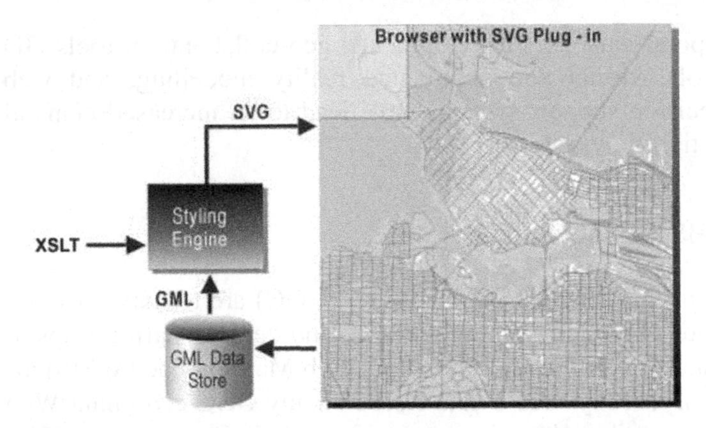

Fig. 19.1. Principle of generating SVG-based maps from distributed databases using Styled Layer Descriptors

poral data to visualize the transition of topographic features over a given period.

19.4.2 Scaleable Vector Graphics

Scaleable Vector Graphics (SVG) is an open, vendor neutral, XML based vector graphics standard for the depiction of resolution and device independent two-dimensional vector and raster data (Fig. 19.1). SVG is a W3C supported format (W3C 2005). For spatial change representation SVG supports animation though start time, duration, and attribute values. It implements animation by supporting the modification of graphic objects colour values, coordinate values, and transformation values. It also permits motion of an object along a path.

SVG may be generated using an XML stylesheet (XSLT) to convert GML to SVG. The advantage of the SVG generation over standard OGC-based WMS lies with the elimination of multiple feature queries to obtain attributes as they are performed by a web map server request. This repetitive querying using web feature server requests is eliminated using SVG since the data has been already downloaded to the client initially. Editing capabilities through web feature server posting of modifications to attributes or geometry is possible using SVG.

19.4.3 Three Dimensional Environments and Virtual Reality

Virtual reality immerses a person into a 3D scene using goggles or other 3D simulating devices and tools. In a temporal 3D environment a computer screen allows for virtual reality type movements but the limitations of viewing a scene on a flat screen mean that the virtual reality environment is not truly immerse.

There are a variety of coding formats, which have been or are being developed to display virtual reality data. Some such formats are Java3D, Virtual Reality Modelling Language (VRML), Extensible Three Dimensions (X3D), and Moving Picture Experts Group 4 (MPEG-4). The formats were conceived to display images over the World Wide Web. A variety of VRML browsers, typically plug-ins for Netscape Navigator® and Microsoft Internet Explorer®, have been developed to display virtual worlds written in VRML. No single plug-in has become a standard though a few have become popular. One particular plug-in -the Cortona VRML Client- stands out because it is widely used, continuously being developed, and is able to read GeoVRML nodes. It has been developed by Parallel Graphics (Cortona VRML Client 2001), is considered fast, powerful and easy to use. Additionally, it is currently supported with new versions which are being developed and released to increase its usability, particularly with Java® and Javascript® Cortona VRML.

VRML is a commonly used format for displaying web-based 3D graphics. VRML is a coding format used to represent geometric objects and behaviours associated with those objects. These behaviours include the ability to react to change in time to create animation, and respond to user interaction. VRML is actually neither virtual reality nor a modeling language but rather a 3D interchange format similar to Hyper Text Markup Language (HTML) that integrates three-dimensions, text, and multimedia to create 3D scenes for viewing on the World Wide Web (Carey and Bell 1999). VRML was first created in 1994, and a second version, VRML 2.0, was released in 1997. It is supported by the Web3D Consortium, which was created to provide a forum for 3D open standards and applications on the web. Web3D Consortium has played a major role in the development of VRML although the Consortium is open to all Internet related 3D technologies. VRML is being replaced by a new 3D file format known as the X3D, or Extensible 3D, which has an encoding format based on XML (Web3D Consortium 2006). The X3D working group within the Web3D Consortium is developing an open source X3D browser using Java3D as its basis since Java3D is an application programming interface using the Java language to draw 3D graphics.

The MPEG-4 format, an object oriented standard, incorporates audio and visual content designed for low bandwidth (Koenen 2002). The powerful nature of MPEG-4 allows it to incorporate VRML objects, which led to the development of the MPEG-4/Web3D working group within the Web3D Consortium. At present VRML or the 3D augmented reality methods are primarily used for the portrayal of the spatial components but they also can be effectively used for the representation of the temporal dimension.

19.5 Applications of Change Visualization

19.5.1 Frame Based Change Depiction using Temporal Aerial Photography and Satellite Imagery

A clear advantage of portraying geospatial data using electronic media is the ability to effectively display the past and present distribution of various geographic phenomena, to analyze the patterns of change and to model the possible future scenarios. Historical maps and aerial photographs are good sources of data to examine the previous distribution of land use, and are a realistic starting point in discussing potential future developments.

One application at Natural Resources Canada selected for representation of temporal change is the historical and spatial evolution of Iqaluit, the capital of Nunavut (http://maps.nrcan.gc.ca/iqaluit). Historical aerial photographs taken at approximately ten-year intervals were assembled to portray the rapid growth of the city over 50 years (1948-2000). They served as a base for creation of historical city maps. The historical photographs and records provided additional information to reconstruct the development of Iqaluit to discover the factors influencing change, explain patterns in development of the city, and provide a cultural and social background using photographs, fly-bys using historical orthomosaics, and 3D renderings of the terrain in the form of animations. Fig. 19.2 shows the user interface created to display historical orthomosaics (internal website only). The animation was created using Macromedia Director. Two orthomosaics were used and the transition between them was visualized by changing opacity over 40 frames. Animation is started by single clicking on the year button and can be run forward and backward.

In another application historical aerial photographs from the National Aerial Photograph Library (http://airphotos.nrcan.gc.ca) and relevant illustrations are being used to examine the history of land use changes and their relationship to change in the transportation modes, which have occurred in response to changing city needs. Transitions in the downtown regions of the City of Ottawa over a span of 82 years (1920-2002) can be analyzed using historic aerial photographs. The interface in Fig. 19.3 shows aerial photographs with a layer control panel on the right. The layers represent eight aerial photos, seven black and whites and the most recent one in full colour. Additional layers show highlighted buildings with opacity allowing

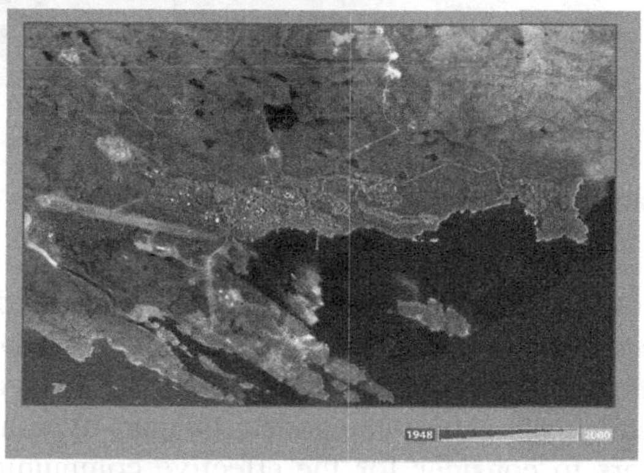

Fig. 19.2. Web based spatial evolution of Iqaluit using user controlled image opacity

viewing the aerial photography underneath. Highlighted buildings can be selected and displayed over selected aerial photographs to visually analyze the change. All of the highlighted buildings are selectable and give access to further information, such as oblique aerial photos with the highlighted building or horizontal views of the building. For earlier years, such as in 1920, buildings are linked to old photographs and postcards. Future plans for interface enhancements are addition or animation and VRML walk-throughs. The interface was created using MapInfo and SVGMapMaker (http://www.dbxgeomatics.com) to visualize spatio-temporal changes shown on orthorectified aerial photography and vectorized land use imagery. Cartographic animation is being used to depict temporal changes using the depiction of a series of maps or images in succession. The success of this technique depends on the ability of users to perceive the change in spatial representation through selected image frames. To do so implies the use of the variables of duration (of each image), rate of change (between images), and order (of images). The animation may be used with the dis-

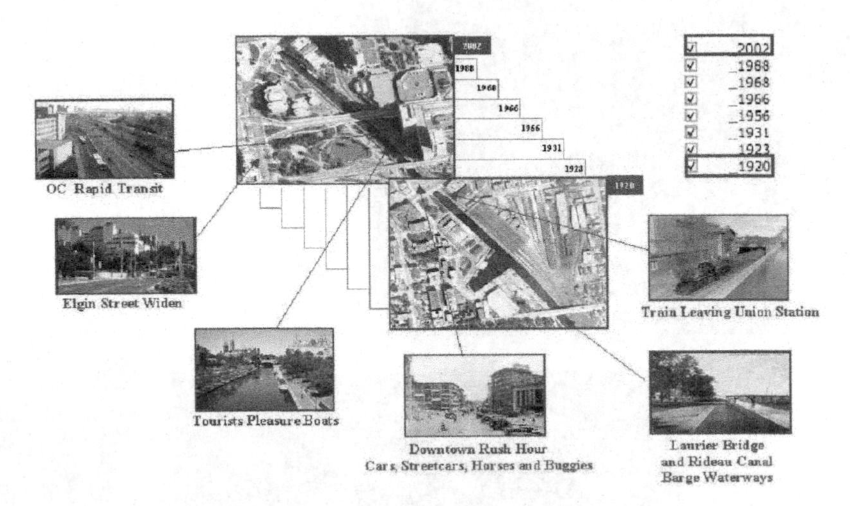

Fig. 19.3. Transition of transportation in Ottawa center

play rate under the user's control or automated as in a slide show with the period between time slice display predefined by the application. A time slice, being temporal spatial phenomenon, is displayed as a discrete spatial frame. The number of feature categories, imagery type and area size are important factors to consider for the effective communication of spatial temporal change. The granularity and density of feature categories selected for the animation of data must be conducive to the effective communication of temporal change.

Monitoring studies of coastal changes contribute to the development of various types of assessments (e.g., impacts, sensitivity, vulnerability, erosion hazard) due to climate change including changes in the sea-level. The project of Reducing Canada's Vulnerability to Climate Change investigates the ocean vulnerabilities to climate change on regional and local scales to provide critical geoscience data to other government departments. This information can be used in assessing climate change impacts and developing adaptation options. The task of this coastal activity is to develop sensitivity and impacts assessments in coastal areas of the Arctic and East coasts of Canada by using earth observation data to provide the necessary spatio-temporal data infrastructure for coastal areas and by developing methodologies for feature extraction and change detection using image data. The study area is the Arctic south-west coast of Banks Island where the settlement of Sachs Harbour is also located. The coastline was extracted from 1961 and 1985 aerial orthoimages, and 2002 IKONOS and

2003 QuickBird satellite orthoimages, respectively. To present the coastal changes, the coastline time series layers are successively displayed over one of the satellite orthoimages, which provides a contextual background. The successive representations of coastlines at different time frames are distinguished using colours and explained in the legend (Fig. 19.4).

19.5.2 Representation of Temporal Data using Dynamic Interactive Scaleable Information Representation

The SDKI's Visualization project is developing the SVG depiction of frame based time snapshots of land use data using OGC WFS technology for the city of Ottawa. A WFS is used to query, access and retrieve temporal referenced data in GML format at the geospatial feature level in real time through the Web for a specific time frame. The retrieved feature level data for the specified time frames are converted to SVG using a XSLT processor and style sheet, for the dynamic web-based display via a web browser using an SVG plug-in.

In addition, the inclusion of raster data into the temporal representation of integrated information for land use for the city of Ottawa is using classified and unclassified Landsat-7 data, roads from the National Road Network, Canada Land Inventory data, historical and current aerial photographs and orthorectified photomosaics. These data will also be integrated for a web-based temporal visualization tool using Scaleable Vector Graph-

Fig. 19.4. Web-based coastal line temporal changes superimposed successively over orthorectified IKONOS satellite imagery using SVG (contains material: ©Space Imaging).

ics. This will permit the visualization and comparison of past states by the combination of feature based vector for known temporal states and raster based imagery for intermediate temporal states.

A prototype using data vectorized from orthorectified photographs was used to create an animation for multiple time slices to examine urban and transportation change for the city of Iqaluit in the Northern Territory of Nunavut. Change can be visualized either by animation or by selecting individual frames. The animation may run automatically by displaying data for a selected time slice consecutively or by allowing users to select an individual time slice or multiple time slices for comparison. Urban features, which have changed during the temporal period are highlighted and their opacities are modified to allow viewing of collocated features. Features whose attribution has modified between time slices are also identified. The change representation was created using five frames, where each frame represents a year between 1948 to 2000. In the animation, the transition from frame to frame was achieved by Macromedia Director's scripting effect called "fast bit dissolve", where the stream path of one map is replaced by the next one. The transition between the individual time frames is highlighted by a change of colour (Fig. 19.5).

19.6 Concluding Remarks

The representation of the spatio-temporal changes using effective visualization techniques improves the inference process by enhancing the cognitive pattern recognition. Concepts for change representation and for their cartographic visualization were discussed. Current technology and several examples for the visualization of spatial changes were presented. The dissemination of change of the geographical information should not only be limited to graphic representation of spatial characteristics but it should also communicate changes in the state information of individual features and changes in their interrelationships as well.

The involvement of the users in the visualization process must also be considered. Interactive cartographic systems not only allow for efficient representation of spatial changes on predefined processes but also provide quick response to dynamic user queries and proactive approaches. They allow the user to decide is own exploratory paths using frame (state) based methods, feature based methods, or a combination of both methods to access, retrieve, process and visualize various distributed temporal geospatial data sources of information during change analysis. The understanding of

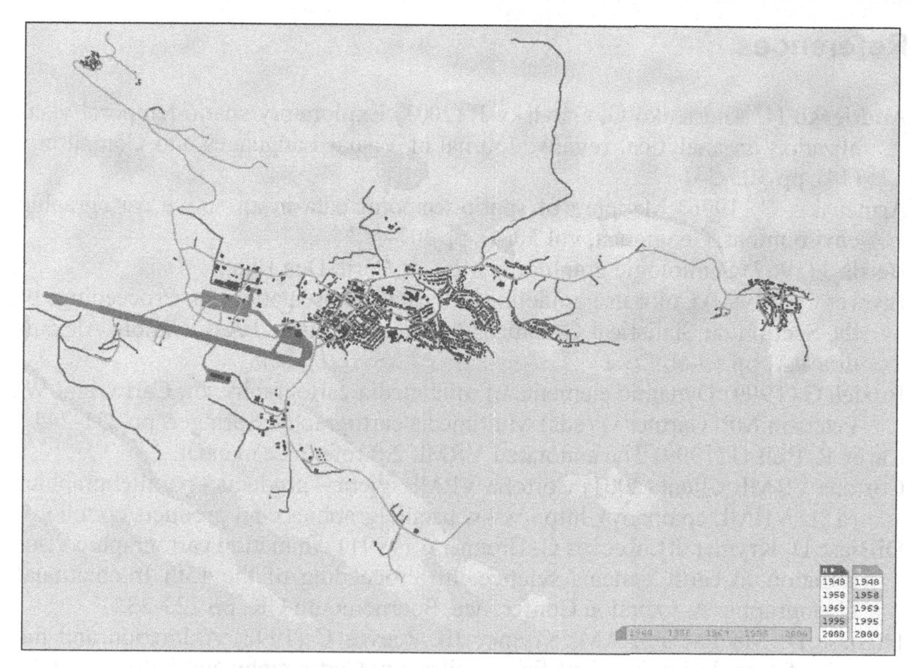

Fig. 19.5. Iqaluit transition using frame based temporal change detection

spatial changes contributes significantly to knowledge-based assessments and decision-making.

The consideration of the basic functionality and requirements of the visualization of spatial change coupled with the capabilities of the web-based multimedia technologies enhances the retrieval and depiction of temporal spatial changes. The remote access of temporally referenced data sources through web feature and web coverage servers will further lead to web-sensor environments for the "real-time" visualization of spatial changes.

Acknowledgements

The authors wish to acknowledge the contributions of Ken Francis in development of interfaces for effective visualization of spatial changes in the evolution of the city of Iqaluit; and of Jean-Pierre Dostaler for the development of visualization of the changing mode of transportation in the city of Ottawa based on the historical aerial photographs.

References

Andrienko N, Andrienko G, Gatalsky P (2003) Exploratory spatio-temporal visualization an analytical review. Journal of Visual Languages and Computing, (14), pp 503-541

Armenakis C (1996) Mapping of spatio-temporal data in an active cartographic environment. Geomatica, vol 50(4), pp 401-413

Bertin J (1967) Sémiologie graphique. Mouton, Paris-Den Haag

Brewer CA (1999) Color use guidelines for data representation. In: Proceedings of the Section on Statistical Graphics, American Statistical Association, Alexandria VA, pp 55-60

Buziek G (1999) Dynamic elements of multimedia cartography. In: Cartwright W, Peterson MP, Gartner G (eds) Multimedia cartography. Springer, pp 231-244

Carey R, Bell G (1999) The annotated VRML 2.0 reference manual

Cortona VRML Client (2001) Cortona VRML client – products - parallel graphics (a 3D VRML company). http://www.parallelgraphics.com/products/cortona/

DiBiase D, Krygier JB, Reeves C, Brenner C (1991) Animation cartographic visualization in earth systems science. In: Proceeding of the 15th International Cartographic Association Conference, Bournemouth UK, pp 223-232

DiBiase D, MacEachren AM, Krygier JB, Reeves C (1992) Animation and the role of map design in scientific visualization. Cartography and GIS, vol 19(4), pp 201-214

Harrower M (2002) Visualizing change: using cartographic animation to explore remotely-sensed data. Cartographic Perspectives, (39) pp 30-42

Koenen R (2002) Overview of the MPEG-4 standard. International Organization for Standardization (ISO)

Langran G (1988) Temporal GIS design tradeoffs. In: Proceedings of GIS/LIS '88, vol 2, pp 890-899

Langran G, Chrisman N (1988) A framework for spatio-temporal information. Cartographica, vol 25(3), pp 1-14

MacEachren AM (1994) Time as cartographic variable, in visualization in Geographical Information Systems. In: Hearnshaw, Unwin (eds), J. Wiley & Sons Ltd, pp 115-130

Open GIS Consortium (2000) Web map service implementation specification, version 1.1.1, OGC-01-68r2

Open GIS Consortium (2002) Styled layer descriptor implementation specification, version 1.0.0, OGC-02-070

Peterson MP (1995) Interactive and animated cartography. Prentice Hall, Englewood Cliffs, NJ

Siekierska E, Armenakis C (1999) Territorial evolution of Canada: an interactive multimedia cartographic presentation. In: Cartwright W, Peterson MP, Gartner G (eds) Multimedia cartography. Springer, pp 131-140

Slocum TA, McMaster RB, Kessler FC, Howard HH (2005) Thematic cartography and geographic visualization. Second Edition, Prentice Hall

Szego J (1987) Human geography: mapping the world of man. Swedish Council for Building Research, Stockholm

Tufte ER (1990) Envisioning information. Graphics Press, Cheshire, Connecticut

Vasiliev I (1996) Design issues to be considered when mapping time. In: Wood C, Keller CP (eds) Cartographic design – theoretical and practical perspectives, Wiley, pp 137-147

W3C (2005) Scalable vector graphics (SVG) 1.1 Specification, http://www.w3.org/TR/SVG/SMIL

Web3D Consortium (2006) X3D overview. The Web3D Consortium, http://www.web3d.org/x3d/overview.html

20 Scalable Vector Graphics Interfaces for Geographic Applications

Randy George

Abstract. Scalable Vector Graphics (SVG) is a w3c recommended standard for xml representation of 2D graphics. As a graphic standard incorporating event listeners it provides an xml standardized approach to interactive web GIS services, accessible through client browsers. Geographic hypermedia is characterized by event driven interactive features. SVG provides a feature granularity not possible using older html image interfaces to GIS services. SVG interfaces connect well with other emerging web GIS services such as the Open Geospatial Consortium (OGC) Web Feature Service (WFS), and Web Map Service (WMS) standards. SVG is a flexible tool for building the gallery of interfaces illustrated in this chapter.

20.1 Introduction

Web based GIS services are still in their infancy. Open standards are now emerging for exposing GIS data stores to internet clients. On the server side, the OGC has published a series of important open standards for publishing GIS data stores to the web. On the client side, SVG enables Internet browsers to view and interact dynamically with the 2D graphics found in many map applications. SVG provides rendering capability for both vector and image data, while event listener attributes can be attached to any feature in the rendered view. Event listeners, such as a user mouse click, can be part of the SVG definition of any geometry feature. This highly granular interactivity leads to the very flexible interface building capability characteristic of SVG. As an open standard, SVG is independent of proprietary commercial interests, and as an XML standard, it is accessible by all the usual XML processing tools such as DOM parsers, XSLT, and XPATH.

Standard compliant service producers make data resources available to consumer clients anywhere in the world. Consumers can be ordinary browsers or aggregating Java J2EE/servlet containers. In this way producer chains which manage a variety of layered spatial data can be merged into a browser where the final rendering takes place. If the end point is rendered in the browser as an image, jpg or png, then there is little interactivity

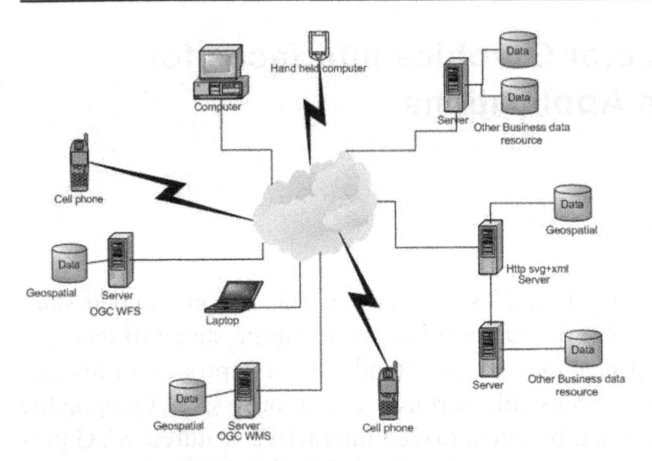

Fig. 20.1. Web GIS Mapping Services

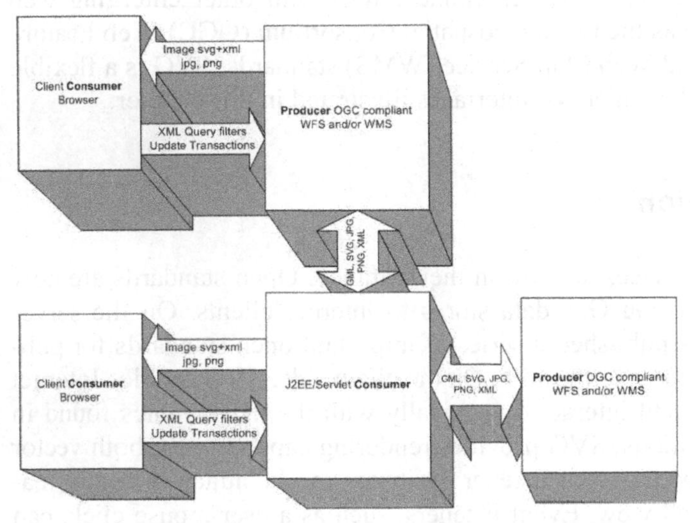

Fig. 20.2. Open standards result in modular web services

available to the client. However, there is a high degree of interactivity available if the endpoint rendering consists of vector data with associated event listeners. SVG combines vector and image data so that the browser interface can make use of the strengths of each type of rendering.

20.2 Examples

The following examples illustrate the variety of interface options open to the SVG developer. It is difficult to convey the dynamic nature of geo-

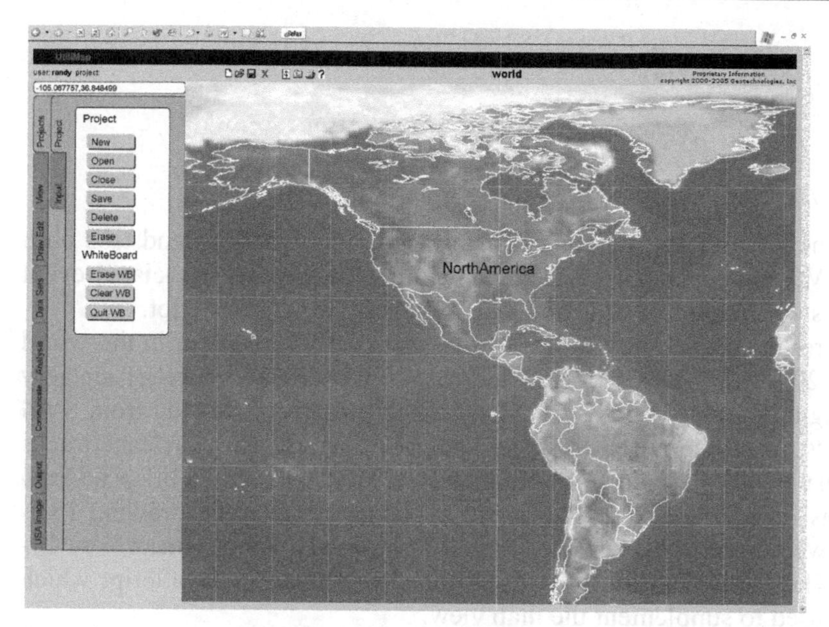

Fig. 20.3. Example of a browser view with both an SVG menu and an SVG map frame

graphic hypermedia in a static media such as the printed page. Most of the interaction occurs through mouse movements and clicks. Mouse movements produce rollover labeling and highlighting affects while clicks produce view changes. Interactive maps can be supplemented with menus and dynamic controls also built from SVG elements.

In Fig. 20.3. the SVG map frame is composed of vector path elements outlining continents with sub group paths delineating individual countries. Events are attached to the continent outlines to allow rollover labeling, highlighting, and click view changes. Under these vectors, as a background, is an image element referencing an external raster jpg of the world. The vector data's strength is in its associated event listeners while the strength of raster jpg is conveying detailed information about land cover in a relatively small amount of data. In this particular snapshot the map view has been zoomed to center around the western hemisphere. Zooming to an individual country would quickly show the weakness of client side raster. Vector outlines would remain sharp and clear while the jpg will quickly degrade unless updated at increased resolution from the server.

Here is an example of the SVG xml element outlining Kuwait:

```
<g id="kuwait" style=" &country;" onmouseover=
"rollover(evt)">
```

```
    <path id="Kuwait" d="M47.95708466,-30.01169395
    L47.95944595,-30.00508308
    ...
    L47.95100021,-30.01377869 L47.95708466,-
    30.01169395"/>
</g>
```

On the left is a simple control menu made up of buttons and tabs built from SVG elements. These are again controlled through associated event listeners which control style and visibility using local JavaScript.

The result of clicking on the North America SVG element in the world of Fig. 20.3 is a view of North America supplemented with relief imagery showing topography, Fig. 20.4. Menu check boxes, constructed from SVG elements, have been used to select additional layers. The hurricane tracks are dynamically derived from national weather URLs. In addition a Draw Tab has been selected on the menu exposing some basic drawing tools which were used to outline the landfall of hurricane Katrina near New Orleans. The Draw Tab tools are local tools developed in JavaScript which can be used to supplement the map view.

This illustrates the use of multiple data sources both from the J2EE producer originally accessed, and additional public URL resources (Unisys 2005). In this instance the hurricane tracks are not published in an open standard format and required custom html scraping to produce the storm

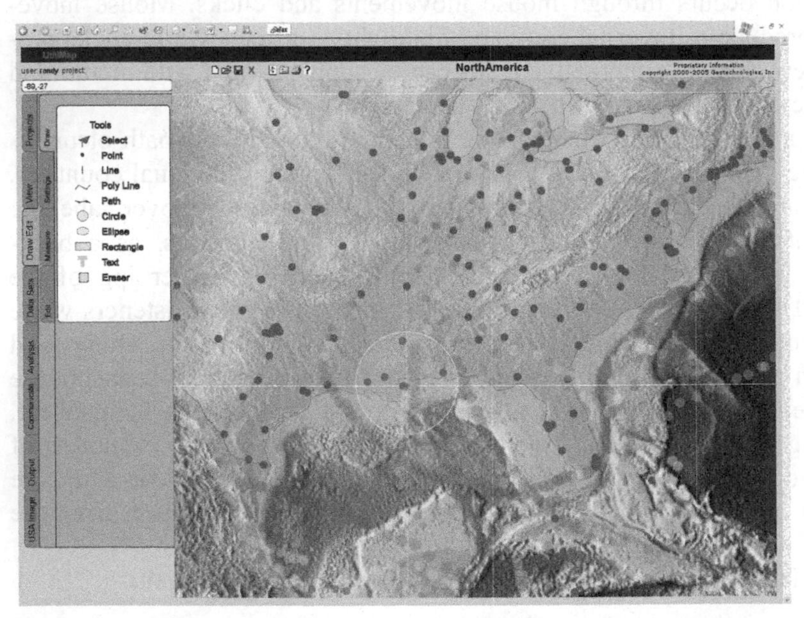

Fig. 20.4. Gulf of Mexico with jpg topography and hurricane tracks overlaid

tracks. Tracks are made up of series of simple circle icons:

```
<circle r="0.5625" id="RITA: Tropical Storm" on-
mouseover="over(evt)" onmouseout="out(evt)"
style="fill:yellow;fill-opacity:0.25" trans-
form="translate(-94.0,-32.1)"/>
```

If emphasis is required to indicate a current position, SVG provides style animation effects to change color and size for a pulsing alert:

```
<circle id="Rita_current" r="0.5625"
    <animateColor attributeName="fill" attrib-
    uteType="CSS" values="red;yellow" dur="0.5s"
    repeatCount="indefinite" accumulate="none" ad-
    ditive="replace" calcMode="linear"
    fill="remove" restart="always"/>
    <animateTransform attributeName="transform"
    type="scale" from="1.0" to="1.25" begin="0s"
    dur="1s" repeatCount="indefinite" accumulate=
    "none" additive="replace" calcMode="linear"
    fill="remove" restart= "always"/>
</circle>
```

Fig. 20.5. shows the result of clicking into North America and United States with a variety of data sets overlaid. In this instance clicking on the icon for Mobile AL has opened an additional window listing the data re-

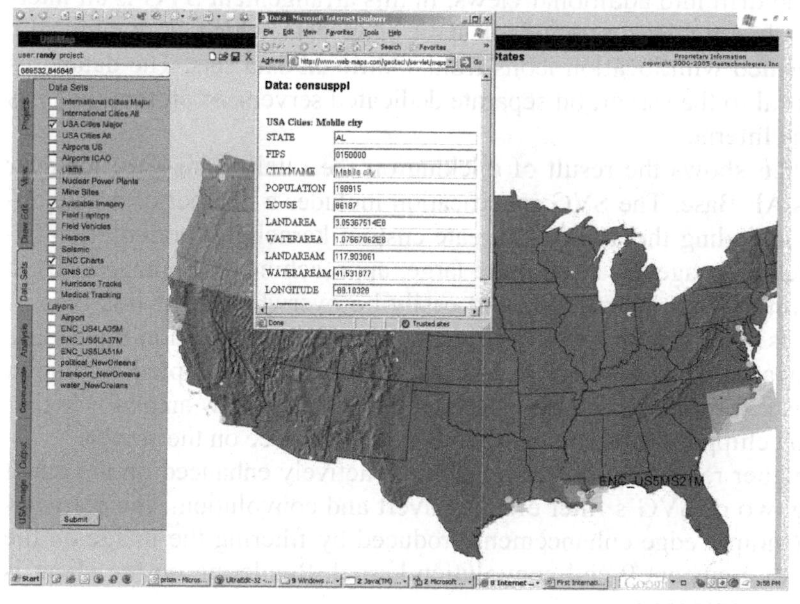

Fig. 20.5. United States with icons showing cities, imagery, and ENC charts

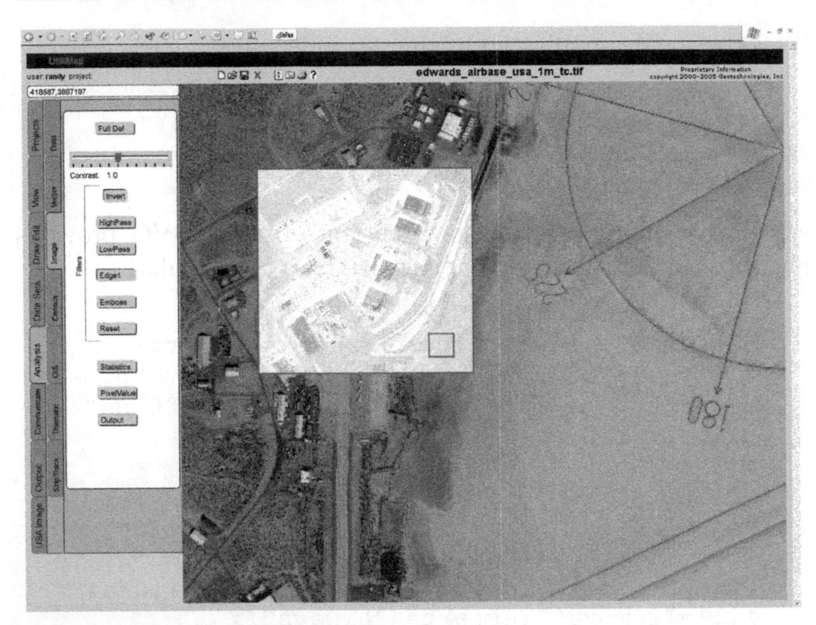

Fig. 20.6. SVG includes imagery capability – Invert plus custom edge convolution

cord behind the selected icon.

All of the icons, state outlines, and ENC chart footprints are active hyperlinks to drill into additional views. In this arrangement SVG is an interface to a hierarchical pyramid of static map data which is dynamically supplemented with location icons from additional data sets. The databases can be local to the server, on separate dedicated servers, or on remote hosts across the Internet.

Fig. 20.6 shows the result of clicking on the satellite imagery icon for Edwards AF Base. The SVG specification includes a number of image operations including the ability to create custom kernels for convolution filters. Satellite imagery can be quite large, approaching 1Gb image sizes in TIFF format. Web interfaces could easily be overwhelmed if this type of imagery is accessed directly. In this case the J2EE application makes use of the Java Advanced Imaging, JAI, api to show a subsampled resolution appropriate for the screen size. The user can then window areas for higher resolution chipping from the massive tiff data resource on the server.

The higher resolution area has been interactively enhanced on the client by using two of SVG's filter effects, invert and convolution. The convolution is a simple edge enhancement produced by filtering the image on the client with a custom 9 cell convolution kernel. Rendering on the client is virtually instantaneous.

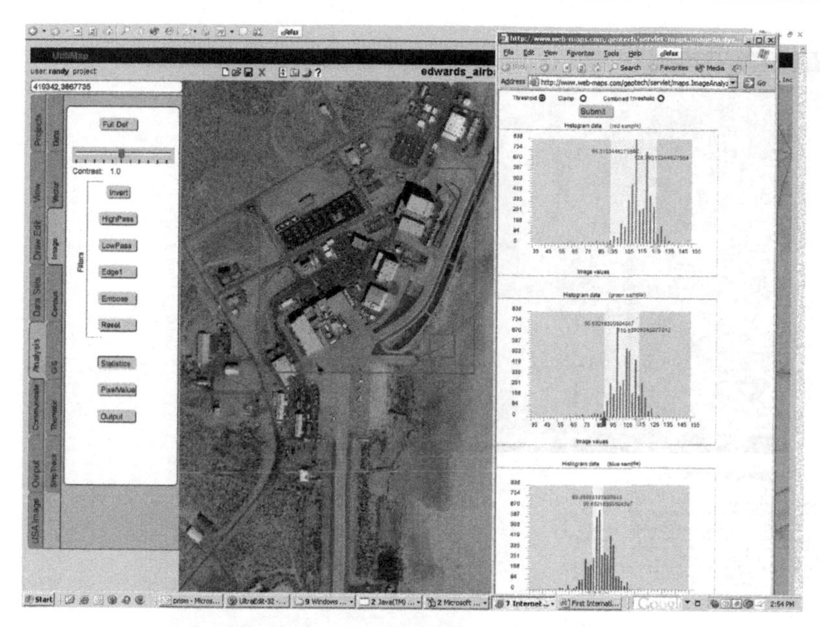

Fig. 20.7. Web interface combining SVG with server side functions

Fig. 20.6 illustrates the use of SVG for dynamic controls. A small window has been outlined for statistical analysis which opens an additional control window. The RGB histograms have been created for the windowed image subset on the server using JAI, and then returned to the client as SVG. The histogram views allow users to dynamically control RGB ranges for clamp and threshold type functions. Once the ranges are set and the type of function selected the server will return the windowed area with the clamp function applied.

Combining the SVG image capability with more powerful server side imagery functions creates a useful web interface without requiring local client side storage of imagery resources. Even though SVG is primarily known as a vector xml language it includes simple image operation capability (Table 20.1).

Table 20.1. Available SVG Filter Effects

feDistantLight	feConvolveMatrix	feMorphology
fePointLight	feDiffuseLighting	feOffset
feSpotLight	feDisplacementMap	feSpecularLighting
feBlend	feFlood	feTile
feColorMatrix	feGaussianBlur	feTurbulence
feComponentTransfer	feImage	
feComposite	feMerge	

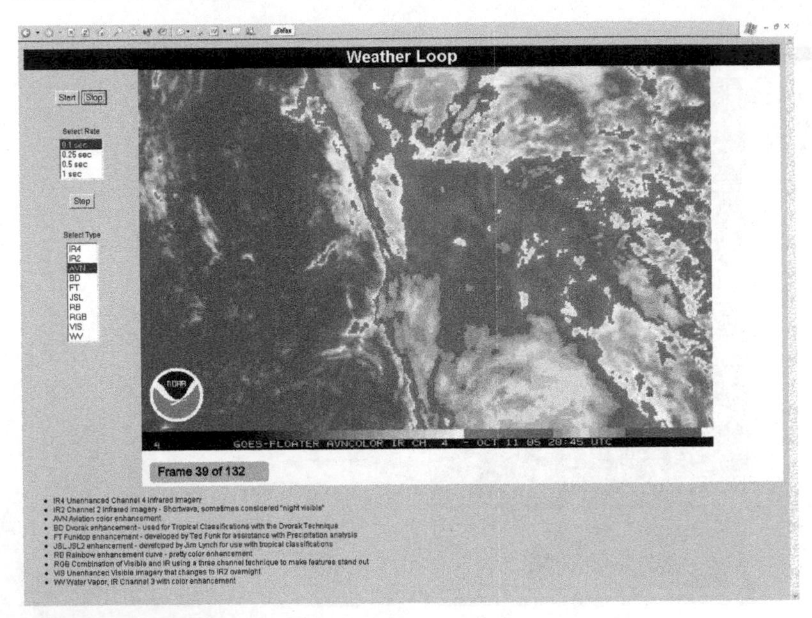

Fig. 20.8. SVG can also be used for a variety of animation affects

As an XML language SVG can act as host to other XML dialects such as Synchronized Multimedia Integration Language, SMIL (2005). This means that most elements can be animated using a subset of the SMIL specification. In addition SVG can be manipulated with client side JavaScript to produce visual affects. Fig. 20.8 illustrates the use of an image stack. Here NOAA (2005) is the source for live imagery in the Gulf of Mexico region during hurricane season. The NOAA images are referenced in a stack of SVG image elements. The visibility of these images is controlled using a simple JavaScript to successively turn on and off images in the stack. This simple trick produces short movie clips based on SVG. Vector overlays can be rendered on top of the stack with additional event listeners. Another approach to video affects, using svg, is to create a film strip image. The successive frames are then passed through an SVG mask element using transform translate functions. In affect this mimics the behavior of a basic movie projector moving successive frames through the projection aperture. These tricks are not useful for anything but short clips. However, subsequent SVG recommendations will support full audio visual references inside svg. For example:

```
<svg xmlns="http://www.w3.org/2000/svg" ver-
sion="1.2"
    xmlns:xlink="http://www.w3.org/1999/xlink"
```

```
        width="420" height="340" viewBox="0 0 420
        340">
        <desc>SVG 1.2 video example</desc>
        <g>
            <circle cx="0" cy="0" r="170" fill="#da4"
            fill-opacity="0.3"/>
            <video xlink:href="sample.avi" vol-
            ume=".8" type="video/x-msvideo"
            width="320" height="240" x="50" y="50"
            repeatCount="indefinite"/>
        </g>
    </svg>
```

Fig. 20.9 illustrates the use of SVG as an interface to an OGC compliant WMS server at the USGS (2005). Selecting a USGS topographic map area on the Colorado map opens a new window showing the results of the WMS query. JavaScript is used to build the WMS query from information in the SVG map. The new window is supplemented with some simple svg controls for moving through the seamless NED relief data and zooming in to higher resolutions. Since the terrain relief imagery is provided by a USGS WMS server, the interface can be built with very little local data.

Here is a sample WMS query for the USGS Seamless NED shaded relief:

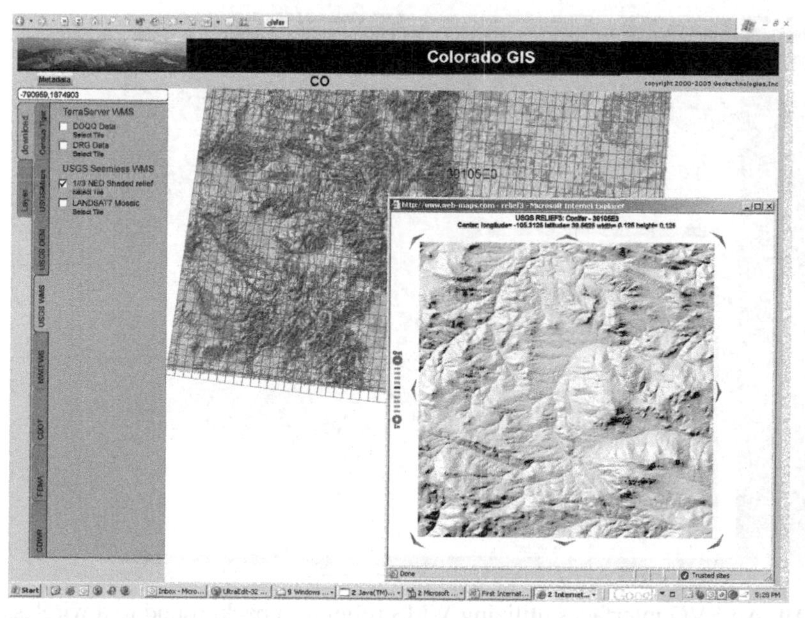

Fig. 20.9. SVG interface to a USGS WMS server for NED relief imagery

```
http://gisdata.usgs.net/servlet/com.esri.wms.Esrim
ap?WMTVER=1.0.0&LAYERS=1/3 NED Shaded Relief&
FORMAT=JPEG&BGCOLOR=0x000000&SRS=EPSG:4326&
SERVICE=WMS
&STYLES=&SERVICENAME=USGS_EDC_Elev_NED_3&request=
GetMap&SRS=EPSG:4326&BBOX=-104.8125,39.375,-
104.6875,39.5&width=1000&height=1000
```

WMS and WFS services can be merged in a J2EE servlet to provide combinations of externally referenced data sources. This is a relatively new resource for GIS and a new way to view GIS. Data sources which have been isolated and locked into proprietary formats are now becoming active internet resources that can be used by a world wide audience.

Fig. 20.10 illustrates how a WMS resource can be merged into user specific interfaces. In this instance the USGS WMS terrain relief is referenced as a useful background to a wireless antenna interface. Antenna coverages are indicated along with customer locations. The shaded relief provides clues for likely wireless access by showing possible terrain masking effects. As the client user zooms into smaller regions of interest more detailed terrain is accessed from the public WMS.

One aspect of SVG that is illustrated by this interface is the ability to custom fit the interface to the user. User interfaces are often more efficient when scaled down to provide only the necessary features and operations

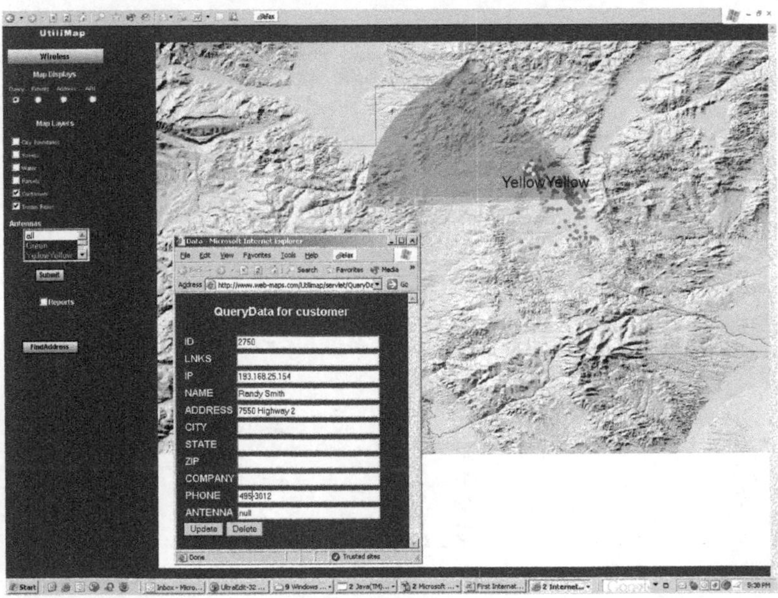

Fig. 20.10. An SVG interfaces utilizing WMS relief as a background to a wireless antenna application

for the job at hand. SVG is a tool for building custom interfaces without resorting to a one size fits all, typical of proprietary web mapping software.

Location based interfaces are growing in importance. Fig. 20.11 illustrates the use of SVG as a monitoring tool for GPS activated Thuraya Satellite phones. As vehicles move through Kuwait City their location is monitored by collecting GPS positions in a database. The SVG monitor interface allows a registered user to view locations from anywhere in the world. Location history is also an option showing the track of a selected vehicle. Adding Ajax techniques allows continuous position polling to move vehicle icons across the satellite imagery background.

20.3 Summary

The sampling of SVG interfaces referenced in this gallery show a wide spectrum of capability available to the developer. SVG is an important XML specification for interacting with both vector and raster data. The SVG specification allows developers to build internet GIS interfaces based on an open standard. These interfaces can be complex or simple. They can reduce the functionality to the subset of capability necessary for a particu-

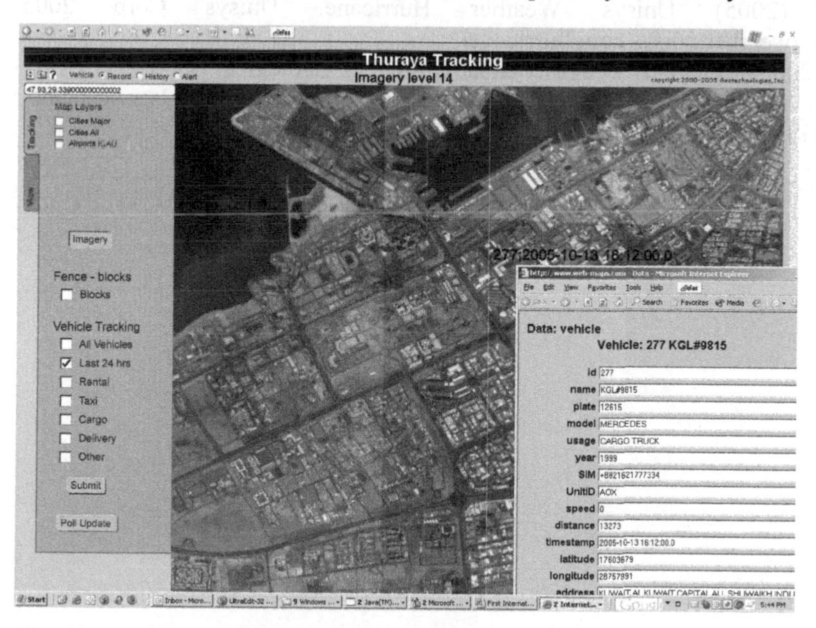

Fig. 20.11. SVG used as a location monitoring interface tracking Thuraya Satellite phone locations in Kuwait City

lar job function. The introduction of GIS producer standards like WMS and WFS have opened up vast data resources to individual web users, but the results of these services must still be rendered in the browser. SVG provides the capability of viewing both WMS and WFS queries from a browser interface anywhere there is internet access.

References

NOAA (2005) NOAA Satellite and Information Services. National Oceanic and Atmospheric Administration, US Government, http://www.ssd.noaa.gov/GOES/FLT/T1/

OGC (2005a) Open Geospatial Consortium web page. Open Geospatial Consortium, Inc., http://www.opengeospatial.org/

OGC (2005b) OGC WFS specification. Open Geospatial Consortium, Inc., https://portal.opengeospatial.org/files/index.php?artifact_id=8339

OGC (2005c) OGC WFS specification. Open Geospatial Consortium, Inc., https://portal.opengeospatial.org/files/index.php?artifact_id=5316

SMIL (2005) Synchronized Multimedia SMIL. W3C Multimedia Activity, W3C, http://www.w3.org/AudioVideo/

SVG (2005) Scalable Vector Graphics. SVG Working Group, W3C, http://www.w3.org/Graphics/SVG/

Unisys (2005) Unisys Weather Hurricane. Unisys Corp 2005, http://weather.unisys.com/hurricane/

USGS (2005) USGS NED WMS capabilities. US Geological Survey, US Government, http://gisdata.usgs.net/servlet/com.esri.wms.Esrimap?servicename=USGS_WMS_NED&request=capabilities

PART V

GEOGRAPHIC HYPERMEDIA: APPLICATIONS AND SERVICES

21 About the Role of Cartographic Presentation for Wayfinding

Georg Gartner, and Verena Radoczky

Abstract. The rapid development of the mobile internet enforces the emergence of location based services. Beside car navigation systems, which are standard equipment to many drivers all over the world today, guiding systems for pedestrians gain more and more importance. While drivers are dependent on tight directions, pedestrians can be provided with more detailed information. Many different possibilities are imaginable in this respect, but not all of them are effective as guiding instructions. In this chapter general considerations about wayfinding tasks in urban environments are discussed and diverse communication methods are investigated regarding their potential as route finding aids.

21.1 Introduction

Within the scope of the project NAVIO (Pedestrian Navigation for Combined Indoor-/Outdoor Environments), which is carried out at the Technical University of Vienna, communication methods for pedestrian navigation systems are investigated and tested in order to efficiently support guiding along unfamiliar environments. Before analyzing various multimedia presentation forms, tests are described that approach the subject the other way around by exploring how users actually gain knowledge about their surroundings and how they represent the obtained information to other people. For this purpose route descriptions and sketches of directions by students of the Technical University were collected and analyzed. Additionally a wayfinding test, where people followed a certain route with the help of different map types, has been carried out in the area of the university. The results of this field test, which give information about efficient map types for pedestrian navigation systems, are inspected and represented in this contribution. Furthermore insights from an extensive literature review about multimedia presentation forms are collected and analyzed. Based on these findings theories about multimedia presentation forms, which could be found in the literature, have been collected and are represented in a subsequent chapter. Here the outcomes and theories of the empiric tests could be mostly verified, confirmed and expanded.

Furthermore the topic "landmarks" was treated in more detail. The integration of active and passive landmarks into a navigation system was analyzed and possibilities on how to realize their application are discussed.

In a final section the major achievements are summarized and discussed.

21.2 Test: Analysis of Route Descriptions and Sketches

In order to obtain information about how people observe and revisualize their environment an experiment that analyzes properties of route descriptions and sketches was carried out. 23 undergraduate students were invited to join an informal test at the Department of Geo-Information and Cartography, where they should take their time and write down 2 different route descriptions and additionally draw sketches of the described paths. The 14 male and 9 female students were therefore asked to describe how they would get from the city centre at 'Stephansplatz' to 2 different auditoriums at the Technical University.

The objective of this test was to get an idea about how mental maps are structured and how people describe these depictions to someone who is not familiar with the area in question (The actual efficiency of these depictions was not tested here and should be explored in a different test). Nevertheless the results of this test can provide information about how people project reality in their minds and can therefore help to find an efficient way to describe routes to people:

- The north-orientation of mental maps seems to be rather important to many people, probably because it facilitates the comparison to street maps. Indoor maps, on the other hand, are not necessarily heading northwards. Presumably some people seem to save them separately in their minds and recall them from the perspective they would have when entering the building (see Fig. 21.1). But, as this test shows, indoor maps do not have a very important status anyway. Most people tend to rely on the logical structure of buildings and on efficient signage and hence neglect to explain indoor paths in detail. None the less this does not necessarily mean that indoor navigation systems should be ignored completely. Many people do not read signs because it is easier for them to ask for directions (Muhlhausen 2002), and even if they do rely on signs, we do not know, if they are distributed by logistic means. Faulty sign design can cause navigation problems in unfamiliar environments. Therefore it is practical to offer another navigation aid than solely signs.
- Another interesting outcome of this test concerns the individual view of different people on the same thing. Even though all participants of the

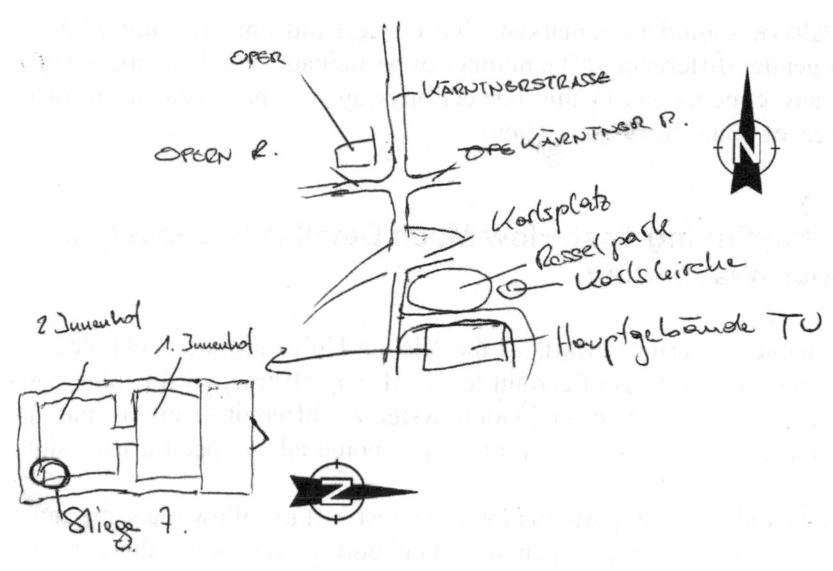

Fig. 21.1. The outdoor map on the top is heading northwards whereas the indoor map (bottom left) is heading west.

test were first-year students of the same faculty and therefore could be defined as one particular user group concerning way-finding abilities in the area of the Technical University, they still have an individual mental map of that area. About 40 % of the students decided to go to the Auditorium Maximum via the main university building even though it is a lot longer than the direct way, but the other 60 % used very individual paths to get to that target. Presumably humans tend to rather use chunks of well-known routes than experimenting with shortcuts, but their familiarity with an area is based on personal experiences and not the same for people of the same user group.

– The same holds for the use of landmarks. Apparently a lot of landmarks used in the descriptions derive from personal preferences. A little café or an Italian restaurant is probably just noticeable to someone who has been in there before, and shoe shops might only be of interest to people who enjoy shopping. However, there are some landmarks that are evidently outstanding to most people, like churches, the Opera House, theatres or market places, thus they should be included in every way description. Moreover places, where a change in direction is required, should be handled as landmarks. This conclusion can be drawn from the students who participated in the test, because they instinctively described decision points in more detail than any other location along the way.

Finally it should be remarked, that the test did not give any evidence about gender differences. The number of participants would be too small to draw any conclusions in this respect anyway, but not even insignificant differences could be observed here.

21.3 Wayfinding Test: How Much Detail is Necessary to Support Wayfinding

This chapter describes efforts at the Vienna University of Technology to prove the effectiveness of certain levels of map abstraction for route communication in pedestrian navigation systems. Different levels of map abstraction are investigated regarding their potential in specific user situations.

All candidates who participated in the test had to follow the same route, merely the map source, which was their only guide during the trip, was chosen coincidentally. That way a quantitative analysis could be achieved by measuring the time needed to orientate at decision points and counting the amount of errors during the task. By asking subjects to draw sketch maps of the visited area and by comparing the results of participants who followed different map sources, conclusions about their effectiveness when generating cognitive maps can be drawn qualitatively.

24 people participated in the test, 11 of them evaluated a schematic map (Fig. 21.2a), 10 were guided with the help of a simplified street graph (Fig 21.2b) and 3 people had to find her way with the help of a typical city map (Fig. 21.2c).

Recapitulating we can say that even though some tricky paths had to be crossed most people got on quite well with the used map. The required time for the walk was somewhere between 20 and 30 minutes, only one person, who nearly got completely lost was on his way for about 45 minutes. Each candidate was accompanied by the test instructor, who was not allowed to help the person in any way, but followed him/her, stopped the time needed and took notes of unusual and surprising behaviour:

- At the meeting point in front of the Opera House in Vienna, the map was handed out to the participants and their current location and orientation was explained to them.
- About half of the candidates accidentally left the route at different spots, but all except one realised the mistake within 2 minutes and could get back on track no matter which map type their test was based on.
- Surprisingly 14 out of 24 people used underpasses and subways in order to overcome traffic or bad weather conditions even though the used

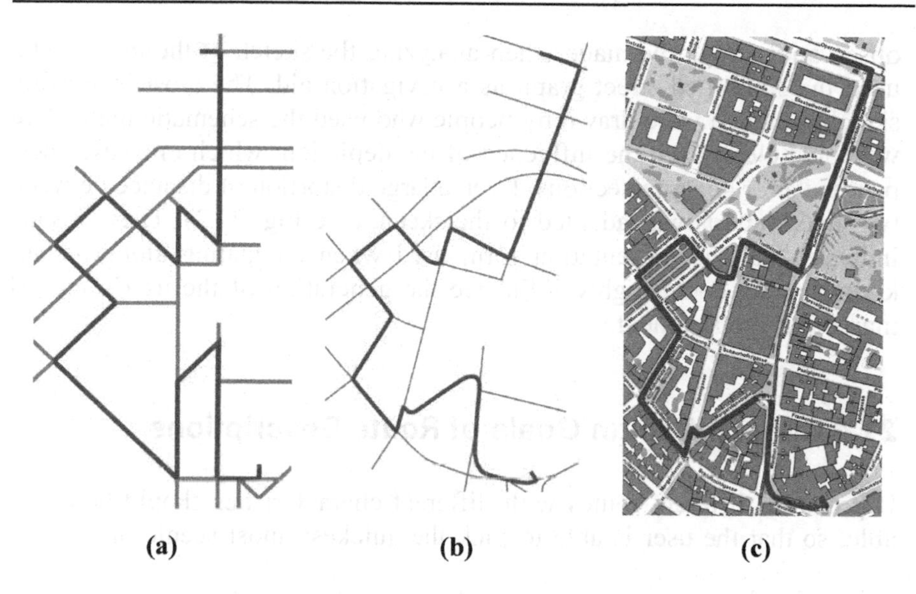

Fig. 21.2. (a) Schematic map; (b) simplified street graph; (c) city map.

maps did not give any information about them. Some of them did know these subways beforehand and others have never seen them before in their lives. It was remarkable to find out, that none of them had orientation problems the moment they left the underground system.

- 22 participants turned and rotated the map in their hands in order to adjust the map alignment to their current walking direction. Only 2 men, one of them a local and the other a foreigner, left it northbound.
- People who are familiar with the area in question did not have any major difficulties with the schematic map as a presentation form once they got used to the unusual depiction. Only one tourist, who has never been in that area before, had major difficulties to follow the route. 2 other foreigners, who tested the street graph and the topographic map, stayed on the right track without any problems.

Subsequent to the way-finding task, subjects had to draw a sketch of the visited path. As expected the sketches of the local people were quite precise even though most of them navigated along the route with the help of the schematic map, and were all pointing northwards. The outlines of the foreigners, on the other hand, were all pointing southwards, which was the starting direction, and as expected lacked a lot of information. When looking at the sketches of the tourists, who did not know the area at all, we noticed that even though a lot of information and essential turns in direction were missing, most of the corners were rounded and many different angles were used to visualise intersections, just like in the original map. Similar observations could be made when analyzing the sketch of the tourist who used the simplified street graph as a navigation aid. The most interesting sketches though were drawn by people who used the schematic map. Here we could clearly see the influence of the depiction, which only uses perpendicular or 45° intersections. Even a large distortion of distance between two intersections was adapted to the sketch (see Fig. 21.3). These results indicate, that the presentation form used when navigating along an unknown route, could highly influence the generation of the user's mental map of the environment.

21.4 General Design Goals of Route Descriptions

If possible, different routes with different characteristics should be available, so that the user is able to pick the quickest, most scenic or shortest

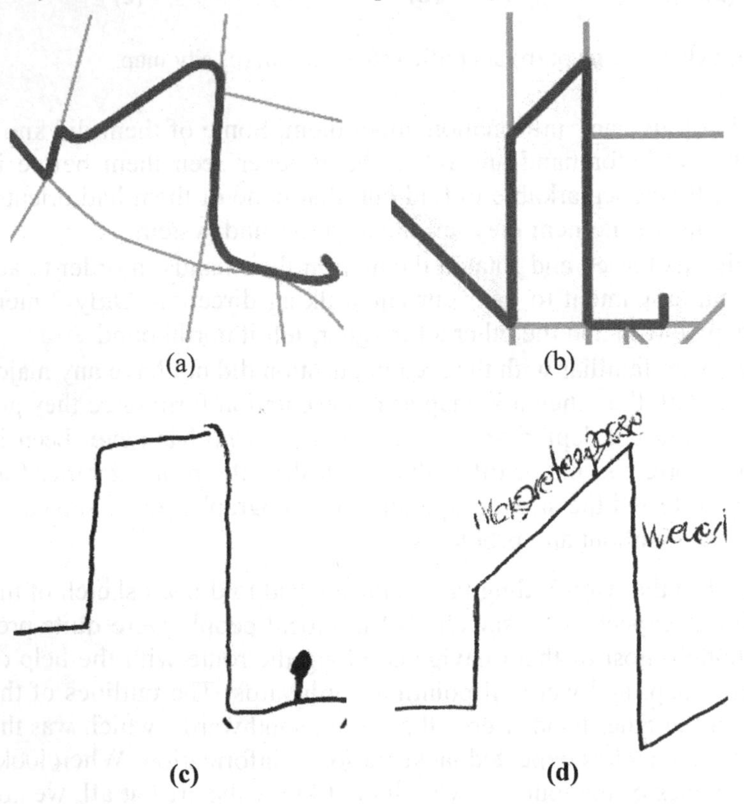

Fig. 21.3. (a) Street graph of the segment; (b) schematic map of the segment; (c) distorted sketch of the segment; (d) distorted sketch of the segment.

route. That way the system does not need to differentiate between user groups, but the user can pick the most appropriate path himself. Nevertheless not too many options should be offered because that might lead to confusion.

When looking at the route description for the first time, the user should be able to get an overview of the whole route at a quick glance (Radoczky 2003, Hohenschuh 2004). This could be realized by displaying a survey map, where the entire route is visualised at small scale. Only when the user actually starts moving, the scale could be adapted to the current situation. Additionally information about the length of the route and the expected travel time could help the user to decide, whether to move along the suggested path or not. Furthermore the start and the end point of the route should be clearly marked, so that the current and the future location can be easily identified (Agrawala and Stolte 2000). When moving along the way, decision points, where a change in direction is needed or could happen accidentally, should be clearly indicated, especially if there are no signs available in the area. In case there are no landmarks existent at critical points, distances between nodes could be mentioned. However, usually the individual length of segments is not required, because humans are generally quite bad in estimating distances. Only in indoor environments this information could be of value, because distances are usually short and assessable.

21.5 Presentation Forms

Rating the efficiency of presentation forms for pedestrian navigation system and comparing them in order to find the optimal communication medium is a rather difficult task that can never lead to a simple conclusion. Many aspects influence the result of such an investigation: the user himself, his experiences and personal preferences, the complexity of the respective area, the available technical resources, the availability and quality of datasets, and so forth. Therefore no universally valid propositions can be made - only rough guidelines can be defined. Moreover it is important to mention, that redundancy reveals to be one of the most important properties a navigation system should consist of and therefore various presentation forms should be used simultaneously (Belke and Rehm 2005, Michon and Denis 2001). Yet exaggerations should be avoided, because even though singular information could cause irritation, too much information can also slow down decision making (Klippel 2003).

21.5.1 Maps

The basic visualisation element when communicating routes should always be a map (Reichl 2003, Thorndyke and Hayes-Roth 1982, Radoczky 2003, Kray *et al.* 2003, Reichenbacher and Meng 2003, Dransch 1997). Even though nowadays new techniques like virtual and augmented reality, animation and many other multimedia tools are imaginable, cartographic representations are still the best form of giving an overview of an area. This acknowledgement applies to all sorts of user situations, like day and night time, fog, rain or snow, quiet and noisy environments, as well as to different user groups, like tourists, natives or business travellers (Reichl 2003). This result implies that even though maps can differ dramatically from the perceived structure of a spatial environment, they can help the user to get a good overview of the area. Precision does not seem to play an important role in the navigation process, as long as essential topological information can be extracted and distortion is not disproportionately high (Dennett 1969, Agrawala and Stolte 2000, Evans 1980, Carstensen 1991, Moar and Bower 1983, Klippel 2003).

Instinctively people typically consult maps when they have to find their way through an unfamiliar environment, and when using conventional paper maps as navigation aids, humans tend to twist and turn the map in order to facilitate way-finding and avoid mental rotation (Zipf and Jöst 2004, Radoczky 2003). In that way, a previously northbound map can be adapted to personal requirements, and turning points can be viewed as simple left and right turns. Some tests showed that even with the help of digital maps in navigation systems, the efficiency of way-finding tasks is generally higher when an egocentric map view is used.

Many different types of maps are available today. They are designed in different styles and contain different levels of detail. Usually a lot of information is displayed on a comparably small sheet of paper, which could hinder information extraction when reading the map. But how much detail is actually needed to guarantee wayfinding success? In some situations schematic maps, like the famous London Underground map, contain enough information to easily find a certain destination, but when moving through a city as a pedestrian, these depictions might not contain enough detail to stay on the right track. On the other hand, topographic maps might include a lot of detail that is not needed by the user. Since neither topographic maps nor schematic maps seem to be ideal for communicating route information, because of their overload and respectively their lack of information, a medium of both seems to be the obvious solution. For that reason we suggest to switch to a highly simplified graph which is derived from a topographic map and add certain features (like street names, parks, landmarks, zebra crossings, monuments etc.) that are essential for the wayfinding task in question.

21.5.2 Verbal Information

In general the visual is superior to the textual human memory. Nevertheless different contents can only be described verbally and not with the help of pictures and spatial layouts of environments can be transmitted remarkably well solely by language. Especially for navigation systems verbal aids can be a major mode of communication when describing routes, as long as simple and clear language is used (Fontaine and Denis 1999). A major difference can be found when analyzing audio and textual information:

Oral route directions should be more concise and tight, because too much detail may pose a problem for the receiver (Lovelace *et al.* 1999). The main advantage of oral information though is the independence of the display that could be hidden in a pocket. The user can walk around without being distracted by looking at the device, which could be a major advantage in the dark or when weather conditions are bad.

With written route directions, on the other hand, the potential of overload on the receiver is less a problem and so longer route directions may be given which can be reread at any time. The advantages of written text are also quite clear: even when the user walks along a busy street or moves in some other noisy environment, he/she will be able to read the text (Reichl 2003). On the other hand, darkness and weather conditions can make it impossible to read written text on a small screen.

Fig. 21.4. Written route directions.

Recapitulating we can say that both, written and audio instructions do not demand a lot of technical resources and are therefore easy to implement (Kray *et al.* 2003). The main disadvantage of verbal instructions though is the fact, that they purely concentrate on the communication of the route and do not convey an overview of the whole area to the user like a visual presentation.

21.5.3 Images

An image can be a very important representation form in a navigation system, even though it is not really valuable as a navigation aid (Radoczky 2003). Users need a lot of time to compare reality with the image, before they walk in the viewed direction. This could be very annoying when being in a hurry, which is why photographs are not used for giving directions. The usage of photorealistic images is only advisable when describing landmarks, and even here we should make sure, that this information can actually help the user to stay on the right track. Therefore photography could be an optional choice that people, who are not familiar with the environment and who are not in a hurry, can choose to gain additional information about a certain object. Beside conventional images panorama views are imaginable as additional information. They provide a better overview of the environment and contain more information than traditional photographs. Unfortunately not many people are familiar with this type of presentation, which can cause confusion and insecurity.

21.5.4 Videos

Videos havc similar properties as pictures: objects and streets can be di-

Fig. 21.5. Image as navigation aid.

rectly compared and identified with reality. The main disadvantage though seems to be the speed of the movement in the film, which does not give a lot of time to watch everything in detail (Radoczky 2003).

Altogether the overall potential of videos as route information aids can be rated as very low (Reichl 2003). However, videos can be used as optional information about landmarks and sights.

21.5.5 3D Presentation Forms

Nowadays new techniques allow different types of three-dimensional presentation forms. Still most of them are not particularly relevant until now because they are not fully functional, require too many resources or are too complicated to imply on a large scale. Nevertheless a few possibilities should be mentioned:

Until today the most common way to include 3D models in computer visualisation is the VRML mode. The Virtual Reality Modelling Language (VRML) is a standardised language that can be viewed with the help of a plug-in and enables the depiction of three dimensional scenes. When used in pedestrian navigation systems the user could view the environment on screen from his/her current point of view. Still the efficiency of this depiction can be rated as rather low, because users are not used to this presentation form and are therefore irritated by it (Rakkolainen *et al.* 2000, Kray *et al.* 2003).

A more reasonable alternative for pedestrian navigation systems is augmented reality, where subjects need to wear special glasses that display additional information to the user whenever needed. As soon as this method functions well enough to be commercialised, it could replace conventional systems that work with the help of a mobile device.

Overall we can say that map mode is more familiar to most people and therefore the use of 3D models, which are also very time-consuming in the production process, is not likely to be adapted in pedestrian navigation systems in the near future.

21.6 Landmarks

A landmark is understood as an orientation point in space that helps humans to find their way in an environment (Klippel 2003, Michon and Denis 2001). Typically it bears a visual characteristic that is unique in the immediate environment or it holds a distinguishing function or purpose or the landmark is located at a central and salient location (Raubal and Winter

2002, Elias 2002, Sorrows and Hirtle 1999). The more distinctive an object is the more relevant it is as a landmark. In route directions landmarks seem to be even more essential than the mentioning of street names.

There are several classifications of landmarks:

A visual landmark shows special visual characteristics, a cognitive landmark is an object with outstanding meaning and a structural landmark is important because of its function or location within an environment (Sorrows and Hirtle 1999). Furthermore we can differentiate between local landmarks, which are only visible from a small distance, and global landmarks which are distant landmarks such as mountain peaks, towers or even the sun (Steck and Mallot 2000, Lynch 1960). Another category which can be defined deals with personal experiences and memories with a specific place and is called "Emotional Landmarks". An Emotional Landmark is originated in a distinctive emotional life-event that associates external landmarks with autobiography and thereby forms an internal reference point (Oakley *et al.* 2005). This type of landmark is very individual and therefore rather hard to acquire. Emotional Landmarks do usually not have a distinctive character. Nevertheless an unobtrusive building can be a very important orientation point to somebody whose best friend lives in there. But when it comes to emotions about areas in a city where people feel uncomfortable and unsafe (or areas which are extraordinarily beautiful with an appealing atmosphere), we can find that these subjectively observed areas often overlap.

The most interesting distinction of landmarks though concerns the function as a positioning tool for the navigation system: An active landmark (as opposed to ordinary passive landmark) realizes when the user moves within its reach and builds up a spontaneous radio contact to the user via an air-interface (see Fig. 21.6). That way the system does not need to locate itself all the time, but receives position information from outside. Another main advantage of this concept is the fact that active landmarks do not need to fulfil classical landmark criteria. Even if the object is entirely insignificant, it can be equipped with a sender and support the navigation system (Brunner-Friedrich 2003). For that reason active landmarks can be located at decision points, in narrow streets where GPS systems fail to deliver accurate position information or in areas, where surrounding objects look very alike and lead to confusion. When thinking about where to place active landmarks, it is important to consider, that they should not be too close to each other, especially if positioned indoors. Signal overlap could cause problems in the intersection area. The actual range of the signal depends on the used air-interface: Bluetooth (10-100m), WLAN (100-300m), IrDA (1-3m) or a proprietary radio interface are possible solutions. The connection itself could be implemented in two ways: either the active

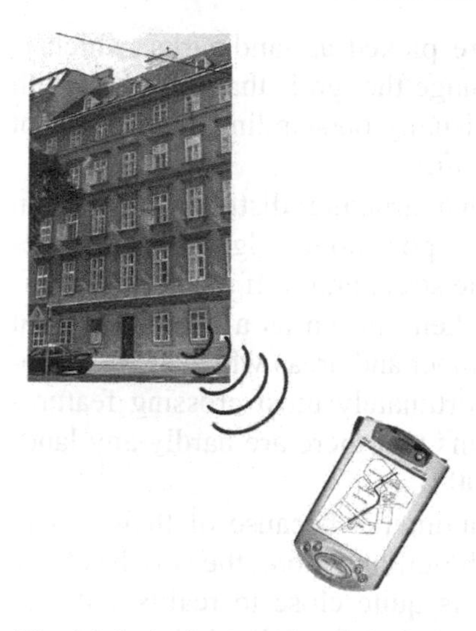

Fig. 21.6. Active Landmark builds up a radio connection with the device.

landmark constantly transmits signals or the handheld device of the user searches for the nearest active landmark.

An extension of Active Landmarks could be realised by a ubiquitous network. Beside only receiving coordinates of the current location, a ubiquitous landmark could also send additional information about the area or describe the object in more detail. Furthermore the concept of ubiquity includes, that these points of interests are interconnected with each other and can therefore exchange information among each other and with the device of the user. That way the provided information could be individually tailored to the users needs. This concept enables a revolutionary opportunity for navigation systems of any kind. Within the last few years a lot of research and development has taken place concerning Location Based Services, which could now be supplemented and expanded with the help of ubiquitous methods, and maybe in the future they could even be replaced. Yet research is still in the early development stage that still requires many new challenges.

21.6.1 Derivation of Landmarks

Six different methods of landmark derivation have been analyzed and tested:

- The first method is based on the concept that landmarks are outstanding because of their content (for example public buildings). All objects that

have special names in the database are picked as landmarks, which is quite easy to fulfil. The main disadvantage though is that buildings with special functions are not necessarily visually outstanding and might not be recognised by users as such (Elias 2002).
- Method 2 is based on the assumption that structural distinctiveness often equals visual distinctiveness. For this purpose node, edge and region degrees are automatically calculated in the street graph. If a place degree is higher than the average degree, it is then chosen as a landmark. That way places where more than 4 streets meet and areas which serve as obstacles are chosen as landmarks. Unfortunately most crossing features have a similar node degree which means that there are hardly any landmarks available (Raubal and Winter 2002).
- In method 3 objects are chosen as landmarks because of their visual, semantic and structural attraction. Each building along the way has to be examined, which is why this method is quite close to reality but also very time consuming and therefore hard to realize (Raubal and Winter 2002).
- Method 4 is quite similar to method 3. Here landmarks are also chosen by walking through an area and inspecting buildings in more detail, but unlike in the last method, there is no objective basis to the procedure. Landmarks are solely chosen in a subjective way by the examiner (Brunner-Friedrich 2003).
- In method 5 the visibility of outstanding points is analyzed with the help of a 3D model. This works very well for global landmarks, but could be difficult for local landmarks in case texture is not attached to building blocks. Another disadvantage is the necessity of a detailed 3D model and the time-consuming work on it (Achleitner *et al.* 2003).
- Method 6 is adapted to the user's needs. In a first step route descriptions by pedestrians are analyzed and landmark types (e.g.: traffic lights) are extracted. Afterwards methods are derived on how to extract these types out of the database. Unfortunately some categories are usually not included in the database which makes results of this method incomplete (Brunner-Friedrich 2003).

Additionally, if a high density of landmarks has been achieved, they could be divided in two classes: landmarks with high or lower importance. According to several tests (Fontaine and Denis 1999, Michon and Denis 2001, Lovelace *et al.* 1999), route descriptions usually include most landmarks and decision points near the start and near the end of the route. Therefore objects in these areas could be classified as highly important landmarks, and objects somewhere along the route could be of lower importance and could therefore be left out to avoid information overload (Brunner-Friedrich 2003).

After the specification which landmarks are to be used in the route description, their visualisation in the map should be decided. Again different methods are imaginable (Elias 2002):

- The object is marked with an arrow
- The object is coloured
- The object is emphasised and highlighted by enlarging
- All other objects are simplified in their geometry apart from the target object
- Less important objects are merged
- The object is presented with a self-explanatory symbol

21.7 Conclusion

The search for an optimum way to present route information to pedestrians is a never-ending task that depends on many different aspects. Different users have different preferences, and even the same users could vary their priorities depending on the respective situation. Nevertheless the project NAVIO demonstrates, that some principles are valid for the majority of people:

1. It is advisable to offer different (but not too many) routes with different characteristics (shortest, most scenic,…), so that the user can adapt the route to his current situation himself. Sometimes it is helpful to include chunks of well-known routes to guarantee wayfinding success.
2. Landmarks are indispensable when describing routes. They have to be included in every navigation system, unfortunately there is no derivation automatism available yet, that is neither incomplete nor too time-consuming to implement. For that reason Active Landmarks, that do not need to fulfill classical landmark criteria, should be implemented. As soon as the user moves within their reach, the Active Landmark builds up a connection and sends information to the user.
3. When deciding which presentation form to use, it is advisable to consider, that the medium could possibly influence the generation of the user's mental map which could again influence wayfinding success. Besides a map, which should always be the basic visualization element, it is advisable to implement one or (at the most) two additional communication methods. Verbal information is the most obvious communication method and could therefore be used orally or in a written format. Tech-

nical resources for text are rather low and therefore easy to implement. Images and videos are impractical as navigation aids, but are useful as additional information about landmarks, and 3D formats are negligible for the time being, because users are generally unfamiliar with these complex formats and because they are very time-consuming to produce. Augmented reality, on the other hand, could soon displace conventional systems.

Acknowledgement

This work was supported by the FWF (Austrian Science Fund), project P16277.

References

Achleitner E *et al.* (2003) Dimensionen eines digitalen Stadtmodelles am Beispiel Linz. In: Schrenk M (ed) Proceedings of 8th Symposion on Info- & Communication Technology in Urban- and Spatial Planning and Impacts of ICT on Physical Space, Vienna University of Technology

Agrawala M, Stolte C (2000) A design and implementation for effective computer-generated route maps. AAAI Symposium on Smart Graphics, March.

Belke M, Rehm M (2005) Im Anfang war das Wort. Research at the University Bielefeld, Germany. In: Sprache, Computer, Roboter, http://www.uni-bielefeld.de /Universitaet/ Einrichtungen/Pressestelle/dokumente/ fomag/fomag25.pdf

Brunner-Friedrich B (2003) Modellierung und Kommunikation von Active Landmarks für die Verwendung in Fußgängernavigationssystemen. In: Proceedigns of AGIT 2003, Salzburg

Carstensen K-U (1991) Aspekte der Generierung von Wegbeschreibungen. IWBS Report 190, Wissenschaftliches Zentrum, Institut für Wissensbasierte Systeme

Dennett DC (1969) The nature of images and the introspective trap. In: Dennett DC, Content and Consciousness, Routledge & Kegan Paul, London, pp 132-141

Dransch D (1997) Computer-Animation in der Kartographie. Theorie und Praxis, Springer-Verlag Berlin Heidelberg

Elias B (2002) Erweiterung von Wegbeschreibungen um Landmarks. In: Seyfart E (Hrsg), Publ. d. Dt.Gesellschaft f. Photogrammetrie und Fernerkundung, Potsdam, vol 11, pp 125 - 132

Evans GW (1980) Environmental cognition. In: Psychological Bulletin, vol 88, pp 259-287.

Fontaine S, Denis M (1999) The production of route instructions in underground and urban environments. In: Frewka C, Mark DM (eds) Proceedings of COSIT'99, LNCS, vol 1661, Springer, Berlin, Heidelberg, New York

Hohenschuh F (2004) Prototyping eines mobilen Navigationssystems für die Stadt Hamburg. Diploma thesis. Department of Computer Science, University Hamburg, Germany

Klippel A (2003) Wayfinding choremes: conceptualizing wayfinding and route direction elements. Doctoral dissertation, Department of Mathematics and Informatics, University of Bremen

Kray C, Elting C, Laakso K, Coors V (2003) Presenting route instructions on mobile devices. In: Proceedings of the International Conferences on Intelligent User Interfaces, Miami, FL, USA, pp 117 – 124

Kray C, Elting C, Laakso K, Coors V (2003) Presenting route instructions on mobile devices. In: Proceedings of the International Conferences on Intelligent User Interfaces, Miami, FL, USA, pp 117 – 124

Lovelace KL, Hegarty M, Montello DR (1999) Elements of good route directions in familiar and unfamiliar environments. In: Freska C, Mark DM (eds): Spatial information theory: cognitive and computational foundations of geographic information sciences, Proceedings of COSIT, Berlin: Springer-Verlag, pp 65-82

Lynch K (1960) The image of the city. MIT Press, Cambridge, MA

Michon P-E, Denis M (2001) When and why are visual landmarks used in giving directions? In: Montello DR (ed) Spatial information theory, Lecture Notes in Computer Science, vol 2205, Springer, Berlin, pp 243-259

Moar I, Bower GH (1983) Inconsistency in spatial knowledge. In: Memory and Cognition, vol 11, pp 107-113

Muhlhausen J (2002) Wayfinding is not signage – but there's more. In: Signs of the Times (magazine), June 1st

Oakley K *et al.* (2005) Emotional landmarks – an alternative way to support wayfinding. To be published.

Radoczky, V (2003) Kartographische Unterstützungsmöglichkeiten zur Routenbeschreibung von Fußgängernavigationssystemen im In- und Outdoorbereich. Diploma thesis, Department of Cartography and Geo-Mediatechniques, Technical University Vienna, Austria

Rakkolainen I, Timmerheid J, Vainio T (2000) A 3D city info for mobile users. In: Proceedings of the 3rd International Workshop in Intelligent Interactive Assistance and Mobile Multimedia Computing, Rostock, Germany, pp 115-212

Raubal M, Winter S (2002) Enriching wayfinding instructions with local landmarks. In: GIScience 2002, Lecture Notes in Computer Science, Springer, Berlin

Reichenbacher T, Meng L (2003) Mobile Kartographie – ein Annäherungsversuch an ein neues Forschungsthema. In: Kartographische Nachrichten, vol 1, Fachzeitschrift für Geoinformation und Visualisierung

Reichl B (2003) Potential verschiedener Präsentationsformen für die Vermittlung von Routeninformation in Fußgängernavigationssystemen (FNS). Diploma

thesis. Department of Cartography and Geo-Mediatechniques, Technical University Vienna, Austria
Sorrows ME, Hirtle SC (1999) The nature of landmarks for real and electronic spaces. In: Freksa CD, Mark M (eds) Spatial information theory: cognitive and computational foundation of geographic information science, Lecture Notes in Computer Science, Springer Verlag, Berlin
Steck SD, Mallot HA (2000) The role of global and local landmarks in virtual environment navigation. Teleoperators, vol 9, pp 69-83
Thorndyke PW, Hayes-Roth B (1982) Differences in spatial knowledge acquired from maps and navigation. In: Cognitive Psychology, vol 14, pp 560-582
Zipf A, Jöst M (2004) User expectations and preferences regarding location-based services – results of a survey. In: Proceedings of the 2nd Symposium on LBS and TeleCartography, Vienna

22 Wireless Campus LBS: Building Campus-Wide Location Based Services Based on WiFi Technology

Barend Köbben, Arthur van Bunningen, and Kavitha Muthukrishnan

Abstract. This chapter describes a project that has started in spring 2005 at the University of Twente (UT) in cooperation with the International Institute for Geo-Information Science and Earth Observation (ITC) to provide Location Based Services (LBS) for the UT campus. This LBS runs on the existing Wireless Campus system that provides the whole 140 hectare University grounds with WiFi based internet access. The project serves as a testbed for research activities as well as an infrastructure to develop practical use cases upon. The former includes research into wireless LAN positioning techniques, into context awareness of ubiquitous data management systems, and into data dissemination for LBS and mobile applications. A first use case was to provide the participants of SVGopen2005, the 4th Annual Conference on Scalable Vector Graphics (August 15-18, 2005) with a location system (called FLAVOUR) to help them navigate the conference locations and locate fellow attendants.

22.1 The Wireless Campus at the University of Twente

In June 2003 the "Wireless Campus" was inaugurated at the University of Twente (UT), allowing cable–free internet access to staff and students anywhere on campus. University of Twente is a young university in the Eastern part of The Netherlands. It employs 2,500 people and has over 6,000 students. On its campus, the university has 2,000 student rooms. The university campus is situated between the cities of Enschede and Hengelo, near the Dutch–German border.

Spread over the 140–hectare campus 650 individual wireless network access points have been installed, making it Europe's largest uniform wireless hotspot. Anyone with a PC, laptop, PDA or other WiFi (wireless fidelity) enabled device can access the university's network and the internet from any building, the campus park and other facilities without cabling.

For education, the WLAN improves the flexibility and independence of time and location. This powerfully facilitates new ways of teaching. The new bachelor's programme Industrial Design, for example, now provides

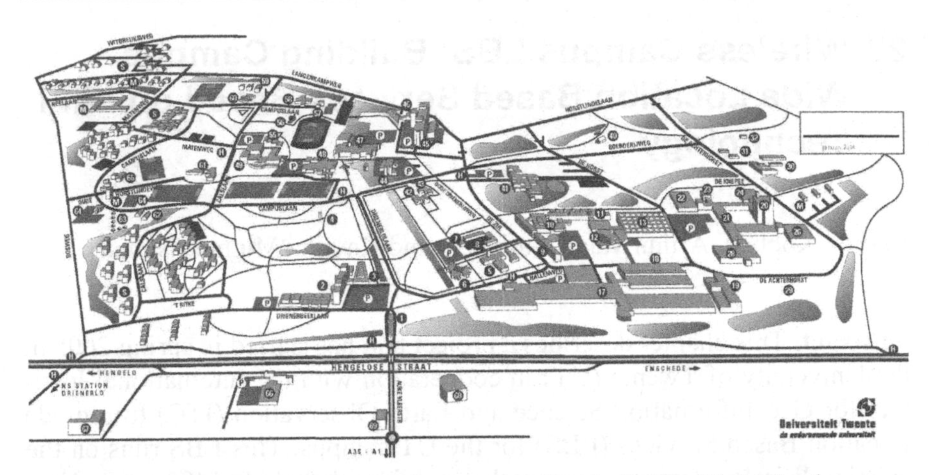

Fig. 22.1. The University of Twente campus in Enschede, The Netherlands.

its students with a laptop, to make use of all possibilities including high-performance CAD software. Students of all programmes use the so–called TeleTOP digital learning environment. These new teaching concepts also enable a more flexible use of teaching rooms.

University of Twente's Wireless Campus aims at a broad range of research and applications of wireless and mobile telecommunication. The UT wants to use the WLAN in cooperation with the adjacent Business and Science Park. Therefore this B&SP is being covered by access points as well. Furthermore, a project has just started in cooperation with the municipality to install further access points to also cover the downtown area of Enschede.

Research projects investigate the technology and the applications of wireless and mobile communication in several ways, mostly in cooperation with industrial and other knowledge partners. The Wireless Campus has become a 'testbed' for wireless and mobile applications. The major part of this research takes place at the Centre for Telematics and Information Technology (CTIT) and the research institute MESA+. Both are key research institutes of the University of Twente. CTIT is an academic ICT research institute of the University of Twente. It conducts research on the design of advanced ICT systems and their application in a variety of application domains. MESA+ is an institute that conducts research in the fields of nanotechnology, microsystems, materials science and microelectronics.

The wireless network facility was made possible with financial support of the Dutch Ministry of Economic Affairs and has been built in cooperation with IBM Netherlands and Cisco Systems. It consisted in first instance mainly of access points that use the 802.11b wireless networking standard, offering a data transfer speed of 11 megabits per second for most users.

However, upgrading the entire network to run using the new 802.11g standard, providing data at speeds up to 56 megabits per second, is an ongoing effort.

22.2 Positioning using WiFi technology

Using WiFi technology for positioning is just one of the many wireless techniques available for positioning of mobile users (others are eg. GPS, Bluetooth or Infrared, and mobile telephony). There are three basic methods for determining the location of users (Kaemarungsi and Krishnamurthy 2004): (a) triangulation that requires at least three distinct estimates of the distance of a mobile device with a WiFi receiver from known fixed locations, (b) using the direction or angle of arrival (AOA) of at least two distinct signals from known locations and (c) employing location fingerprinting schemes. In indoor areas, the signal will almost always be reflected from various objects (like walls) and because of this multipath environment, techniques that use only triangulation or direction might not be very reliable. Location *fingerprinting* refers to techniques that match the fingerprint of some characteristic of the signal that is location dependent. The fingerprints of different locations are stored in a database and matched to measured fingerprints at the current location of a receiver. In WLANs, an easily available signal characteristic is the received signal strength (RSS) and this has been used for fingerprinting. But the RSS is a highly variable parameter and issues related to positioning systems based on RSS fingerprinting are not understood very well. The big advantage of RSS-based techniques is that we can use the existing infrastructure to deploy a positioning system with minimum additional devices. It is far easier to obtain RSS information than the multipath characteristic, the time or angle of arrival, that require additional signal processing. The RSS information can be used to determine the distance between a transmitter and a receiver in two ways. The first approach is to map the path loss of the received signal to the distance travelled by the signal from the transmitter to the receiver. With the knowledge of the RSS from at least three transmitters, we can locate the receiver by using triangulation.

In the Wireless Campus LBS project the positioning component is part of a wider PhD research into a variety of positioning techniques for LBS. This WiFi based component will build upon an earlier test done in 2004 for two specific buildings on the University campus. In this project, called "FriendFinder" (Bockting 2004), a prototype client-server architecture was built, where the client program on the mobile device determines its location with respect to the Access Points (APs) by determining the RSS-s and comparing them with data about the APs that are in a server-side database. This database stores in the first place the location in XYZ of all APs inside the two buildings chosen, their BSSIDs (the unique identifier of an AP), and their antenna signal strength. Furthermore, maps of the buildings are stored for use in the Graphic User Interface (GUI) of the client application.

The client application first buffers the RSS measurements because not all APs are detected in any single scan. Then it detects probable faulty measurements and deletes them. The accepted measurements are then put through a filter that calculates their centroid. Now the client has a first estimate of its position. Further filtering takes place, using among others standard deviations and maximum likelihood calculations, to get a better estimate of the position and the final estimate is determined by so-called "iterative multilateration". In this technique a clients position, with its estimated inaccuracy, is used by other clients as a reference frame. In that way all nodes use each others information to jointly improve the accuracy of the positioning. An important part is played by further filters that implement a learning effect from the stored positioning history of the application to achieve further improve the accuracy.

By using the XML–based Instant Messaging protocol "Jabber", the client applications can communicate and relay their positions to each other and show them in the GUI by placing symbols on the building maps mentioned earlier.

Tests have shown that the average positioning accuracy this first prototype could reach was just under 5 meters (4.6m), for non-moving devices. The system provides the user with an estimation of the current positioning accuracy. One of the research tasks for the Wireless Campus LBS project will be to reach better accuracy of positions. For that, a more precise determination of the locations of the APs and their properties is needed, covering this time the whole UT campus.

22.3 Mapping the Access Points

For the FriendFinder project mentioned above, only a limited number of the Access Points (APs) have been used. As no geoscientists where involved at that stage, their positioning was done in a rather improvised way. The height of the APs especially was a problem, it was determined only by estimate and with respect to the building's ground floor height. In this limited project that was not a big problem, as only one building was involved, but for the larger project the elevation differences between the buildings (more than 5 meters, which is a lot for the Dutch!) will have to be taken into account.

The 650 individual wireless network APs that have been installed are currently only indicated on paper maps, one map per floor, of the individual buildings of the University. These are print-outs from CAD–drawings ("blue–prints") maintained by the Facility Management Services that have a high level of detail, but they are not georeferenced and thus have a local, arbitrary, coordinate system that's basically just 'paper coordinates'. Furthermore, the location of the APs has been indicated haphazardly by hand-drawn symbols at the time of installation of the devices.

Therefore the first task, starting February 2005, has been the digital mapping of the AP locations in a geodatabase. In order to do this, it was decided to digitise all locations using GIS software and digitally georeferenced versions of the CAD-drawings. The georeferencing was achieved by transformation of the CAD drawings, using control points from an overview map of the whole campus that is available in the Dutch national coordinate system "RijksDriehoeksstelsel" (RD). First test have determined that it is possible, when using simple first order transformation, to achieve RMS errors of less than 0.1 meter.

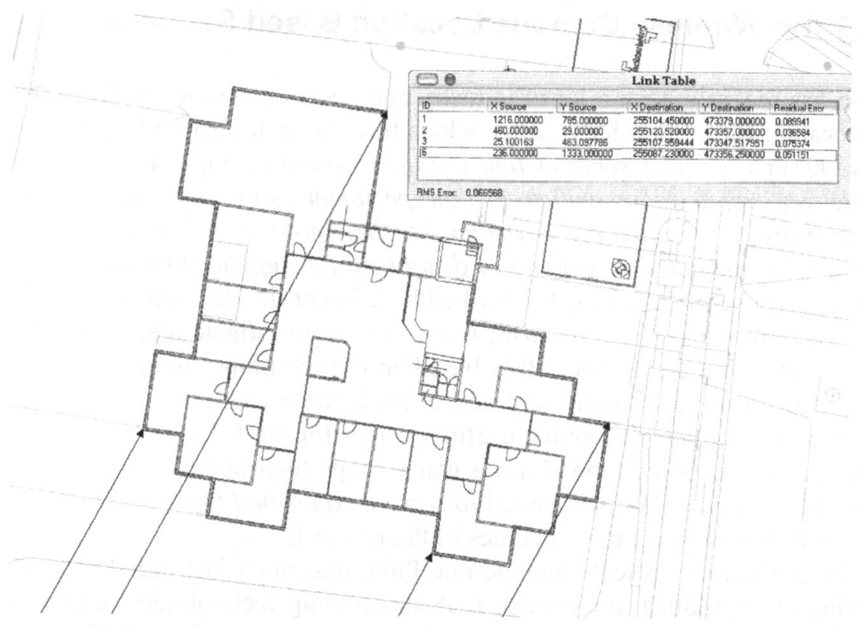

Fig. 22.2. Screenshot of the GIS used to map the APs. It shows one of the building CAD drawings (darker colours) after georeferencing on the UT overview map (light colours). The arrows show the control points, the link table depicts their unreferenced coordinates and their equivalents in RD, as well as the residual errors (all in m).

For all buildings a base elevation will also be determined in meters above NAP (the Dutch vertical datum) by combining the campus map with the Actual Height model of the Netherlands, a detailed elevation model of the whole country made by airborne laser altimetry, which has a point density of minimal 1 point per 16 square metres and a systematic error of 5 centimetres maximum (AHN site: http://www.ahn.nl/english.php). In order to get precise location measurements, it was deemed necessary to physically visit all APs and use a laser measurement device to determine the relative location of the AP antenna with respect to the elements of the building present in the CAD drawings (walls, floors, windows). The height of each AP, measured from the floor or ceiling, will be combined with a determination of that floor or ceiling's height from the base elevation of the building. By combining all these relative measurements with the georeferenced maps a precise XYZ location has been determined and put into the geodatabase. The added bonus is that all APs have been checked and additional attributes were gathered, such as antenna type, antenna connection length for estimating signal loss, etcetera.

22.4 The Wireless Campus Location Based Services

Their has recently been a lot of industry and research activity in the realm of Location Based Services (LBS), which have been defined in Urquhart *et al.* (2004) as *wireless services that use the location of a (portable) device to deliver applications which exploit pertinent geospatial information about a user's surrounding environment, their proximity to other entities in space (eg. people, places) and/or distant entities (eg. destinations)*.

The purpose of the project described here is not the development of *the* or even *a* Wireless Campus LBS, but rather to investigate and set up the infrastructure necessary for LBS's based on it. It combines input from several research projects with the practical application of new as well as established techniques to provide useful services for the UT campus population. The research mentioned has a wider scope then just this project: the Wireless Campus LBS is intended *to serve as a testbed for* the research as well as *to benefit from* the outcomes of the research.

These research projects include one PhD, described in Muthukrishnan, Lijding *et al.* (2005), on various LBS positioning technologies, that will look, among other things, into improving the accuracy of the WiFi positioning. To achieve this, the research investigates the positioning algorithms, the filters and methods used, and also the effects of signal-reflecting obstacles on the measurements. These obstacles, such as walls

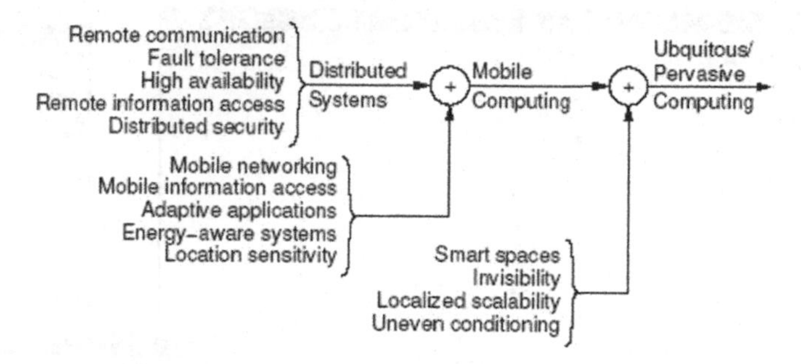

Fig. 22.3. Evolution from distributed to ubiquitous computing (reproduced from Satyanarayanan (2001)).

and pillars, are included in the geodatabase and could therefore be accounted for in the positioning algorithm. Another area of further research will be the self-learning abilities of the system, that should theoretically make the positoning more accurate over time.

Another PhD concentrates on the impact of context awareness on ubiquitous data management. This research deals with consequences that the evolution from *distributed computing*, via *mobile computing* to *ubiquitous* or *pervasive computing* (as shown in Fig. 22.3) is having on data management issues. Context-awareness is thought to be a major requirement for computer systems to be ubiquitous. In a recent paper resulting from this research (van Bunningen *et al.* 2006), a design is presented of a context-aware data management supporting platform. One of the important characteristics of context, and therefore of the supporting platform, is spatial information, and the spatial context information provided by the Wireless Campus LBS will be used in implementing said platform. Another factor of context for any system is the (un)certainty of the information it provides, and providing the user with relevant information about that uncertainty will also be part of the CampusLBS services.

On the client-side of the system, ongoing research on data dissemination for LBS and mobile applications (Köbben 2004) will be concentrating on the Wireless Campus LBS as a testbed for adaptive, task-oriented delivery of mapping information to mobile users.

22.5 Test at SVG Open 2005

The first use case test of the Wireless Campus LBS was to provide the participants of a conference held at the UT grounds this summer (August 15–18, 2005) with an LBS to help them navigate the conference locations and

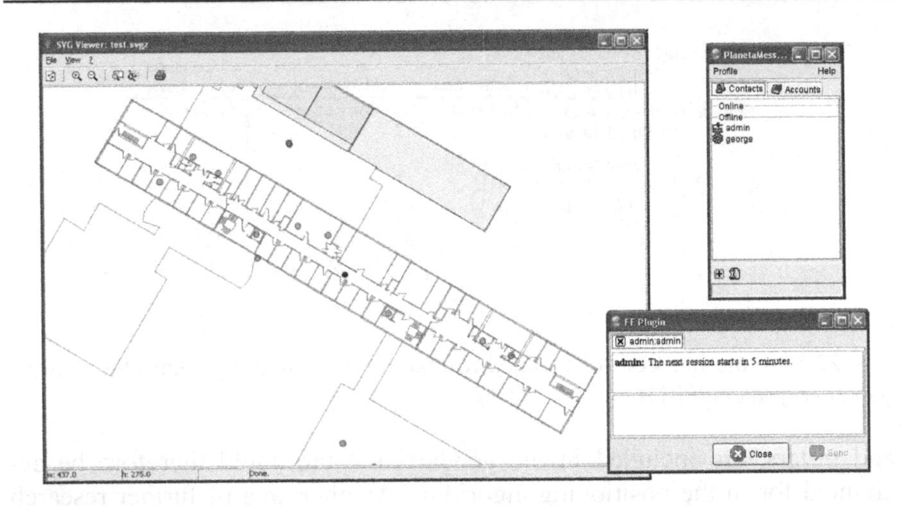

Fig. 22.4. Graphical Interface of FLAVOUR.

locate fellow attendants. This conference, SVGopen2005, the 4th Annual Conference on Scalable Vector Graphics (SVGopen site: http://www.svgopen.org), was deemed to be a good testbed as it drew a crowd of some 170 people from 20 countries all over the world, from a very wide field of applications: electronic arts & media, geospatial sciences, information technologies, computer sciences, software developers, Web application designers, etc. They share an interest in Scalable Vector Graphics (SVG), the W3C open standard enabling high-quality, dynamic, interactive, stylable graphics to be delivered over the Web using XML. Most of them are technology–oriented and there is a high degree of interest in, and ownership of, mobile devices.

The application built for testing by the participants has been called FLAVOUR (Friendly Location-aware conference Assistant with priVacy Observant architectURe). Services offered by FLAVOUR can be categorized into:

- *Pull* services, in which location of attendants play an important role as the attendants' request will be replied by the system on the basis of their whereabouts. Examples of pull services offered are:
  - Finding your own location and that of fellow attendants;
  - Locating resources available in the infrastructure such as printers, copiers, coffee machines etc.
- *Push* services, in which individual and bulk messages are sent to the attendants. This enables the attendants to:

– Be notified about important events by conference organizers;
– Communicate with their contacts, i.e., colleagues, friends, etc.

The architecture, described in more detail in (Muthukrishnan, Meratnia *et al.* 2005), is based on a Location Manager, which provides services using the Jini platform (Jini platform: http://www.sun.com/jini). Each Location Manager registers with the Jini Lookup Service to offer the location of the user it represents. Interested users can look up the service and subscribe to the location of a given conference participant. This is done using publish-subscribe mechanism. The Location Manager uses a privacy policy to decide if a client is allowed to subscribe to the location of its owner (publisher). It also publishes to all the subscribers relevant changes in the location of its owner.

The Jini architecture also provides other kinds of services, such as the message board to which every conference participant can subscribe. The message board is used by the conference organization to publish changes in the schedule, information related to the social events, etc. Participants can also use the message board to make announcements to the other participants, as for example asking about lost objects, or to chat.

The graphical depiction of the maps and the location of the users is done in SVG, providing vector graphics in high graphical quality with a small memory and file footprint.

The tests at SVG Open 2005 were relatively succesful: Most conference participants experimented with the localisation features of the system. The messaging and friend-finder functions were used to a lesser extent. Various extensive interviews have been held with test persons and also written feedback was collected. The localisation functionality worked quite reliably, although the accuracy was varying quite a bit over the various conference locations. In the computer science building the results were clearly better then in the main conference halls. The tests still have to be analysed further, but the most obvious reasons are the non-optimal configuration of access points and the fact that the database of these access points still was incomplete at the time of testing.

22.6 Outlook

The implementation of the Wireless Campus LBS described in this chapter has only just started. But as it builds on the solid foundations of the well-established infrastructure of the Campus-wide WLAN at the University of Twente, and has had a successful pilot in the FLAVOUR tests at SVG

Open 2005, we expect that it will be put into use and expanding relatively quickly in the coming years.

Probably the most exciting aspect of the project is the fact that it provides the opportunity for a very diverse group of people from quite different disciplines to contribute to a technical infrastructure that can serve as a testbed for their respective researches, and at the same time has the potential to become a useful everyday feature for mobile users at the University Campus.

References

Bockting S (2004). FriendFinder project, localisatie met wireless LAN. University of Twente, Enschede

Kaemarungsi K, Krishnamurthy P (2004) Modeling of indoor positioning systems based on location fingerprinting. In: Proceedings of INFOCOM 2004, the Conference on Computer Communications. IEEE, pp 1013-1023

Köbben B (2004) RIMapper - a test bed for online risk indicator maps using data-driven SVG visualisation. In: Proceedings of Location Based Services and TeleCartography. Institute of Cartography and Geo-Media Techniques, Wien, pp 189-195 http://kartoweb.itc.nl/kobben/publications/RIMapper_paper_Vienna2004.pdf

Muthukrishnan K, Lijding ME, Havinga P (2005) Towards smart surroundings: enabling techniques and technologies for localization. In: Proceedings of LOCA2005 – co-allocated with the 3rd International Conference on Pervasive Computing. Munich, Springer Verlag, p 11

Muthukrishnan K, Nirvana M, Koprinkov G, Lijding M, Havinga P (2005) SVGOpen conference guide: an overview. In: Proceedings of SVG Open 2005. Stichting SVG Nederland, Enschede, http://www.svgopen.org/2005/papers/SVGOpenConferenceGuideAnOverview/SVGOpenConferenceGuideAnOverview.pdf

Satyanarayanan M (2001) Pervasive computing: vision and challenges. IEEE Personal Communications, IEEE, vol. 8(10-7)

Urquhart K, Miller S, Cartwright W (2004) A user-centered research approach to designing useful geospatial representations for LBS. In: Proceedings of Location Based Services and TeleCartography. Institute of Cartography and Geo-Media Techniques, Vienna, pp 69-78

van Bunningen A, Feng L, Apers PMG (2005) Context for ubiquitous data management. In: Proceedings of the International Workshop on Ubiquitous Data Management (UDM2005), IEEE, Tokyo

23 Developing Web-GIS Applications According to HCI Guidelines: The Viti-Vaud Project

Jens Ingensand

Abstract. A variety of different projects makes it possible to develop full-functional web GIS-applications. Nowadays most of these projects focus on one specific part of a system (e.g. a map-engine, scripts for data-visualization etc). Although certain GIS projects concentrate on the development of web-interfaces, some problems arise (e.g. consistency-problems) when these projects are compiled and modified to fit a certain context. Furthermore the development of a system according to Human Computer Interaction (HCI) guidelines involves the participation of the end-user at almost all stages of the process. This chapter describes the adaptation of HCI guidelines during the development of a specific web-GIS application, based on open source-GIS projects: an interactive system for wine cultivation in the Swiss canton of Vaud. Based on this work, some design suggestions for developers of web - GIS projects are presented.

23.1 Introduction

An increasing number of people can today interact with geographic data through the internet. The use of this medium as an interface for geographic information systems (GIS) implies that GIS are being opened for a wider public who not necessarily are experienced GIS-users. This development is part of a democratization process of geographic information; most common examples for this process are online mapping-systems (e.g. Swissgeo: http://www.swissgeo.ch, or Map24: http://www.ch.map24.com). Making geographic information available for everybody puts the term "usability of GIS" in another light as most standard desk-top-GIS have been designed to support only one experienced user at a time (NRC 2003). For these systems literature about and studies regarding the usability is available. Brewer (2002) but for web-GIS specific theories and methods are rare.

In this chapter I will describe the development of a web-GIS application with a Human-Computer Interaction (HCI) perspective. HCI is a research field that occupies with the usability of computer systems and the development according to user requirements. The system that I have developed is a prototype of an interactive system for winegrowers in the Swiss canton of Vaud. The system is not only a pure mapping system for the consultation of online-maps, but also a tool to input spatial information.

The outline of this chapter is the following: In section 23.2 the relevance of the HCI perspective within the development of web-GIS will be discussed in order to clarify the context. Section 23.3 describes the development of a specific web-GIS application. In section 23.4 suggestions and ideas for further developments, based on the experience from the development of the prototype and also from the evaluation of this prototype, will be presented. Finally section 23.5 describes conclusions and future work.

The goal of this chapter is to discuss the importance of HCI for web-GIS applications and to present my ideas for developers in order to improve the usability of such applications.

23.2 Related Work: HCI Aspects for the Development of Web-GIS

HCI is a field of research that emerged in the 80's through a combination of different disciplines: computer sciences and applied social and behavioral science (Carroll 2001). Today there are different theories and methods that help the developer to reach the most important goal: *"The reduction of the users' cognitive efforts when interacting with the system"* (Preece *et al.* 2002). By minimizing the user's cognitive efforts, the use of the system becomes as efficient and pleasant as possible and the usability is maximized. The term usability has been defined as (ISO 1998): *"The extent to which a product can be used by specified users to achieve specified goals with effectiveness, efficiency and satisfaction in a specified context of use."* It is therefore important that the user intuitively can understand what he can do with the system without needing to start with a help file. There are three important elements of HCI theories and methods for the development of web-GIS:

- The user centered system development: this method framework is an essen-
- tial part of HCI. It helps the developer to identify the user and the users
- needs;
- The design of the user-interface. e.g. interface concepts and metaphors, display and interaction techniques, and software development methods.
- Cartographic and webmapping methods.

According to the user centered system development, the development process starts with the identification of the users (Fig. 23.1), e.g. in which situation the users would use the application and what it could help the users with. Interviews and inquiries are examples for strategies or methods to identify the user.

Further, the contact with the users must result in a specification of the users' requirements. The specification is a list of the desired functionalities, but should also contain some reflections about the design and the technologies that can be used during the development. The next step is the development of one or more prototypes. These prototypes can finally be evaluated and validated. If the users aren't satisfied with the final prototype the development process starts again with a new analysis of the users needs (Preece *et al.* 2002).

According to Gould & Lewis (Gould *et al.* 1997) a user-centered approach to system development and design should be preferred in all situations. There are two main reasons for this: End users are experts on their work and therefore the only ones that can describe it, and end users are the ones that are most suitable for testing and evaluating prototypes and systems that are developed for them. Further system developers often believe that they can find the perfect design for a system from the first try, but in reality, a good design involves continuous iterations of the design-model described.

There is a difference between the interaction with a standard desktop-GIS application and the use of a web-GIS application: the latter shouldn't require knowledge of a GIS-expert. A standard GIS-application usually requires the use of specific knowledge because standard GIS are not devel-

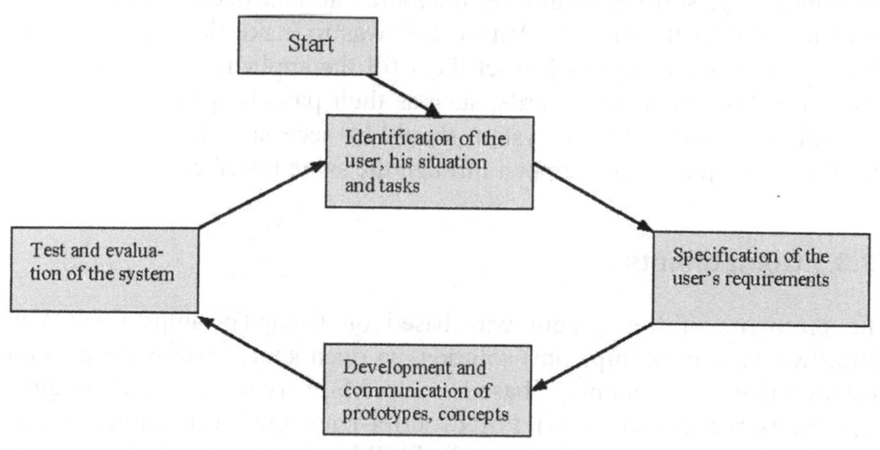

Fig. 23.1. The user-centered system development (Preece *et al.* 2002)

oped for a specific purpose or for a certain kind of data. The interaction with a web-GIS application implies that the user can only modify and visualize the accessible data within the framework that the developer defines (Räber and Jenny 2001).

As the map is the central part of any web-GIS - interface, it is also the main element of interaction. For users that are not familiar with GIS, the interaction with computer maps is relatively new, as paper maps had been dominating. Map design is also highly subjected to human perception (MacEachren 1995). The reality usually contains much more information that can be visualized on one single map and certain purposes require a certain amount of abstraction. Therefore it is necessary to generalize and abstract the reality by omitting less important things and summarizing similar things. Map representations should further match the user's cognitive map– the user's perception of a certain context (Tezuka *et al.* 2001).

23.3 Development of a Web-GIS Interface According to HCI Guidelines

23.3.1 Context

The goal of the project "Viti-Vaud" at the GIS Lab at EPFL was to develop a prototype of a web-GIS application for the winegrowers in the Swiss canton of Vaud, using open source web-GIS components and HCI theories and methods. This project was part of a project called "terroirs viticoles" (wine soils) (Pythoud and Caloz 2001) where different layers containing e. g. soil types and the microclimate had been gathered. Thus one main idea of the project "Viti-Vaud" was to make these layers available for all winegrowers and to let them fill the application with other information about their wine-yards, such as their parcels spatial extent, harvest information etc. As the system should be accessible through Internet, also the data-input should happen through the same interface.

23.3.2 Components

The prototype of the system was based on CampToCamps CartoWeb (http://www.camptocamp.com) solution, an open-source web-visualization system that is mainly based on MapServer (a map-engine, http://mapserver.gis.umn.edu), PhP (a script-language for dynamically creating html-pages, http://ch.php.net), PhPMapScript (the link between MapServer and PhP, http://www.maptools.org/php_mapscript), and PostgreSQL/ PostGIS (a spatial database-system, http://www.postgresql.com and http://postgis.refractions.net). Dynamic PhP scripts and the protocol XMLRPC (http://www.xmlrpc.com) connect the different parts of the system. Java (http://java.sun.com) is used to visualize the map and to permit the input of spatial data (Fig. 23.2).

In the standard CartoWeb application a user interface had already been partially implemented, e.g. a navigable map (in two versions - as html version and as a Javaapplet (the Rosa Java-applet, http://www.maptools.org/rosa) Further a part of the interface showed all available layers and permitted to choose the layers to be displayed. An overview-map showed the position of the main map and enabled the user to change the main map's position through clicking on it.

23.3.3 Requirements

In order to adapt and enhance the system, the development process began with an analysis of the user through interviews and enquiries. One first goal was to list and specify – both functional and user requirements for the project.

The main functional requirements for the system were:

– The consultation of interactive maps (including the possibility to query

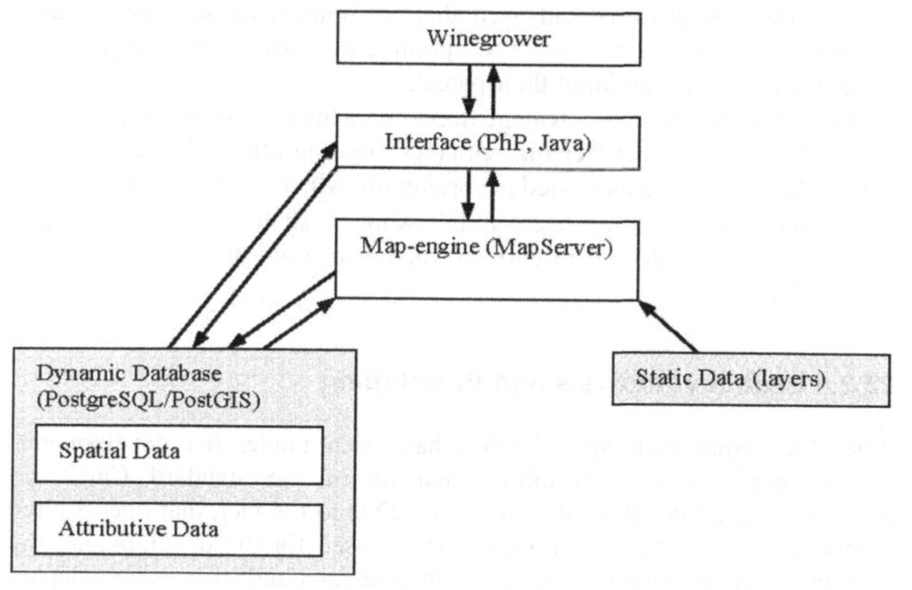

Fig. 23.2. Functional model of the System

objects);
- The input of spatial data and attributive data directly through the web-interface.

To find out about the users requirements, some initial questions were formulated:

- How familiar are the users with computers?
- Do they use computers to record data about their wine-cultivation?
- Do the users have a connection to the Internet?
- Do the users have experience in searching spatial information on the Internet?
  Which data are the users interested in?
- Which data the users would agree to share?

Based on these questions, an inquiry was formulated and sent out to 35 winegrowers in the region (28 sent back the inquiry). During the same period, interviews with a researcher at the federal wine-research institute were conducted. These sources permitted to establish requirements. The most important ones were:

- An optimization according to the most common systems. (e.g. almost all winegrowers were using Microsoft Windows and Internet Explorer)
- Optimization of the spatial input function: The inquiry showed that only very few winegrowers had used their computers to draw or to paint, therefore it seemed necessary to optimize the part of the system where the winegrowers can input their parcel;
- Data classification: the winegrowers were more interested in certain data from the project "terroirs viticoles" than in other. Hence the available data should be classified according the winegrower's interest;
- Restricted usage: some users didn't want to share all their data, thus they should be allowed to choose whether or not other users could see their data.

23.3.4 Usability Analysis and Prototyping

After the requirement-specification had been made, the development-process began with a usability analysis of the standard CartoWeb (http://www.camptocamp.com) interface. Due to the fact, that open-source components from different projects where used for the development, the usability of these components had to by analyzed and, if possible adapted to the given context. For this analysis Ben Shneiderman's golden rules (Shneiderman 1998) are a good starting point. They uncovered some problems which probably could have caused problems for the user. Below I give two examples of such interaction-problems:

The first problem was not only revealed in CartoWeb's (http://www.camptocamp.com) compilation of open-source software, but also in other projects, based on the same components. The user's interaction with the map was different than with the surrounding elements. Clicking on the map automatically recharged and updated the map, but e.g. choosing a different theme required to click the button "reload map" (Fig. 23.1a). An option would have been to omit this button.

The second issue that had been discovered is a very good example of programmer-centered system development: the option allowing switching between the html-mode of the interface and the java-applet mode. A button named "Java mode disabled – click to enable" (Fig. 23.1b) with a wizard-figure indicates these choices. A normal user (and even someone who knows what Java in a computer-context is) can impossibly figure out what this button means or what it does. It doesn't say anything about the choice to zoom in using a rectangle. In this case it would have been helpful to just change the icon to e.g. a rectangle with a zoom-lens and to write "rectangle zoom mode" under the icon.

After interaction-problems in the basic application were recognized, the development process continued with the implementation of the prototype:

Due to a large amount of available layers, these ones had to be categorized. In many visited web-GIS (e.g. EcoGIS, http://www.ecogis.admin.ch), themes had often been categorized using a tree-structure. Through a "+"-sign, different parts of this tree-structure could be evolved. This structure unfortunately had the disadvantage that the tree could grow quite tall, and that the amount of information could be confusing. For the development of the prototype, the different layers had been categorized into different groups of layers. These groups were visualized in different frames that the user could switch between.

One further user requirement was to restrict the access to a part of the system where the winegrower could add his parcels as winegrowers some-

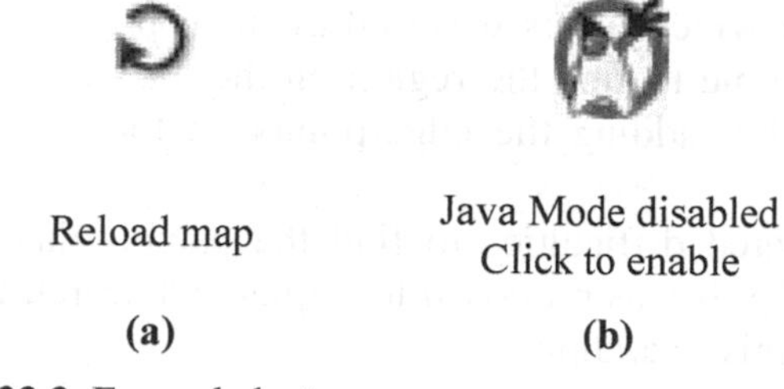

Fig. 23.3. Example buttons.

times don't want to share their data. Therefore the interface was enhanced with a login segment. In this new section, the winegrower could add/modify and delete his data. One difficult part in this section was to create a module where the winegrower could share the spatial extend of his parcels through drawing them directly on the map. This action is a standard action in most of all common GIS, but for the development of web-GIS, few modules exist. For all data that had been input, the creator and owner can choose whether or not this information to be accessible for other winegrowers that are interested in his data.

23.3.5 Evaluation

After the prototype had been implemented, the system was evaluated and validated by four winegrowers. During the evaluation the users were given some tasks that they had to accomplish. Basically the evaluation consisted of some map-navigation exercises, some queries, measuring distances, drawing a sample parcel on the map and adding data to a parcel. The developer was present, but was asked to assist only in case of problems. Main results of this evaluation were:

- The navigation buttons (e.g. zoom-in, re-center, query) were recognized, but sometimes their usage wasn't understood correctly. After a while, users became used to the zoom functions. Especially the function to draw a rectangle in order to zoom in was found to not be intuitive. An idea proposed by one winegrower was to explain the zoom tools better through text fields. One winegrower used the scale choice menu and the arrows around the map to navigate, instead of the navigation tools. Another winegrower wanted to have a panning function instead of a re-center function where the map follows the mouse-pointer;
- All winegrowers encountered problems with the function to measure distances included in the Rosa Java applet (http://www.maptools.org/rosa). It was proposed to display the result of a measurement near the map and not in the status-bar.
- The most difficult task for winegrowers was to draw their parcel on the map. It didn't take much time to find the region on the map where the parcel should be drawn, but adding the edge-points on the map was found to be un-intuitive.
- All winegrowers encountered difficulties to find the parcels that they just had created on the map. It was proposed to implement search functions in order to facilitate this operation.

23.4 Ideas for Web-GIS Developers

In this section I want to present some ideas for web-GIS developers that can be helpful for the development of future systems. These ideas are based on my experience from the Viti-Vaud project and the comparison with related work.

Generally, the development of a whole web-GIS application from scratch would require enormous resources. For a developer, it is therefore necessary to choose existing components and to assemble them into a functioning system. A web-GIS application is therefore in most cases a composition of different projects, systems, modules and specifications. This causes two problems:

- The inclusion of the user in the development process. In order to develop a system strictly according to the user's requirements it is in an ideal case necessary that the user is part of the whole design process. As certain components are already developed partly or entirely, the developer has to compromise;
- The adaptation of HCI-guidelines within this composition. The developer has to choose the design-pattern that can be adapted for the whole system. Usually many components have been implemented with a certain design and certain interaction-patterns. In some cases it can be difficult to change or to adapt these patterns.

There are further some design patterns developers can focus in order to create a usable application:

- *Consistency:* Ben Shneidermans (1998) first rule of good interaction design is "strive for consistency". A consistent application forms a unit where all its elements have the same physical design and the same manner of interaction. This rule is very important for the design of a web-GIS application based on different modules;
- *Feedback, error handling and labels:* The system should always give feedback to all the actions that the user takes. Error messages should be understandable for the user. System-developers should not make the mistake to consider error-messages as a help for debugging the code but as a help for the user to understand why something went wrong. Also buttons and other components such as menus and icons should always give the user an idea what they do. These components can often be easily and quickly modified, and a proper choice of text messages and icons can considerably reduce the users cognitive efforts;
- *Compatibility:* web-GIS are often using "higher" programming and visualization methods such as Flash, Java, Javascript in order to main-

tain functionality, that is not supported by standard html-pages. These programming languages can cause compatibility problems with some operating systems or browsers. In some cases, the user has to install plugins in order to view the data. Compatibility problems can be very annoying for the user and sometimes even inhibit the use of the system. System developers must be aware of this problem, but as always compromise in some cases. One way to compromise is for example to investigate the most common systems and to optimize the web-GIS application to these systems.

Other practical ideas that can help developers during the process are:

- *Contact with the users:* For a system developer, the contact with the users should be an important part of the process. Already a few users can specify usability requirements and during an evaluation reveal many interaction and design problems that programmers maybe never had thought of. In case one user has a problem it should be made clear that his problem, is not his incapability to handle a system but the programmer's problem to make it understandable.
- *Offering the user the possibility to give feedback:* As described earlier, the user-centered system development is a continuing process. Offering the user the possibility to give feedback about the system through forms or email can help the developers to discover problems within their system.

23.5 Conclusions and Future Work

Developing web-GIS application according to HCI guidelines is a research field that hasn't been investigated very much yet. One reason is that the development of web-GIS (on the contrary to standard GIS) is an emerging field of research. Another reason is that it is difficult to define the whole web-GIS system as one product such as a stand-alone GIS. In most cases a web-GIS is a composition of different components, developed by different teams and sometimes for different purposes. Further a web-GIS system always implies multiple users that not necessarily have expert GIS knowledge. This reason validates and justifies the use of HCI theories and methods even more.

The development of the web-GIS application for winegrowers showed that all available components that were integrated in the user-interface already had some sort of given interaction and design patterns. Some components were easier to modify than others. It is therefore important that the developer chooses the components of the web-GIS application to develop not only from a functional point of view, but also from the user's perspective. Further the evaluation of the prototype showed that many users often think differently about the interaction than developers and that it always takes some time to learn how a system works. I think that web-GIS are special in that way that it is the first time, that the interactive use of geospatial information is getting available for everybody. On one hand users have to get used to this interactivity, on the other hand developers have to think of the user and adapt their application to user's requirements. Small changes and adaptations of the user's requirements are quite easy to realize and do not take much time. Such small changes can have a profound effect from the perspective of HCI. On the contrary the complete adaptation without compromises of the user's requirements can imply much more work.

All requirements that the system was based on, were specific for this project. A similar system for another purpose could be an interesting enhancement of the project. However it must be stressed that the requirements for the development of a system are basically of two kinds – functional requirements and user requirements. A system for e.g. hikers would certainly change functional requirements (e.g. the possibility to add a track instead of a parcel, etc), but maybe less general interaction-patterns On the other hand a system for e.g. land surveyors would most likely change interaction-patterns.

The use of open-source components for web-GIS can both be an advantage or a disadvantage for the design of a system. Open-source products are completely adaptable and extensible, but these products are not designed according to user requirements, but rather to functional requirements.

The Viti-Vaud prototype is the cornerstone of a new project at the GIS-Lab called "Réseau interactif en viticulture" (http://lasig.epfl.ch/projets/riv/) (Interactive network for wine-cultivation) that includes federal research institutes, the regional government and winegrower-associations. Within this project a new system will be developed, providing statistical functions and interfaces for all actors involved in the wine-cultivation in the region (e.g. wine cellars, wine-shops, etc).

References

Brewer I (2002) Cognitive systems engineering and GIScience: lessons learned from a work domain analysis for the design of a collaborative, multimodal emergency management GIS, pp 1-3

Carroll JM (2001) The evolution of human-computer interaction. Addison Wesley, http://www.awprofessional.com/articles/article.asp?p=24103

Gould JD, Boies SJ, Ukelson J (1997) How to design usable systems. In: Helander M, Landauer TK, Prabhu P (eds) Handbook of human-computer interaction, Elsevier Science B.V, Amsterdam, pp 231-253

ISO (1998) International Organization for Standardization, 9241-Part 11, http://www.iso.ch, Geneva, Accessed January 2004

MacEachren AM (1995) How maps work. Representation, visualization, and design. The Guilford Press. New York, London

NRC (2003) IT Roadmap to a geospatial future. Committee on Intersections Between Geospatial Information and Information Technology, National Research Council, pp 73-104, http://www.nap.edu/books/0309087384/html/

Preece J, Rogers Y, Sharp H (2002) Interaction design: beyond human-computer interaction. Wiley, New York pp 18-19, 165-170

Pythoud K, Caloz R (2001) Etude des terroirs viticoles vaudois – Rapport d'avancement. Laboratoire de SIG – EPFL, Lausanne

Räber S, Jenny B (2001) Attraktive Webkarten – ein Plädoyer für gute Kartengrafik. Institute of Cartography, ETHZ, Zürich pp 2-4

Shneiderman B (1998) Designing the user interface. 3rd Edition, Addison-Wesley, Reading Massachusetts, pp 75

Tezuka T, Lee R, Kambayashi Y (2001) Web-based inference rules for processing conceptual geographical relationships. In: Claramunt C, Winiwarter W, Kambayashi Y, Zhang Y (eds) Proceedings of the 2nd International Conference on Web Information Systems Engineering, vol 2, IEEE Computer Society, Los Alamitos, California

24 High Definition Geovisualization: Earth and Biodiversity Sciences for Informal Audiences

Ned Gardiner

Abstract. Geographers can play an essential role in interpreting Earth and Biodiversity Science concepts to general audiences. Museums and informal science institutions have begun to use visualizations of geospatial data within high definition video (HDTV) programming to present complex science concepts to audiences of varying ages and varying scientific training. The diversity of experience and understanding among audiences imposes many demands on geographers working with HDTV, for they must simultaneously provide content of interest to novices and experts alike. This chapter introduces essential guidelines for producing HDTV geovisualizations while emphasizing the particular talents that geographers bring to bear on time series visualizations involving geospatial data.

24.1 Introduction

Geovisualization is at the top of research agendas in the public sector (McMahon *et al.* 2005) and among academic consortia (McMaster and Usery 2005) because of its power to help communicate complex relationships among data and, implicitly, the spatial context of such data. The literature on how and why to design and implement geovisualizations is heavily skewed toward applications aimed at expert users. For example, Buckley and colleagues (2005) emphasize applications of geovisualization that focus on analysis, knowledge discovery, data mining, and cognition of high-level information from spatial data. This emphasis is logical because experts who are familiar with conventional means of displaying spatial data can perceive subtle details that are revealed through spatial and temporal patterns intrinsic to geovisualizations. Further, innovations in geovisualization are typically advanced by geoinformatics specialists. Such expert users of spatial data have a great deal to gain by optimizing the methods by which they themselves view spatial and multivariate data. Since geographers have a long-standing interest in map use, design, and interpretation, they continue to push frontiers in both cognitive theory and algorithm development for data processing and display as they are applied in many visualization contexts (see MacEachren 2004).

Sustained effort has led to remarkable advances in commercially and publicly available visualization software. Commercial, off-the-shelf geographic information systems (GIS) and remote sensing image processing software packages provide very powerful tools for visualization in both two and three dimensions. A few examples are listed in Table 24.1 for illustrative purposes; a complete listing would involve most GIS and remote sensing packages. Visualization of spatial data is an important way that developers of such data can evaluate the accuracy or efficacy of their products. Research, decision support, and planning in many disciplines have all increased demand for geovisualizations, but they have also led to innovation in how tools are used and in software development. Geologists have adapted visualization tools to examine outcrop patterns (Thurmond *et al.* 2005), stability and instability of various surfaces on varying slopes (Giardino *et al.* 2004), the three-dimensional structure of aquifers (Artimo *et al.*, 2003), and ocean bathymetry (Schlitzer 2002). Geovisualization has long been applied to ecosystem management for its utility in exploring alternate management scenarios (Kato *et al.* 1997, Schmid 2001, Chertov *et al.* 2002, Twery *et al.* 2005). For example, Chertov and colleagues (2005) used visualization to evaluate effects of varied forest management scenarios upon carbon storage and wildlife habitat.

Experts are not the only users of geovisualization products, however. Because navigation and exploration software programs are available for free download to anyone with an interest in spatial data, people with varying understanding of spatial data processing and interpretation can now access complex spatial data through the Worldwide Web. For example, the United States' National Aeronautic and Space Agency's (NASA's) Worldwind navigation software (http://worldwind.arc.nasa.gov/) and the commercial application Google Earth (http://earth.google.com/) both allow client software to query and export data and imagery from a server via the internet. Worldwind allows one to load and navigate through many types of data that researchers use on a daily basis for spatial analysis tasks, such as rasterized topographic maps and passive optical satellite imagery at

Table 24.1. A few commercial-off-the-shelf GIS and remote sensing software visualization tools.

Company (alphabetical)	*Software Packages*	*Reference URL*
Environmental Systems Research Institute (ESRI)	ArcGlobe ArcScene	www.esri.com
Leica Geosystems	IMAGINE VirtualGIS	gis.leica-geosystems.com
Research Systems, Inc.	ENVI 3D Surface View	www.rsinc.com
The Orton Family Foundation	CommunityViz	www.communityviz.com

multiple resolutions. Google Earth facilitates a very similar set of tasks. Such tools are designed to be easy to use by amateur map users. For instance, an Italian citizen recently identified a significant archeological site in his back yard by using the Google Earth mapping tool (Nature 2005). Another example followed the destructive impact of Hurricane Katrina on the Louisiana and Alabama coastline. Hundreds of homeowners, displaced by flooding and other dangers, used Google Earth to view high resolution digital images of their neighborhoods. The images were purchased by the United States Federal Emergency Management Agency (FEMA) and allowed people to assess personal losses despite lack of physical access to their property (http://www.google.com/earth/katrina.html). In brief, the concerted effort of geographers over the past several decades has culminated in widely available data and software applications, both of which are putting spatial data viewing and analysis capabilities in the hands of users of all skill and experience levels.

By personally seeking out software tools, visualization products, and spatial data, the users described above each engaged in a self-directed, query-based exercise of exploring data and relationships among data. The dominant paradigm among geographers discussing geovisualization is that private engagement with spatial data, as exemplified above, contrasts directly with non-interactive presentation of data, i.e., "communication" (sensu MacEachren 1994). Setting "visualization" and "communication" as polar opposites may inadvertently diminish the importance of geovisualization for informal education in the eyes of the research and applications community who are designing and implementing geovisualizations. Indeed, it is the union of visualization methods and careful design techniques that make geovisualization an effective medium for informal education. There is a great deal to gain from showing all audiences the measurable causes and consequences of fundamental science concepts and processes. Geovisualization presents many opportunities for breaking down cognitive barriers about concepts and Earth processes with tremendous scientific, political, social, and economic importance.

The power of visualization for education of novices is gaining momentum, and not only in the geosciences. The island nation of Singapore, for example, is training its youth in mathematics using a combination of visual learning and traditional, conceptual approaches to the subject (Friedman 2005). Friedman posits that their strategy of employing mixed media has not only proven effective, but it is part of a concerted strategy at the federal level in that country to surpass the perceived inadequacies in math education in the world's largest economy, the United States. For users and viewers of all skill levels, visualization, and spatial data per se, can help people associate tangible, perceivable objects with complex concepts that may be difficult to grasp in the absence of geovisualization products.

Among the greatest challenges facing geographers whose goal is to educate the general public about Earth processes, global change issues rise to the top (McMahon *et al.* 2005). In this context, "global change" refers to land cover change, ecosystem change, variability in atmospheric constituents, alterations of the hydrologic cycle, climatic shifts, and a large suite of factors that have both natural and anthropogenic forcings (Mahoney et. al. 2003) that together form a superset of the issues receiving international attention and concern about climate change (Watson 2001). Despite its significance to people around the world, many people have a limited or simplified understanding of global change. For example, many erroneously equate the phrases "global warming" and "global change," despite the scientific consensus that rising global temperatures are but one of many pervasive factors altering the Earth as a system (Watson 2001), i.e., the functioning of Earth's physical system and the distribution of organisms that live here (Vitousek 1994). The general public, at all levels of scientific background and expertise, need educational media and programming that explain global change concepts. Greenhouse gas emissions, atmospheric warming, climate change, and subsequent alterations of Earth's physical and biological systems are all directly and indirectly related, but each requires concerted study, and, many would agree, political action. Given the large number of direct and indirect effects of anthropogenic global change, the great costs associated with studying and ameliorating such alterations, and the widespread misunderstanding among novices, these issues are justifiably among the grand challenges for educating large numbers of people with varying scientific backgrounds. Geovisualization can play an important and powerful role in this grand challenge.

Informal science education institutions (ISIs) use geovisualization to explain Earth System Science concepts to people with varied backgrounds and levels of scientific understanding. Fig. 24.1 describes the prevalence of geovisualization in an ISI current science program. To be effective for informal education, geovisualization techniques require careful design. Because design and implementation for informal education have not received abundant attention amongst geovisualization researchers, this chapter focuses on innovations that stemmed from successful high definition video (HDTV) geovisualization productions for informal education in Earth and Biodiversity Sciences. Academics and government agencies have identified several priorities to advance the art and science of geovisualization including the following list paraphrased from McMahon and others (2005) as well as McMaster and Usery (2005): (1) use of HDTV display media that provide enhanced presentation qualities in comparison to computer

Science Bulletins (http://sciencebulletins.amnh.org) presents current science topics in Astrophysics, Earth Science, and Biodiversity Science. Geovisualization plays a central role in the latter two topic areas, as indicated by the highlighting in the accompanying figure; shading represents program elements that use geovisualizations. Within all Bulletins are an introduction, a news section, a data visualization, and a feature story. Each introduction lays out the themes of its respective Bulletin as a whole. In other words, the themes that recur repeatedly are introduced here. News sections, updated weekly, present recent events or scientific findings; for Earth and Bio Bulletins, these news stories exemplify Earth or Biodiversity processes. Imagery and map data are integrated to provide 2-D representations of momentary conditions or change through time. Visualization sections render satellite imagery on the whole Earth and at larger map scales; these employ both 2-D and 3-D views, time series data, image morphing, and other techniques that are described in detail in the main text. Feature stories, which focus on scientists conducting Earth and Biodiversity research in the lab and in the field, utilize geovisualizations to help explain concepts as they are verbalized by the scientists themselves.

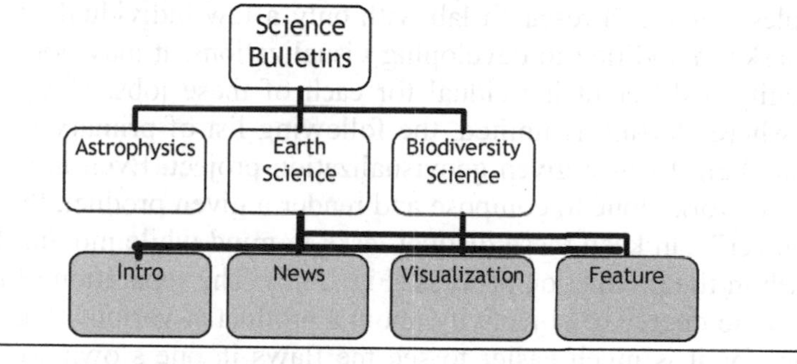

Fig. 24.1. Science Bulletins.

CRT monitors; (2) use of the internet to distribute programming; (3) incorporation of near real-time data from Earth Observation (EO) satellite platforms; and (4) complementing data visualization with interpretation to focus audiences on specific learning goals. The examples that follow are taken from a current science program (see Gano *et al.* 2005) that has addressed each of these goals that are at the frontiers of current efforts in geovisualization research. This chapter describes the foundations for successful HDTV geovisualizations: manipulating geospatial data to explain spatially and temporally dynamic processes. The processing methodologies and interpretation techniques described will be of interest to others who are building similar educational products. In addition, geovisualization experts who are primarily focused on experts' use and interpretation of geospatial data may come to appreciate that informal science educators have advanced some practical innovations for presenting complex data and important scientific concepts.

24.2 HDTV Production and the Role of the Geographer

Buckley and colleagues (2005) emphasize design is an essential element of geovisualization research, for here the connection is made between spatial data and complex, real-world phenomena. Designing and producing geovisualizations for HDTV requires many distinct talents, including production oversight, scientific and artistic interpretation, science writing, and geoinformatics. These roles may be separated among qualified individuals, in which case a production team emerges that can work effectively and efficiently on many projects at once. There is efficiency in such an arrangement because specialization does allow one to bring depth and relevant experience to one's contributions on each project. In many real-world examples, such as a research lab with only a few individuals doing many other tasks in addition to developing visualizations, it may not be possible to identify a different individual for each of these jobs. Nevertheless, in cases where staffing is limited, the following list of primary responsibilities may help focus a given geovisualization project. Even in cases where one must work alone to compose and render a given product, the "lone geovisualizer" can keep these distinct tasks in mind while moving from start to finish in the production process (Fig. 24.2). The separation of roles does allow some degree of objectivity about a product at various stages of completion, for it is much easier to see the flaws in one's own work when a colleague with an equally vested interest in the outcome states clearly that an element detracts and does not add to the overall presentation. This goes for everyone on the production team. Thus, a writer may notice a problem with the way data are presented, or the artist who is composing a project may notice inconsistencies between what the visual data show and what the script says. Each person's opinion is important in the production process, for each individual can view the project with the eyes of an expert or as a novice. Since the objective is to communicate with novice and experts alike, this feedback process (loop in Fig. 24.2) is essential to the success of the final product.

Because geospatial data and science concepts underpin the efforts of the production team, the bulk of the following presentation focuses on the geographer's role in interpreting and providing these raw materials. Image processing and digital cartography are the nuts and bolts of geovisualiza-

Producing informal education media for HDTV requires an iterative approach in which artists and scientists collaborate to devise a story (top left), assemble the necessary data to tell that story (middle left); iterate as indicated (right), and publish a finished product (lower left). At every stage, team members evaluate the overall goals of the final product. In so doing, they may identify weaknesses in the overall presentation or specific elements. In some cases, this leads to a reformulation or refocusing of the artwork or even of the science concepts that can effectively be conveyed by the medium.

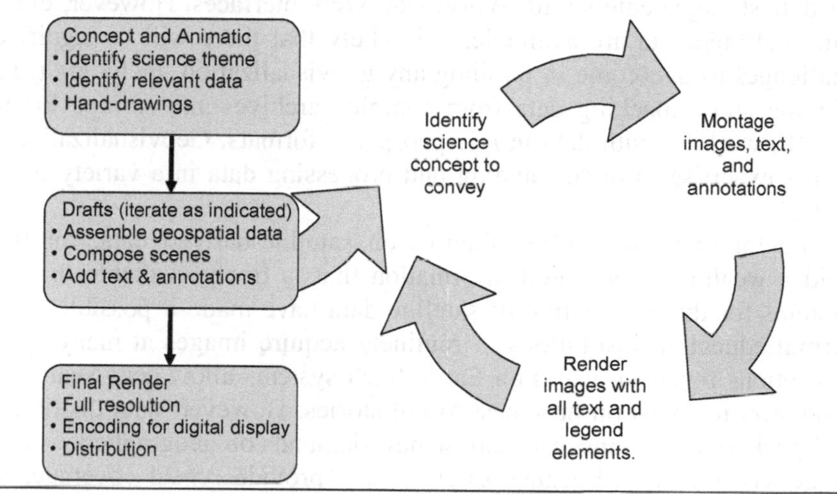
Concept and Animatic
• Identify science theme
• Identify relevant data
• Hand-drawings
Drafts (iterate as indicated)
• Assemble geospatial data
• Compose scenes
• Add text & annotations
Final Render
• Full resolution
• Encoding for digital display
• Distribution
Identify science concept to convey
Montage images, text, and annotations
Render images with all text and legend elements.

Fig. 24.2. The Production Process.

tion. Consequently, there are a number of essential tips and techniques that are applied routinely to provide geographic data to the other production team members. These tasks can be arranged in a number of ways but are presented below from more technical to more interpretive. This ordering is meant to convey the gradual transition from information to knowledge that is the hallmark of effective science communication and synthesis of any kind.

24.2.1 Obtaining and Manipulating Data

Geographers, especially if broadly trained, can help translate complex science concepts into language and spatial representations that general audiences can and will understand. To be effective in describing research of others, it is important to identify well-documented science efforts. If a scientist's work is adequately documented, if data are stored in a format that can be shared, and if metadata are provided along with descriptions of results from analyzing the data, others are more likely to be able to work with those primary science sources in explaining the significance of the work in question to lay audiences. This same reasoning applies to the ability of other scientists to work with one's own research results, so proper data storage and archiving should lead to more rapid advances within specific disciplines. For these reasons and others, the past decade has brought widespread acceptance of metadata standards and the establishment of major data storage centers with Worldwide Web interfaces. However, even if data and metadata are available, it is likely that there will be significant challenges to overcome in building any geovisualization, including (a) retrieving and unpacking data from complex archives and storage formats and (b) working with data in *non*-geospatial formats. Geovisualization requires expertise in understanding and processing data in a variety of formats.

The following discussion focuses on satellite-derived data, for these hold a wealth of geospatial information that is freely available. Existing systems for the distribution of satellite data have made it possible for informal education institutions to routinely acquire images at many spatial resolutions for most places on Earth. Such systems allow geovisualization producers to focus on how best to tell stories. However, data distribution and packaging systems also impose new demands on geographers working in geovisualization. Because satellite data provide repeat coverage, this type of information is vital for portraying the dynamic changes on land, in the oceans, or in the atmosphere. Passive optical remotely sensed data are intuitive to use in presenting spatial data because the detail that can be discerned from false- or true-color images is familiar, even if the synoptic perspective is not. Generating color composites is familiar to geographers, so the following discussion centers on data sources and interpretation techniques for color composite imagery derived from passive optical sensors. The discussion then proceeds to include interpretation of quantitative data derived from remote sensors. False color and true color images can be generated for most of the land's surface at many different spatial resolutions.

The moderate resolution imaging spectroradiometer (MODIS) instruments aboard NASA's Terra and Aqua spacecrafts provide current data for most of Earth's land surface on a daily basis. Using data from the Terra/MODIS sensor, Stöckli and colleagues built a composite image of Earth from Terra/MODIS data collected in 2000. A newer version of this data set, called Blue Marble: the Next Generation, provides a composite image of Earth's surface for every month based on data collected in 2004

(Stockli *et al.* 2006). These data are available via the Worldwide Web (http://bluemarble.nasa.gov/).

To obtain current imagery, swath data from MODIS are available for free download and may be obtained directly from NASA's Earth Observing System Data Gateway (http://edcimswww.cr.usgs.gov/pub/imswelcome/) within hours of acquisition. Using these data requires that one become conversant with the hierarchical data format (HDF; Cohen 2005). Personal communications from a variety of users suggests this format causes a great deal of anguish because specialized software, whether from the public sector or from a vendor, is needed to extract data from HDF files. The advantage of HDF is that data and metadata are stored together. The complexity of HDF is off-putting to many scientists, but, as Fig. 24.3 illustrates, HDF can be conceived of as a container, within which one or more data sets can be accessed. Within a given data set, one may extract values and attributes that are necessary for further data processing and interpretation.

Alternatively, one may obtain pre-processed, georeferenced versions of many images in near-real-time by using the MODIS rapid response system (http://rapidfire.sci.gsfc.nasa.gov/). With spatial resolutions ranging from 250 m to 1 km, these data provide vivid detail. If working within a 1920 x 1080 HDTV format, these data are suitable for spatial extents ranging from roughly 2,000,000 km^2 to 130,000 km^2 (Fig 24.4). Archived Landsat data (e.g., Landsat 7 ETM+ has nominal resolutions of 15 m for the panchromatic band, 30 m for multispectral bands, and 60 m for the thermal bands), are freely available for much of the world via the Global Land Cover Facility at the University of Maryland (http://glcf.umiacs.umd.edu/). This archive includes data contributed by researchers from around the world, as well as EarthSat's orthorectified mosaic of Landsat images that covers Earth's land surface. Focusing on coral reef ecosystems, one web site supplements these freely available data with a global set of Landsat images that covers all coral reefs around the world (http://eol.jsc.nasa.gov/Reefs/).

Higher resolution data are not freely available on a global basis, but such imagery is available from commercial providers in the United States and governments in France (http://www.spot.com/html/SICORP/) and India (http://www.nrsa.gov.in/). A wide variety of data sources are featured in the daily image archive on NASA's Earth Observatory web site (http://earthobservatory.nasa.gov/). Given a study area of a given size and the simple conversions between nominal spatial resolution and screen size (Fig. 24.4), it is possible to determine which satellite data sources are appropriate for a given project.

Compare the HDF file structure to a highly organized storage building. First, one must enter the building. This is analogous to simply opening the file. Before entering, it is not possible to see what is inside. However, once inside, one will find either a swath or a grid or point data. Enter a door with one of those labels by "attaching" to a swath, grid, or point data type. Within this data structure, there are a series of neatly arranged shelves or cabinets; these are "scientific data sets" or "attributes." Each is labeled clearly; by explicitly sending a query, one may determine the binary address of each type of information and reference it by name or by index value. Having read this label, one may pull open the drawer of interest. For example, once a swath is opened and the labels for the data sets within are known, one may extract a data set of interest. More information on NASA's implementation of HDF is available at: http://hdf.ncsa.uiuc.edu/hdfeos.html.

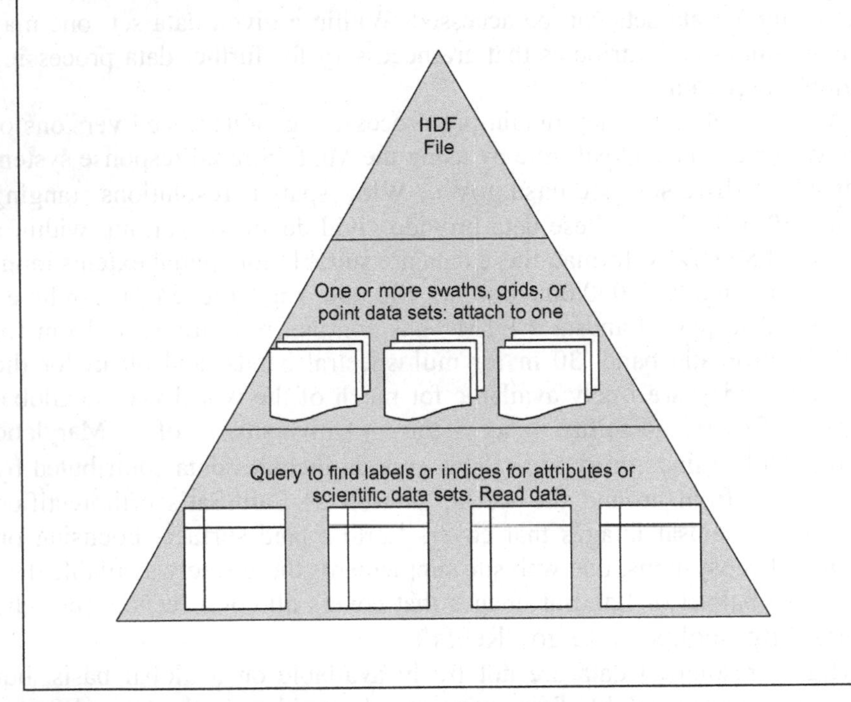
HDF
File
One or more swaths, grids, or
point data sets: attach to one
Query to find labels or indices for attributes or
scientific data sets. Read data.

Fig. 24.3. The Hierarchical Data Format (HDF) in brief.

At the outset of a project, it should be readily apparent whether other scientists have done extensive research developing or using geospatial data for the topic at hand. If such data are available, the process moves quickly to the task of interpretation. Using others' data requires both mutual trust and adequate metadata to incorporate their raster and vector data into one's own project. Interpretation for HDTV informal education geovisualization products requires at least two principal steps: first, one converts any relevant spatial data into a format from which an artist may prepare HDTV compositions; second, one hones the message of a visualization to a few key visual elements and scientific concepts. The first step, exporting geospatial data into formats suitable for artistic refinement, requires that all data be represented on a common two-dimensional frame. The conversions in Fig. 24.4 are useful for exporting vector data as rasters at a common resolution to be overlain on other raster data such as satellite images. Some commercial GIS packages allow one to export map projects into a format that is suitable for manipulation by an artist. For example, ESRI's ArcGIS software will export entire map compositions, with their many layers, into Adobe Illustrator format. This has the advantage of preserving vector attributes for data represented as points and lines. If an already-composed multi-layer format is not available or does not meet one's needs, it is possible to compose and export map layers so that an artist can subsequently combine and manipulate them as coincident raster images. Being able to reverse-engineer a GIS by exporting map layers into a set of images is an important skill to master because artists frequently want to modify line colors and weights using their own software and aesthetic styles. While it is certainly possible to train artists in the use of GIS software, it is not always practical. Following is a methodology that works well for providing spatial data to image composition experts who may not use GIS.

Two key issues immediately rise to the surface when exporting from a GIS: (1) map projection; and (2) map scale. The projection from which one exports must, obviously, match that of his or her imagery. Keeping data in a common projection is simple if careful attention is paid to the metadata of each data layer along the way or if the GIS software being used manages this step seamlessly. These are familiar tasks for geographers and don't require extensive treatment here. Producing data at a common map scale may require a few calculations and care in recomposing multiple data layers in an environment outside of a GIS. It is desirable to match exported layers to the full resolution of a satellite image or other raster. To compose a set of map layers at the full resolution of an image, assume a one:one pixel:pixel ratio on an image:page layout basis. For a 1920 x 1080 pixel layout, determine the default dot-per-inch (dpi) density of the GIS software being used. Many GIS packages use 72 dpi as the default print density, thus implying a page width of 67.7 cm.

$$\frac{1920px}{72\,{}^{px}\!/_{in}} * 2.54cm/in = 67.7cm \tag{24.1}$$

There are several HDTV standards in common usage. For example, 1080i is a version whose native resolution is 1920 pixels x 1080 pixels with interlaced images. This standard allows different display manufacturers to adopt a given pixel density for a wide variety of products. If a 4-m wide screen and a 1-m wide screen both have resolutions of 1920 x 1080, they may both display HDTV content produced using the 1080i format, provided the overall hardware configuration will support 1080i. Because display devices can vary in size, it is not useful to think in terms of absolute map scale but rather strictly in terms of scale ratios. Keeping this in mind, a few conversions are frequently used to optimize the matching of raster resolutions with maps displayed on an HDTV screen (assume the 1920 x 1080 screen size for the time being):

- Max. extentx = Resolution x 1920
- Maximum extenty= Resolution x 1080
- Maximum Extent = Pixel Resolution x (1920 x 1080)

Thus, scale ratios are used to match the raster resolution with the display, or the portion of the display, being used.

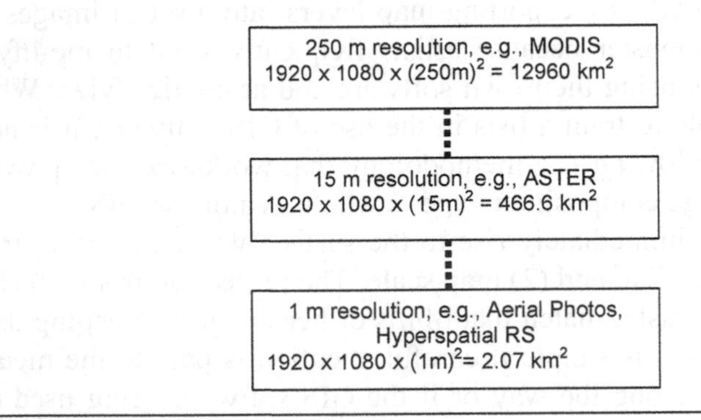

Fig. 24.4. HDTV (1080i), Pixel Resolution, and Display Scale.

For satellite data at a nominal 30-m resolution, a map scale of 1:83,039 is implied:

$$\frac{0.677m}{(1920px * 30m/px)} = \frac{1}{83{,}039} \tag{24.2}$$

If vector data are displayed at this scale and exported as a raster layer, the exported image will overlay with the 30-m resolution satellite imagery. Different mapping software packages manage exporting differently, and many have become very sophisticated about exporting data at a desired output resolution. In such a case, it is possible to define the spatial extent of a map composition and simply adjust the output image size so that it perfectly matches the spatial extent and resolution of satellite imagery. Nevertheless, it is worth understanding the relationship between map scale and image resolution for cases in which one must join multiple map layers into a single art project without the benefit of these useful built-in software functions. Once the scale inherent to an art project is defined, one may export any number of layers as flat images and recompose the map project by re-assembling all of the coincident map layers.

24.2.2 Interpreting and Animating Maps in Two Dimensions

Having obtained a color composite image of the appropriate resolution and having overlain appropriate ancillary data, it is possible to address the artistic and aesthetic issues inherent in clearly presenting one or more science concepts to a general audience. When the output medium is a video that offers no interactive capability, each message and each visual element must be deliberately chosen and easy to digest. Therefore, organizing a sequence of images and interpretive statements or annotations involves paring down the number of data sets that can be meaningfully shown or the number of science concepts that can be clearly presented. Slocum and colleagues (2003) underscored the need for simplicity in their summary about the effectiveness of cartographic visualization. They emphasize that the ability of programmers and geographers to manipulate and present complex data sets may far outstrip the cognitive abilities of most viewers to grasp what they are watching. Working with producers, writers, and other artists who focus on communicating to audiences with varied backgrounds helps to tame the tendency to present complex spatial information. Professionals who work with media such as film understand the need to keep attention fixed upon a clear, understandable concept. The following section lists a few simple rules of thumb that are very useful in composing maps and imagery for HDTV display.

Provide very simple geographic reference points so that the location of the image or map may be unambiguously determined. One solution to the need for geographic reference is to provide a continuous zoom from the Earth as a globe or on a flat global projection down to the focal area of interest. Another method is to highlight a point on a global representation of the Earth, then cross-fade or expand an image from that point. A third is to provide a reference map in the corner of an image for the entire time it is on screen. While most familiar to cartographers, the latter option is not favorable because it utilizes screen area. When a screen is filled with map data, legends, and ancillary information, it is difficult for viewers to quickly assimilate what is being presented.

Moving from global to local perspectives is an effective way to make geospatial data relevant. Showing the Earth as a sphere or on a flattened projection provides not only geographic common ground, but it emphasizes that there are processes that occur globally but that have local manifestations. Moving to an area of interest then provides a more local example of a process with global significance. Images that illustrate explicit examples of organisms or land areas that are affected by a process further emphasize the impact of any given process in specific locales.

If thematic or quantitative data are to be shown on screen, it is necessary to introduce the elements of the legend in a very deliberate way (for example, see Science Bulletins 2004a). A convenient way to do this is to use annotation or a line of script to emphasize that a new map element is being presented on screen. For example, if presenting data about habitat loss in the trinational region of Cameroon, Central African Republic, and the Republic of Congo, one might state, "...This land-clearing activity is displacing native lowland gorillas." At this point, the phrase "land-clearing activity" could be parsed from the sentence and moved to the side to indicate the color on a map legend associated with land-clearing. This legend would appear at the same time a shaded polygon showed the region on the ground where such land-clearing was taking place. By using dynamic map elements in concert with text, their cumulative impact is greater than if a single element is introduced. Everything on screen should appear for a reason and at a time that makes the most sense for communicating a specific idea.

The images and ideas presented build upon one another toward a focused message or statement. This "single effect" is difficult to achieve with spatial data because there are always multiple causes and effects for a given pattern on the landscape. One approach is to indicate that, while many factors are at play, there is one specific activity or change that is important for the point being made. One might list a variety of reasons for land cover change in the upper Congo River Basin, but if a story is about road building due to logging, the map data and imagery on display will show the roads themselves. While other land cover changes continually take place, the roads would be emphasized and their effects on local vegetation and habitats would be described. Subsequent images might show the influence of roadways on light regimes in a forest or movement of trucks into wilderness areas, or text might reinforce how roads lead to changes to ecosystems in the study region.

These guidelines stem from practical experience presenting data in two dimensions (see examples at http://sciencebulletins.amnh.org/bio/archive/). The process of composing map data for HDTV is similar to standard cartographic representation of data and map information. The principal differences are: (1) map information must be simplified; and (2) camera movements are simulated using graphics rendering software. The latter requires that spatial data be output to a format that is useable by the artist setting up the lighting, camera movement, and scene composition.

24.2.3 Mapping in Three Dimensions

All of the considerations mentioned above implicitly deal with data presented in two dimensions, the simplest case of visualizing spatial data according to Dorling (1992). Many software packages are available for rendering images on surfaces and shapes, thus making it possible to dynamically change the shape and appearance of spatial data. This facilitates a number of techniques for connecting viewers' attention between global and regional views or between the Earth as a sphere and the Earth as represented on a flattened projection. A spheroidal rendering of the Earth is arguably the most easily understood representation of the planet. On a sphere, continents are in proportion to one another, distances along great circle routes are well-represented, and directionality is preserved. Yet a sphere is not always the best shape on which to show data. A flattened perspective is needed in order to view the entire planet at one time. This is a useful perspective for seeing movement of clouds, aerosols, stratospheric gases such as ozone, ocean currents and temperature patterns, and other physical parameters that are measured from space. Showing the geometrical transformation between spheroidal and flattened perspectives not only maintains continuity visually, but it also reinforces the idea that a flattened perspective is inherently a distortion of Earth's features as one observes distortions affecting the shapes and sizes of land masses, ocean basins, and other Earth features during the morphing process.

Draping imagery over elevation profiles is more familiar to many than morphing between Earth as a sphere and in a flattened projection. Draping imagery on digital elevation data is well-known within the field of geography, and this capability has been implemented in most remote sensing image processing software packages. Existing remote sensing and GIS software packages, however, are not adept at producing sequences that are well-suited to HDTV. Animation software is better suited to HDTV because the camera, lighting, frame rate, rendering resolution, elevation offsets, vertical exaggeration, histogram stretching, timing, and a multiplicity of other parameters are each adjustable with more specificity than is typically available in commercial GIS and remote sensing image processing software. For example, as one of these factors is adjusted, the low resolution raster images may be used. Once the camera, lighting, movements, and so forth are established, full resolution images can be substituted within the animation software layout environment before rendering a final product. Although this type of capability may be offered in commercial GIS and remote sensing software environments, they have not been proven robust for HDTV image sequence development. Furthermore, animation software packages such as Maya (http://www.alias.com/) have additional uses in an HDTV setting, for they may be used to build custom animations or renderings based on artwork.

As with two-dimensional data, using art production software to render imagery in three dimensions requires that all data layers are in a common projection and must overlay perfectly before an artist can incorporate them into an animation project. The latitude/longitude grid system is useful for artists and geographers to exchange data because the simplicity of this representation of the Earth makes it relatively simple for three-dimensional geometry to be mapped within an animation package. Furthermore, it is very easy for the geographer to provide individual satellite scenes or map projects that overlay neatly into a spherical latitude/longitude grid. A system that has been repeatedly useful for representing data at all scales, from global to local, is to grid the Earth into a set of tiles of manageable size, e.g., 45 degrees longitude by 22.5 degrees latitude. Most regional study areas fall within one of these 64 tiles of the Earth's surface. Because the geometry of mapping each of the tiles to the spheroidal Earth is established once, the same geometry can be applied to a subset study area by defining the top left and lower right boundaries of the study area within one of the 64 tiles. Having done this, a higher resolution image or map composition can be given to an artist. This composition must also be projected to geographic coordinates, but it can be at a higher resolution than the 45 degree by 22.5 degree tile. Subsequently, the artist can substitute the portion of the larger tile with the map composition when the camera zooms in past the scale at which the larger tile's resolution begins to degrade visibly. Animation software is designed to allow for the substitution of data at these different resolutions, provided care is taken to define the geometrical constraints that characterize how two data sets of different resolution fit together.

24.2.4 Global Earth Observation Data Streams

Earth Observing satellites provide many terabytes of data about Earth's land, oceans, and atmosphere every day. These data are often associated with missions whose histories span many decades, thus opening avenues for demonstrating Earth's spatial and temporal variability. The MODIS and Landsat programs, briefly mentioned above, provide optical data that can be paired with observations from similar sensors with legacies that span more than three decades.

For example, MODIS provides sea surface temperature products that are comparable, though not identical, to sea surface temperature as measured by Advanced Very High Resolution Radiometer (AVHRR). Examining variation in sea surface temperature through time gives the viewer an immediate appreciation of the seasonal variability of Earth's thermal budget: one can truly see the cooling and warming of sub-equatorial waters in both hemispheres. Further, this warming is out of phase with the peak in solar irradiance at Earth's surface. In order to demonstrate deviations from expected temperature patterns, the long time-series of data available from AVHRR is used to compile a "climatology record" (http://podaac.jpl.nasa.gov/cgi-bin/dcatalog/fam_summary.pl?sst+clim) which in this context means the average temperature recorded on a 1 km x 1 km grid that covers the planet. Deviations from this expected temperature are measured in degrees Celsius, offering a Sea Surface Temperature Anomaly (SSTA) data set. These data are colorized from blue (representing below-average temperature) to red (representing above average temperature) and visualized on the spherical Earth. This technique allows climatologists to study how decadal and other multi-year patterns might alter the observed temperature profile in a given year. It is also the technique used by El Niño researchers to identify warmer than average temperatures (El Niño events occur when at least a 0.5 degree Celsius positive SSTA is observed for at least three months in the 3.4 region) or cooler than average temperatures (La Niña events occur when a 0.5 degree negative SSTA is observed for at least three months in the 3.4 region) in the equatorial Pacific.

In their animation of global sea surface temperature, Science Bulletins (2004b) employs image morphing between two- and three-dimensional views of Earth, side-by-side comparisons of conditions at the surface of the planet, and time series animations rendered on a three-dimensional sphere. The visualization begins by showing a time series of temperature measurements from the MODIS instrument aboard Terra on a flattened sinusoidal projection. A color scale and legend portray values from -2 to 35 degrees Celsius. The most obvious pattern is the seasonal signal mentioned above, and this is explained in the text. The script goes on to explain that El Niño events occur when temperatures in the equatorial Pacific are warmer than average. The colors are desaturated, and an arrow points to the "warmer than average" region during the 2002-2003 El Niño event. The data set is then cross-faded to SSTA, whose color scale is deliberately altered to show deviations from average temperature. Some generalized global consequences of El Niño climate events are described and annotated on the spinning globe. Ongoing variations in equatorial Pacific temperature anomaly are compared to the extreme El Niño of 1997-1998, which appears more intense and vast than the 2002-2003 event. This comparison is done via side-by-side views of Earth that are clearly labeled with the observation dates (January 1998 and January 2003 for globes on left and right, respectively). The sphere showing data from January 2003 is shown with continually advancing data values so that the viewer can appreciate both the extent of the January 1998 anomaly and the relative quiescence of subsequent SSTA observations.

Image processing for the SSTA visualization required hundreds of global sea surface temperature data sets, one for each day of MODIS observation, to be read and written to tagged image file format (tiff) as colorized images representing either sea surface temperature or temperature anomaly information. These data were then recombined into 8-day running averages of temperature so that data voids due to clouds could be minimized and also to reduce data volume. These procedures were written in Interactive Data Language (IDL; http://www.rsinc.com/) based on unpublished work by Jesse Allen, scientific programmer at NASA's Goddard Space Flight Center.

24.2.5 Image Processing for Global and Regional Visualizations through Time and Space

Code and techniques were augmented greatly to produce another Earth Science visualization about the origins and fate of carbon monoxide (CO) in the troposphere (Science Bulletins 2004b). The morphing and visualization techniques in this visualization were very similar to those described for the sea surface temperature visualization which had proven effective in relating the geographic and temporal distribution of physical measurements on the surface of the Earth. Carbon monoxide is monitored by the Measurement of Pollution in the Troposphere (MOPITT) instrument on the Terra satellite and are collected within 22 km x 22 km pixels in a 640-km swath. The MOPITT sensor covers the planet approximately once every seven days. Its relatively small footprint required individual swaths from seven consecutive days to be combined into running averages representing global patterns. However, since CO is transported and transformed in the atmosphere over this length of time (David Edwards, personal communication), data were averaged within 0.5 x 0.5 degree pixels over each 7-day period so that average conditions were better represented in the visualization. Cloud cover is persistent over many areas of the Tropics. Clouds do not eliminate the presence of CO, however, they merely obscure the observable diffraction of radiation by CO gas. For visualization purposes, average CO concentrations within 3-degree x 3-degree pixels were calculated in order to fill regions lacking CO data due to persistent cloud cover. These averaged and filled data were combined in a time series visualization that balances the need to represent scientifically accurate results with an aesthetically viewable product (Science Bulletins 20004c). A few patterns are worth noting. In the northern hemisphere, very large plumes of CO arise over northeastern Asia and over much of the upper latitudes. These plumes persist over the course of several weeks and can be clearly seen moving with sub-Arctic air masses (Fig. 24.5). Another striking feature is the outgassing of CO from biomass burning in tropical regions where fire is used to clear agricultural areas on a multi-year rotation. These plumes travel with prevailing easterlies from Africa to South America along the Equator (Fig. 24.6).

Some data sets are stored and distributed in formats other than the HDF. For example, ozone data from the Ozone Monitoring Instrument (OMI) aboard NASA's Aura satellite are stored as ascii files representing ozone measurements on a 1.25 x 1 degree rectangular grid. The National Snow and Ice Data Center (http://www.nsidc.org/) stores sea ice concentration data in a binary format that must be read and converted to an image format that is viewable by common software packages. After reading these data, processing steps are sufficiently similar to the examples above that they do not require additional treatment. However, the variety of data storage tech-

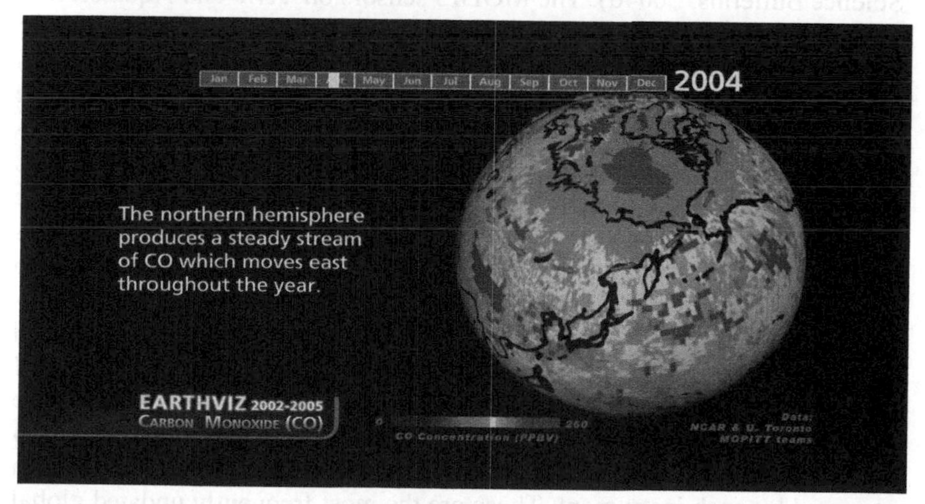

Fig. 24.5. Carbon Monoxide (CO) plumes in the troposphere across the Northern Hemisphere. Time-series sequences of images such as this show that large CO plumes arise in central Asia and persist as they circumnavigate the Arctic Circle.

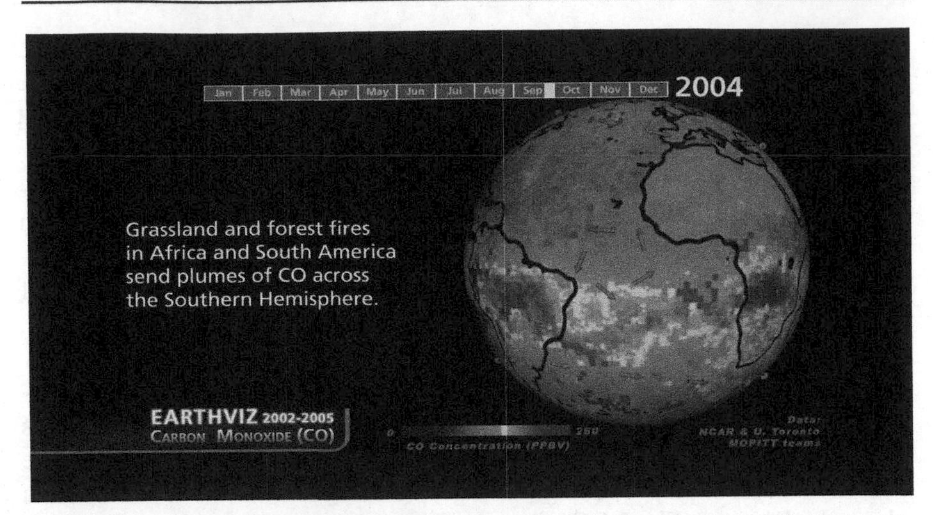

Fig. 24.6. Carbon Monoxide from incomplete biomass combustion fills the troposphere in the Southern Hemisphere. Time series visualization demonstrates that winds transfer plumes between Africa and South America.

niques used by satellite remote sensing missions does require the programmer to be flexible and conversant with a variety of methods for reading and processing data.

An innovative image processing technique was developed to visualize wildfire spread in California during the Fall of 2003, which is often the peak of California's annual fire season (Jentoft-Nilsen and Gardiner 2004; Science Bulletins, 2004d). The MODIS sensors on Terra and Aqua offer at least four observations over California every day, so this sensor can be used to monitor fire spread, even during a rapid fire event. Yet MODIS does not have sufficient spatial resolution to closely observe landscape changes that occur over tens to hundreds of kilometers, i.e., within relatively small spatial extents. In contrast, Enhanced Thematic Mapper Plus (ETM+) data cover a given patch of ground roughly every two weeks. The less frequent revisit time renders this sensor unsuitable for detecting fires while they are spreading, but its higher spatial resolution allows one to view the effects of wildfire spread by either mapping burn scars or examining pre- and post-fire vegetation patterns. The visualization entitled "Fire" (Science Bulletins 2004d) combined the high temporal frequency of MODIS with the higher spatial resolution of ETM+. NASA provides fire detection and thermal anomaly products from MODIS (named MOD14 and MYD14 for Terra and Aqua products, respectively) for every swath observed by each instrument. These are the most frequently updated global fire products available, and they are distributed freely. The algorithm written for this visualization interpolated the locations of fire occurrences between MODIS observations by using a modifiable parameter for the distance between subsequent fire detections to identify which fire detections were modeled as having spread contiguously and which were treated as new ignitions. Fire detections in subsequent time intervals that exceed the distance parameter were mapped as new detections, while those under the distance limit had their locations interpolated to the new time step. The interpolated path of the fire showing the growth of a fire as it spread across a landscape was obtained by finding the shortest path connecting active fire pixels between successive image acquisitions. Interpolated pixels were calculated at time intervals of 30 minutes and were filled based on the presumed intensity and rate of travel of the fire in question. To improve the aesthetic quality of the time series data, random shapes were assigned to observed and interpolated fire pixels. This temporal interpolation algorithm was complemented visually by displaying the results on a satellite image whose spatial resolution was an order of magnitude greater. Animations of fire spread based on 500-m MODIS data were overlain upon nominal 30-m data from ETM+; the shape made by the path of the fire was used to reveal a post-fire image overlain upon a base image acquired before the fire ignition occurred. The process of revealing underlying imagery was accomplished by coding the interpolated fire path as an alpha channel that was multiplied by the post-fire Landsat image and superimposed over the MODIS image. Interpolation and animation composition were performed using IDL (http://www.rsinc.com/) in combination with Adobe (http://www.adobe.com/) Photoshop and After Effects software packages.

A very similar technique was successfully adapted to simulate the growth of Phoenix, Arizona between 1912 and 2000 (Gardiner and Jentoft-Nilsen 2005). As for many long-term studies of land-cover and/or land-use change, urban and agricultural areas were known for seven discrete years during the 1912-2000 time period. Linear interpolation was used to estimate agricultural and urban areas for the months between dates for which land cover estimates were available. Previously generated land cover maps were buffered by a constant distance and stored in computer memory until the accumulated areas for a given land cover class were equal to those predicted by linear regression for a given time interval. At this point, a map of predicted land cover was written to disk. Growth was constrained to match the outlines of the next date for which land cover data were available. The final product, an HDTV piece entitled "Urban Sprawl" (Science Bulletins 2005), was composed at 30 frames/second using one month per frame, or 2.5 years per second as the chosen frame rate. Acevedo and Masuoka (1997) noted that linear estimates of area do not produce the smoothest possible visualizations. However, a variety of alternative growth models were fit to available data, namely logistic growth and two different exponential models. None was able to accurately predict each of the seven dates for which data were available. Therefore, lowess regression was used so that each of the seven dates would be accurately represented in the visual presentation. Once the visualization reached 1975, the date for which a Landsat MSS image could be obtained (special thanks to Ron Beck, USGS, for data provision), the map view changed from thematic data to a false color image of Phoenix. The urban area was outlined in red, and that red boundary grew while encroaching into the agricultural landscape surrounding the city. This showed the expansion of urban land use into previously agricultural areas. As the red boundary expanded, an ASTER scene from 2004 was revealed. Thus, the spatial and temporal interpolation techniques developed for the "Fire" visualization proved very effective also for mapping urban sprawl through time. These methods have general applicability where both spatial and temporal interpolation are required.

24.3 Summary: The Geographer's Role in Producing HDTV Geovisualizations

Because geospatial data are useful for explaining concepts in many disciplines, this chapter introduces a set of problems and solutions when using spatial data in informal education products. Examples begin simply and become more complex; the purpose of examples shifts from primarily technical (in the case of obtaining and formatting data) to more interpretive (in the case of generating spatio-temporally interpolated data). In short, more sophisticated image processing techniques are required to convey more complex phenomena. Special attention is given to registering raster and vector data within two-dimensional software packages that do not natively support geospatial formats but that may be more familiar to artists, whose talents are particularly helpful in presenting data to general audiences. The principles described for working with two-dimensional data also apply to preparing data for three-dimensional rendering software. While some rendering software environments do support geospatial data formats, many of those favored by artists do not. Moving between georeferenced and non-georeferenced image formats requires one be able to accurately tile and nest image data within non-geospatial image formats. The examples in this chapter illustrate convenient methods for exchanging raster data with artists who may then render such data in two or three dimensions. Many examples point out the applicability of satellite imaging data at different nominal resolutions for presenting data at distinct map scales and display resolutions. Because satellite data provide repeat coverage at a variety of spatial resolutions, guidance is provided for accessing global earth observation data streams, particularly from NASA. To portray spatially and temporally dynamic processes, several of the visualizations described here were built from creative, innovative interpolation techniques that could be applied to other visualization projects. These examples provide a starting point for other geographers seeking to present complex spatial data for audiences with varied levels of understanding about the Earth as a system. On the whole, the examples of geovisualization described here required innovation and competence with geospatial data manipulation. Professional geographers may be pleasantly surprised that they can continue to learn about the data and concepts that underlie their own discipline while contributing to informal science education.

Acknowledgements

The work described here was funded through ongoing support from the National Aeronautics and Space Administration (NASA); this manuscript was funded under award NNG05AG25P from NASA. The author is particularly grateful to the Center for Remote Sensing and Mapping Science at the University of Georgia for hosting him as a visiting scientist. The following individuals (listed alphabetically) have significantly contributed to Science Bulletins' Earth and Biodiversity Science public education efforts over the past several years; their mutual support encourages a high quality product and uncommonly pleasant work atmosphere: Alyson Abriel, Laura Allen, Bill Bourbeau, Al Duba, Arlene Ducao, Steve Gano, Michael Hoffman, Catherine Jhee, Rosamond Kinzler, Ilias Koen, Shay Krasinski, Jason Lelchuck, Timon McPhearson, Leigh Morfoot, Len Siegried, Bancha Srikacha, Joanne Teo, Vivian Trakinski, Ben Tudhope, Tania van Bergen. The following NASA scientists and personnel as well as Science Systems and Applications, Inc. contractors have provided vital collaborations for the efforts described herein: Jesse Allen, Wayne Esaias, David Herring, Marit Jentoft-Nilsen, Claire Parkinson, Rob Simmon. David Edwards at the University Consortium of Atmospheric Research generously advised the appropriate use of MOPITT CO data. Scott Goetz and Nadine Laporte of the Woods Hole Research Center have provided data and feedback about land cover mapping visualizations in the U.S. and Africa, respectively. Map data for the Phoenix metropolitan statistical area come from the Central Arizona, Phoenix Long Term Ecological Research site.

Ron Beck of the United States Geological Survey provided Landsat-2 MSS data for Phoenix.

References

Acevedo W, Masuoka P (1997) Time-series animation techniques for visualizing urban growth. Computers & Geosciences, vol 23(4), pp 423-435

Artimo A, Makinen J, Berg RC, Abert CC, Salonen VP (2003) Three-dimensional geologic modeling and visualization of the Virttaankangas aquifer, southwestern Finland. Hydrogeology Journal, vol 11, pp 378-386

Buckley AR, Gahegan M, Clarke K (2005) Geographic visualization. In: McMaster RB, Usery EL, A research agenda for geographic information science. CRC Press, Boca Raton, pp 402

Chertov O, Komarov A, Andrienko G, Andrienko N, Gatalsky P (2002) Integrating forest simulation models and spatial-temporal interactive visualisation for decision making at landscape level. Ecological Modelling, vol 148, pp 47-65

Chertov O, Komarov A, Mikhailov A, Andrienko G, Andrienko N, Gatalsky P (2005) Geovisualization of forest simulation modelling results: a case study of carbon sequestration and biodiversity. Computers and Electronics in Agriculture, vol 49, pp 175-191

Cohen A (2005) HDF-EOS library user's guide for the EMD project, Volume 1: overview and examples. Technical paper 170-EMD-001, revision 3. Raytheon Company, Upper Marlboro, pp 65

Dorling D (1992) Stretching space and splicing time: from cartographic animation to interactive visualization. Cartography and Geographic Information Systems, vol 19(4), pp 215-227

Friedman TL (2005) Still eating our lunch. New York Times, September 16, A-27.

Gano S, Kinzler R, Trakinsky V (2005) Science bulletins: cross-media publishing of current science stories. In: Trant J, Bearman D (eds) Toronto, museums and the Web 2005 Proceedings. Archives & Museum Informatics. http://www.archimuse.com/mw2005/papers/gano/gano.html, Accessed March 31, 2005.

Gardiner N, Jentoft-Nilsen M (2005) Urban sprawl visualization for Phoenix, AZ. Technical Proceedings, Annual American Society of Photogrammetry and Remote Sensing (ASPRS) Conference, Baltimore, CD-ROM

Giardino M, Giordan D, Ambrogio S (2004) GIS technologies for data collection, management and visualization of large slope instabilities: two applications in the Western Italian Alps. Natural Hazards and Earth System Sciences, vol 4, pp 197-211

Jentoft-Nilsen M, Gardiner N (2004) Visualizing large wildfire spread via temporal interpolation and spatial resampling of MODIS fire occurrence data. Technical Proceedings, Annual American Society of Photogrammetry and Remote Sensing (ASPRS) Conference, Denver, CD-ROM

Kato Y, Yokohari M, Brown RD (1997) Integration and visualization of the ecological value of rural landscapes in maintaining the physical environment of Japan. Landscape And Urban Planning, vol 39, pp 69-82

MacEachren AM (1994) Visualization in modern cartography: setting the agenda. In: MacEachren AM, Taylor DRF, Visualization in modern cartography. Pergamon, Oxford, pp 1-12

MacEachren AM (2004) How maps work: representation, visualization, and design. Guilford Press, New York, pp 513

Mahoney JR, Assrar G, Leinen MS, Andrews J, Glackin M, Groat C, Hohenstein W, Lawson L, Moore M, Neale P, Patrinos A, Schafer J, Slimak M, Watson H (2003) Strategic plan for the U.S. climate change science program: a report by the Climate Change Science Program and the Subcommittee on Global Change Research. Climate Change Science Program Office, Washington, pp 202

McMahon G, Benjamin SP, Clarke K, Findley JE, Fisher RN, Graf WL, Gundersen LC, Jones JW, Loveland TR, Roth KS, Usery EL, Wood NJ (2005) Geography for a changing world, a science strategy for the geographic research of the U.S. Geological Survey, 2005-2015. USGS Circular 1281. Sioux Falls, pp 76, http://geography.usgs.gov/documents/gcw/index.html, Accessed October 4, 2005

McMaster RB, Usery EL (2005) A research agenda for geographic information science. CRC Press, Boca Raton, pp 402

Nature (2005) Enthusiast uses Google to reveal Roman ruins. Nature, vol 437, pp 307

Schlitzer R (2002) Interactive analysis and visualization of geoscience data with Ocean Data View. Computers & Geosciences, vol 28, pp 1211-1218

Schmid WA (2001) The emerging role of visual resource assessment and visualisation in landscape planning in Switzerland. Landscape and Urban Planning, vol 54, pp 213-221

Science Bulletins (2004a) Land cover in the Congo. http://sciencebulletins.amnh.org/bio/v/congo.20040818/. Accessed October 4, 2005

Science Bulletins (2004b) Global sea surface temperature in earth bulletin http://sciencebulletins.amnh.org/earth/v/sst.20050401/. Accessed October 4, 2005

Science Bulletins (2004c) Global CO emissions in earth bulletin. http://sciencebulletins.amnh.org/earth/v/co.20041001/. Accessed October 4, 2005

Science Bulletins (2004d) Fire in bio bulletin. http://sciencebulletins.amnh.org/bio/v/fire.20040616/. Accessed October 4, 2005

Science Bulletins (2005) Urban sprawl: Phoenix in bio bulletin. http://sciencebulletins.amnh.org/bio/v/sprawl.20050218/. Accessed October 4, 2005

Slocum TA, McMaster RB, KesslerFC, Howard HH (2003) Thematic cartography and geographic visualization, 2nd edition. Prentice Hall, Englewood Cliffs, pp 575

Stöckli R, Vermote E, Saleous N, Simmon R, Herring D (2006) True color Earth dataset includes seasonal dynamics. EOS 87: 49/55

Thurmond JB, Drzewiecki PA, Xu XM (2005) Building simple multiscale visualizations of outcrop geology using virtual reality modeling language (VRML). Computers & Geosciences, vol 31, pp 913-919

Twery MJ, Knopp PD, Thomasma SA, Rauscher HM, Nute DE, Potter WD, Maier F, Wang J, Dass M, Uchiyama H, Glende A, Hoffman RE (2005) NED-2: A decision support system for integrated forest ecosystem management. Computers and Electronics in Agriculture, vol 49, pp 24-43

Vitousek PM (1994) Beyond global warming: ecology and global change. Ecology, vol 75, pp 161-1876

Watson RT (ed) (2001) Climate change 2001: synthesis report. International Panel on Climate Change (IPCC). Cambridge University Press, Cambridge, pp 3

25 Mobile Geographic Education: The MoGeo System

Jerry Mount, David Bennett, and Marc Armstrong

Abstract. Recent developments in technology such as mobile computing devices, global positioning systems, and wireless networks provide avenues for the development of innovative educational alternatives to traditional computer laboratory exercises. We have created a mobile geographic education (MoGeo) system that places students in the environments they are studying while retaining access to network accessible knowledge repositories and the laboratory instructor. Students can simultaneously gain first-hand experience in the field coupled with planimetric and analytical views created by GIS software. This chapter demonstrates features of the MoGeo system and provides an example of how it is currently being used in the GIScience curriculum in the Department of Geography at The University of Iowa. The framework provided by this system is easily extensible.

25.1 Introduction

Geographic Information Science (GIScience) education focuses on geographic processes that occur at multiple temporal and spatial scales (Armstrong and Bennett 2005). It is often necessary to use varying levels of abstraction to understand how these processes work. These abstractions rely on data structures, techniques, and methods that reduce real-world phenomena into a form that can be stored, manipulated, analyzed and visualized using a computer. Abstraction is unavoidable because we cannot collect or store an infinite number of potential geographic observations. These abstractions, however, make GIScience education difficult because students are usually removed from the environments they are studying. GIScience educators can use developments in information technology to provide multiple modes of presentation, enhance the efficiency of knowledge transmission, allow exploratory examination of the geographic phenomena, and help understand the ways in which we model that phenomenon.

Web-based mapping, virtual environments, multimedia cartography, and interactive displays link images with associated information allowing a user to examine trends in a map display and to associate aspatial information with spatial phenomena. The viewer can also make cognitive linkages between visually explicit trends in a display and knowledge they already possess. Educational systems, designed to present information to students in an attempt to foster critical modes of thinking, often do not take full advantage of the developments in technology and new information presentation techniques. Although many courses in higher education have adopted web-based methods for class management, including posting of lecture notes and grades, more can be done to enhance the learning experience using new technology and methods. The purpose of this chapter is to illustrate the use of experimental, field-based hypermedia as a teaching tool for GIScience.

25.1.1 Background

Our mobile geographic education system (MoGeo) started as a small project designed to examine the potential of using field-based education to teach GIScience. The fundamental idea was that abstract modeling concepts (e.g., topology) would be more clear when students are embedded in the environments they are being asked to represent digitally. In this case, traditional field-based activities are augmented through the use of technology. The system is implemented using personal digital assistants (PDAs) equipped with global positioning system (GPS) receivers and wireless network interface cards (NIC), server-side spatial databases, internet mapping applications, and field-based course exercises. The result is a locationally and contextually aware educational system that is able to monitor student activities and provide each student with an individualized learning experience based on their preferences and performance.

The MoGeo system uses both field-based and traditional classroom environments. The field-based activities complement traditional teaching approaches by reinforcing very specific educational objectives. Successive field-based exercises may be designed to build upon knowledge obtained in previous exercises (classroom or field-based) and, therefore, the sequencing of specific goals is important. The objective of each activity is clearly defined. Performance, therefore, can be evaluated and, if a problem exists, remedial activities can be explored. For example, if a student is answering direct questions concerning some geographic phenomenon that they are near to, and the system determines their level of understanding is sufficiently low, then it can provide alternative questions or provide access to related knowledge repositories on the internet. In addition, students can create personal profiles that contain their preferences for various factors that may be applicable to field-based activities in the MoGeo system. For example, a student may prefer to receive audio cues when they approach a location where they must complete some task. Individual needs, such as physical disabilities, can be addressed in a similar manner through adjustments to each lesson. The ability to adapt to the individual provides a dynamic educational framework based on the location and context of the student. This framework can also be easily adapted to other disciplines such as archaeology, environmental sciences, geosciences, or planning.

Although many of the lessons encourage individual completion of goals, the MoGeo system can offer group learning activities as well. In this case, students must cooperate to complete lesson goals. This mode also encourages experimentation because individuals can be considered parameters in a spatial analysis. For example, the location and elevation of each student (derived from GPS) distributed on a landscape could be used in an interpolation analysis designed to model the landform of a study area.

25.2 Educational Framework

The MoGeo system was developed to provide *in situ* learning that places students in the environments they are studying. It creates a unique approach to student experiences by providing multiple representations of the natural environment including the first-person perspective, the view using orthographic imagery, and the abstracted environment through analyses on that environment. The combination of first-hand experience with the environment (the experiential), the representational perspective of the environment through imagery, and the abstracted environment provided by spatial and statistical analyses provides a complementary overall perspective of the environment and the models through which the environment is studied. A student can simultaneously examine all three perspectives of the environment they are studying. Fig. 25.1 shows a conceptual model of the additive effects of the various perspectives.

The MoGeo approach provides the following advantages over traditional field-based methods:

- an individualized learning context
- flexible and dynamic development environment
- access to online knowledge repositories and the instructor
- access to computationally intensive GIS analyses
- experimentation with immediate feedback on changes in parameters

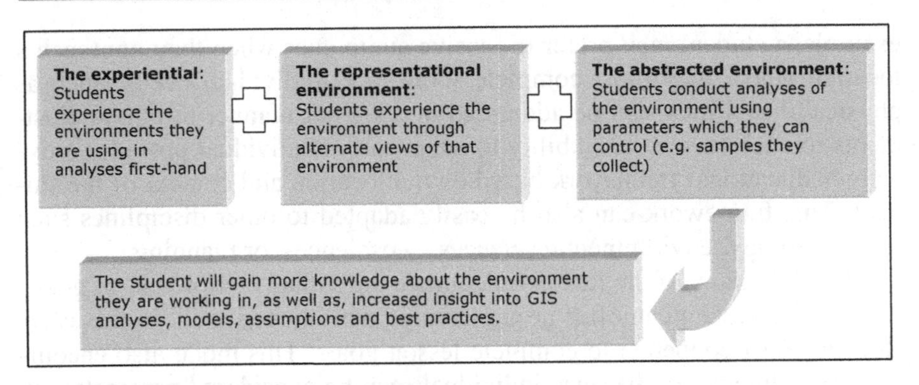

Fig. 25.1. Environments within MoGeo.

In the MoGeo system we have explicitly defined educational objectives and methods for each lesson. The educational objectives provide the instructor with a list of goals and a framework for measuring the effectiveness of teaching methods. Fig. 25.2 shows the process involved in designing MoGeo lessons starting from the choice of an educational goal. This goal is the general, overarching purpose for the lesson. Teaching students about the proper use of a GPS device is an example of an educational goal. An important decision made during the lesson design is whether the goal is to teach students about the ways in which we study the environment (e.g. GIS modeling, statistical analyses) or to teach them about the environment itself. The first approach focuses on the methods by which GIScience can capture, store, model, analyze, or present information. In this case, the environment is merely a background for the educational activity. For example, the lesson might focus on concepts such as interpolation methods, generalization, or sampling error. The lesson would be designed to illustrate features of the method, assumptions, concerns and best practices. The second approach would focus mainly on domain specific education, the environment in this case, and less on the ways we study it. In this mode the instructor could use a "guided tour", or directed observation, approach that takes the student to important features of the environment where further information is presented using a collection of software methods.

After the educational goal has been defined, the instructor will choose the educational mode. This decision, whether to tailor the lesson to individual or group learning, structures the development process by helping the instructor determine which field-based activities are appropriate to use in the lesson. Field-based activities are tasks completed by the students that are designed to teach the subject. Section 25.2.1 defines the three education modes available in the MoGeo system and section 25.2.2 provides further information on field-based activities.

25.2.1 Educational Modes

We use three main educational modes in the MoGeo system. The first, individual learning is focused on the individual student through activities designed to be completed by one person. The second, group learning is focused on group activities and goals that may be reached through cooperative efforts. The third mode, data collection, uses the system as a data collection device with an indirect educational goal. Although students will learn about the environment and sampling strategies, the focus of this approach is to gather data. These observations may be used in subsequent lessons, however, so there is learning involved in this approach as well.

Individual Learning

In this mode the student is asked to conduct individual tasks that demonstrate features of an environment or the way we study that environment. The subject of the activity is based on the lesson goal defined by the instructor. For example, if the class is an introductory level GIScience class, then student activities might focus on ways in which the environment is modeled. This might include activities such as sampling. The environment is used to reinforce assumptions, conditions, and parameters which serve as inputs into a geographic analysis, and serves as part of an "experiential feedback" in the field-based exercise. Rather than sitting in a computer laboratory and using data given by the instructor, the student is asked to conduct an exercise in which they become a "locational parameter" as they sample the environment using a GPS receiver or other attached sensor. The involvement of students in all aspects of a geographic analysis will help them more fully understand considerations involved in this process. For example, a student may be asked to create a digital terrain model of an area using sample elevation points of their choice. After the area is sampled and

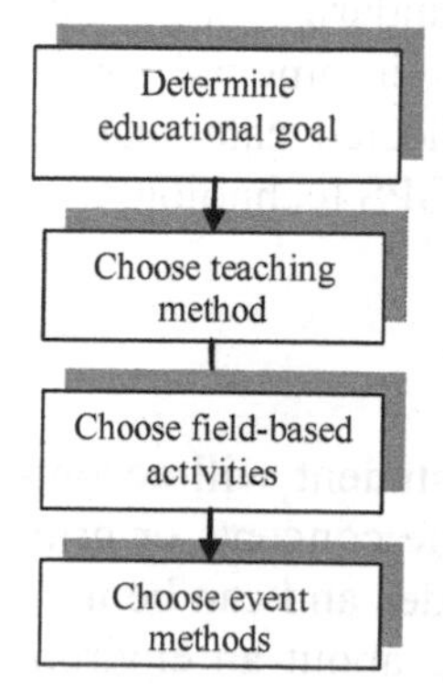

Fig. 25.2. Lesson design.

subsequent interpolations completed, the student may compare their results with analyses based on high-accuracy surveys of the same area to determine how they deviate from the "truth". If the error is sufficient the student may choose to resample and try again. In addition, when significant error is determined by the MoGeo system it may offer the student more information about sampling or interpolation processes. Engaging the student in all aspects of a field-based geographic analysis, coupled with testing of their results and access to knowledge repositories, enhances traditional computer laboratory methods.

Cooperative or Group Learning

In addition to individual lessons the instructor can create and assign cooperative or group lessons. The tasks can be larger in scope because each student is required to complete only part of the global set of tasks. A group of students should be able to cover a larger area in a given amount of time than an individual working on the same goal. Each member of the group has access to the collected data of the entire group so individuals may make recommendations to other students about best practices. The group learning environment is also partitioned into two different educational methods as with the individual learning environment. Group activities can focus on learning about a specific knowledge domain like an ecosystem, or focus on the methods by which we study that environment. Most, if not all, of the field-based activities that can be used in individual learning can be used in group learning.

Data Collection

As the name implies, this approach is designed mainly to assist in collecting data for an environment. The MoGeo system can assist, however, by reinforcing sampling strategies, denoting collection boundaries, or guiding students to locations. It can also warn the student of environmental conditions that may affect the precision or accuracy of collected data such as proximity to buildings and other problems when using GPS technology.

25.2.2 Field-Based Activities

Field-based activities are the general tasks that each student will accomplish during a lesson. They are designed to demonstrate concepts or principles of the lesson in individual or group learning modes and can be used to teach geographical methods and present knowledge about an environment.

Problem Solving

The goal of this activity is for a student to solve a geographical problem in the field. The instructor presents the problem, for example a way-finding exercise, and the student attempts to solve it using their understanding of the problem, examples given by the instructor, previous experiences, or knowledge gained by accessing network knowledge repositories available in the field. The student can also initiate direct communication with an instructor through the wireless network to gain more information about a particular task. In addition, the instructor can monitor each student's location, and their performance in the assigned tasks. If a problem is detected, then the instructor can initiate a conversation with the student to offer assistance, deliver context specific support material, or make changes to the tasks. The analysis of performance can be conducted in real-time or post-hoc to determine whether students completed each task successfully. When students deviate significantly from an expected outcome then additional information about a task or an optimal solution as determined by the computer can be provided. The instructor can also use student activity data collected from multiple students and across semesters to examine systematic problems experienced by students and improve the design of course activities.

Simple-Complex Tasks

Simple activity tasks require students to conduct a series of activities but no direct problem-solving is involved. The GPS accuracy and precision lab described below in section 25.4 is an example of a simple activity task. The goal, in this case, is to show students the relationship between accuracy and precision, the impact of environmental factors like building proximity or tree cover, the number of fixed satellites used to determine a location, and the calculated positional dilution of precision (PDOP) compared with a graphical or statistical distribution. Simple activities can also be combined into a series of tasks designed to be part of a larger goal. In addition, simple activities can be combined to develop increasingly complex tasks. The complexity of the tasks can be adjusted based on the performance of the student or the expectations of the course.

Quizzes and Testing

Field-based testing is a locationally and contextually dependent evaluation technique that relies on direct student questioning through dialog boxes. This is similar to a "pop quiz" but the questioning occurs, for example, when students enter within a predetermined proximity of an observable phenomenon that is the subject of a question. A tree identification lesson may require students to visit selected trees on a campus and when they enter within 5 meters of a tree they are presented a series of questions that focus on that particular species. If students do not perform adequately then alternative questions can be offered, or additional forms of information can be provided. In this activity students may access online knowledge repositories to answer specific questions. Students gain the experience of, for example, examining a type of tree in its natural environment while retaining access to virtual information repositories that can provide additional information if required or desired. The instructor can select questions which focus the student's attention on features that are important for the course.

Directed Observation

In this activity the students are directed to certain locations that exhibit a phenomenon of interest. As they approach these locations they are asked to observe a phenomenon and its environmental setting. Important features of the subject area can also be included as contextually dependent spatial triggers that draw the focus of the student to these key areas. The instructor can direct students to key features that help them understand the phenomenon by building on knowledge gained from previous tasks. For example, in a course studying soil erosion students may be required to travel to various locations that provide examples of different erosion processes that have been discussed in lectures. As they examine the impact of soil erosion they are provided with more information based on their proximity to important sub-features of the phenomenon. Students will receive "pop-up" messages related to environmental factors, such as distance to streams, extreme slopes, or a description of the soil at a given location. The instructor may organize this additional information as a series of ordered spatial triggers that must be visited in the correct sequence to facilitate learning.

"Gameplay"

This activity embeds the learning experience within a game environment designed to present geographical concepts or illustrate environmental features. The goal is to provide an informal learning experience where the emphasis is on learning through games and experimentation. For example, the instructor can create a geocaching game where students have to inductively solve a geographic mystery using a GPS and clues provided by finding secret stashes of information. The game could use physical or virtual objects that must be discovered by a student to successfully complete the lesson. The gameplay also can be configured to provide more revealing hints if a student appears to be performing poorly.

Combination

The last form of activity is a combination of one or more of the techniques presented previously. For example, a lesson may use directed observation of some environment followed by a series of direct questions based on the observation experience. Rewards may also be given based on student performance. For example, a popup movie might provide feedback which congratulates the student on an excellent performance. The combination of these various teaching techniques provides a powerful toolkit for developing innovative field-based geographical education.

25.3 Technological Framework

This section illustrates the technological features of the MoGeo system. The first two parts, event triggers and events, describe key software features and their role in education while the last part briefly describes the hardware used in the MoGeo system.

25.3.1 Event Triggers

Event triggers initiate software actions that are designed to enhance student education by providing a response to an initiating stimulus. The instructor designs event triggers to demonstrate important concepts when certain conditions are met. The triggering conditions and the response events are chosen by the instructor during the lesson design phase. An event trigger is the evaluated condition under which an action occurs based on Boolean operations (e.g. AND, OR, NOT, and XOR operators). There are four basic classes of event triggers. A spatial trigger causes an event to initiate when defined spatial relationships occur. Spatial relationships, in this context, include any of the relationships (e.g. contains, touches) defined by Egenhofer and Herring (1990) or any subsequent extension of these. An example would be a spatial trigger that is activated when a student enters within a certain proximity of a location. A contextual trigger is initiated when certain aspatial conditions are met. For example, a trigger may be activated when student performance on a quiz exceeds a certain threshold. A temporal trigger executes when temporal conditions are met. For example, if students are supposed to finish the lesson by a certain time, then an event is triggered when that time period ends. A composite trigger is constructed using a combination of the other kinds of triggers. For example, a composite trigger could employ spatial, contextual, and temporal trigger conditions and be activated only when students enter a certain location at a certain time. If only one condition is met, then the trigger would not be activated.

The triggering conditions can range from simple to complex evaluations. An example of a simple trigger equation is:

```
if (userfinished == TRUE) {
  call statsfunction();
  printf("Your score is in the " + scoreVar +
          "percentile\n");
} else {
  call donothing();
}
```

The syntax of the event trigger will change depending on where the evaluation occurs. For example, the preceding code might be used when the system communicates with a COM-compliant object such as a statistical analysis using S-Plus. Event triggers can also be evaluated using structured query language (SQL) queries:

```
Select * from table where X > 10
```

If the returned table is empty then the condition is FALSE. The query can be made more complex by adding AND, OR, XOR, NOT, BETWEEN, LIKE conditions to the equation:

```
Select * from table where X > 10 AND Y < 10
Select * from table where X > 10 AND Y < 10 AND
   Z > 100
Select * from table where X > 10 AND Y < 10 AND
   Z BETWEEN 100 AND 200
Select * from table where X > 10 AND Y < 10 AND
   USERNAME LIKE 'Jo%'
```

Fig. 25.3 shows the decision process involved in triggering an information response when a student finishes a lesson.

25.3.2 Events

When trigger conditions are evaluated to TRUE, then an event occurs. Events are the resulting actions based on the initiating conditions (i.e. the event triggers) and these can change the system state for some variable or variables. The launched event is contextually dependent on such factors as

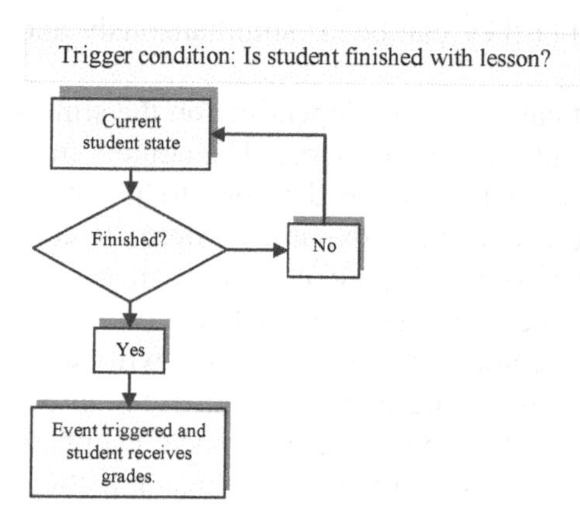

Fig. 25.3. Simple context trigger decision process.

the goal of the lesson, whether the activities are individual or group learning, and the field-based techniques being used. Generally, events can be classified into push or pull situations through which information is requested, transmitted, presented, modified, or stored. Multiple event modes can be used simultaneously, or serially, to respond to an initiating situation. For example, if a student has a question about their current activity they can initiate direct communication with the instructor using a Voice over IP (VoIP) application. The instructor may, in response to a request, push information to the student through links to webpages, or verbally explain the activity to the student, or both.

The cardinality of the relationship between the initiating stimulus and the response is determined by the instructor. In a one-to-one relationship the stimulus (trigger) is directly tied to the response (event). For example, when students complete their assigned work they are presented a webpage showing their score on the assignment. A stimulus in a one-to-many relationship triggers more than one response event. A many-to-many relationship is one in which a series of triggers must be activated to initiate a set of events. This situation may commonly occur in group activities where students collaborate on a common goal. Events can also occur serially or in parallel based on the design of the lesson and the intended goal. GIS analyses, for example, may occur serially as data is processed or transformed before it is used as input in subsequent parts of an analysis. In this case, a script may be used to trigger geoprocessing events in the proper order. Events occurring in parallel are not necessarily dependent on each other as in serially occurring events. For example, a spatial trigger may cause the server to push a movie to the client's PDA, update a table in the database, and trigger a sound on the instructors PC. All of the events are triggered by the one spatial trigger but they can occur simultaneously and independently.

The content delivered by an event can also vary depending on the stimulus and it can be tailored to the student, group, or course. The content may be static, or pre-determined by the instructor such as direction to a website containing more information about the subject, or dynamic, where the content is shaped by current conditions, such as student performance on a quiz. In this case, the content of the event may be constructed "on-the-fly" based on those conditions. It should be noted that not all events will be directly observable by the student. Many of these events will occur in the server and the student may never know they occurred. Logistical events in the database, such as the creation or deletion of tables, can occur without student input or knowledge.

There are many different types of events that can be initiated by system triggers. These can be grouped into two main categories based on whether the events occur on the client-side or server-side.

Client-Side Events

The effects of client-side events are usually experienced directly by the student. These events include the following types:

- Multimedia
  - Audio
  - Video
  - Animation (interactive or non-interactive)
  - Images
- Communication
  - Voice over IP (VoIP) call
  - Audio/video conferencing
  - Instant messaging
- Information access
  - Web site access
  - Access to stored documents (e.g. PDFs, Word documents, spreadsheets, GIS data)
- Course management
  - Quizzes and tests
  - Data collection/sampling
  - Queries

In the MoGeo system multimedia events are designed to transmit information to students using audio, video, animation, or imagery. These events are used to provide background information or to provide additional information on a subject if desired by the student. Although these usually involve simple viewing by the student, some forms, such as animation, can be interactive to engage the student more directly. Flash and SVG animations, for example, can be interactive or passive.

Communication methods are designed to transmit information between the student and instructor or the student and the computer. These methods provide additional information to the student, initiate corrective feedback based on performance, or provide alternative options for completing assignments. The application of these feedback methods can be static, where an event is directly tied to a communication method, or dynamic, where ancillary contextual data is used to determine the proper method for the current situation. They can be interactive, in the case of certain animations, queries, or messaging; or non-interactive, as in audio or video playback. Information feedback can be initiated by the student, instructor, or computer. The event can be initiated as part of a trigger or by conscious effort by the student or instructor. Potential types of communication methods include audio and video conferencing, animation, direction to online knowledge repositories through web-browsers, and instant messaging.

Course management events are designed to manage practical aspects of the education system, such as taking quizzes, sampling, and database queries. These are a series of methods designed to collect or query information. These can be interactive, such as a quiz, or non-interactive, as in GPS tracking events which occur automatically without user input.

Server-Side Events

Server-side events, on the other hand, are often not directly experienced by the student in the field. These are often processes associated with the background functioning of the MoGeo system. These include the following types:

- Database activities
- GIS analyses
- Media server activities (webpage/video and audio streaming)

Database events are any activities involved in the management of MoGeo data. These events include the creation/deletion of tables or databases, data modification, queries, and the creation of database views. GIS analyses are server-side events designed to manipulate spatial data. The MoGeo system uses ESRI's ArcObjects to provide GIS functionality to students in the field or computer laboratory. The GIS functions have access to all of the data stored in the MoGeo database, including the current location of each student, and they can access external databases as well. Computer languages, like Python and Visual Basic, which have access to the GIS objects, allow "on demand" GIS processing so students in the field can initiate an analysis, view the results, change a parameter, and re-run the analysis. The server is also responsible for media streaming when that event is called or pushed to the client.

25.3.3 Technology

Although many improvements have occurred in mobile computing, limited processing power and battery life still plagues ubiquitous computing. Due to these constraints it was important to design a system where most of the computation occurs on the server-side. Fig. 25.4 shows the basic diagram of the MoGeo system emphasizing the role of components that are involved. The field-based side of the system is light-weight with the PDA only providing location determination, input through the keypad, and information display. In the MoGeo system the students perform tasks in a field environment which serves as the geographical background for the lesson. Each student is assigned a PDA for use in the current lesson and they will complete all tasks through the PDA interface.

A database server (e.g. Microsoft's SQL Server) is a component of the MoGeo system. In addition to the routine storage and retrieval of data, the server initiates data transformation services and GIS analyses. This is accomplished through SQL, C, C++, Python, JavaScript or VBScript languages. In this process component object model (COM) compliant objects such as ESRI's ArcObjects, Windows applications, or statistical methods in the S-Plus statistical package are called. The SQL Server can also trigger events like spatial analyses using Transact-SQL, data transformation services (DTS), stored procedures, or extended stored procedures. The database server can be anywhere on the network as long as it is reachable by the clients. It is the common repository for all data collected, stored, analyzed, and presented in the MoGeo system.

Another important part of the system is the wireless network which runs on the 802.11b protocol using the wired equivalent privacy (WEP) encryption protocol. When we started this project we used the wireless infrastructure already in place at The University of Iowa, but nothing prevents the use of other network protocols. We can also set up a wireless network in remote environments using a laptop as the database server and a wireless access point powered over Ethernet (PoE). The laptop can be connected to a car battery using a power inverter for prolonged field use. In remote environments, internet resources might be unreachable; however, satellite

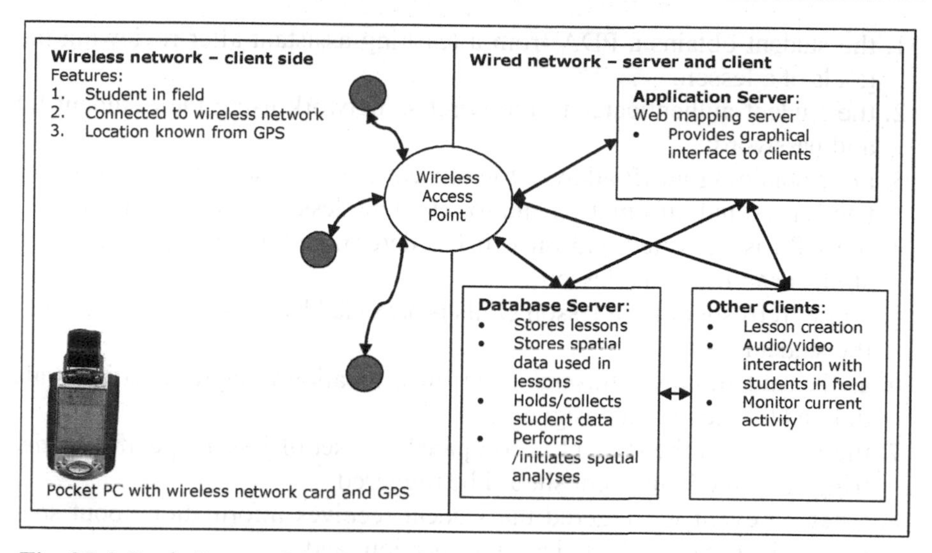

Fig. 25.4. Basic System Diagram.

and wireless internet service providers are making this problem obsolete. Instructors can load pertinent information onto the laptop server to provide additional knowledge resources to students when the internet is not available.

The last hardware component of the system is the PDA, and although it is fairly light-weight in terms of its computer processing power it does provide essential services. It allows us to determine a student's location at a particular time and the wireless network connects the student to online knowledge repositories, spatial data servers, and the instructor. It is the combination of location, aspatial context, access to data, extensibility and connectivity that make the MoGeo a useful tool in our courses.

25.4 Example

Although the MoGeo system is a mesh of applications, protocols, programming languages, and hardware it is quite simple to operate. We have been using various versions of the system for over a year and have experienced few problems. Most of the problems were due to network dropouts or loss of GPS signal. Students have adapted to the new technology rapidly and seem to enjoy the associated lessons.

The following is an example of the steps involved in accessing and using the system:

1. the student obtains a PDA from a teaching assistant after reviewing the goals of a lesson;
2. the student authenticates to the wireless network using their student ID and password;
3. after obtaining an IP address the student authenticates to MoGeo to retrieve personal information and to set up the lesson to be completed;
4. the GPS is activated and once it has determined a current location the student can begin the lesson;
5. a brief refresher of the lesson goals is provided by information pushed to the student;
6. as the student moves through space their location is captured at intervals defined by the lesson designer;
7. the location of the student is compared to a set of lesson-specific spatial triggers to see if an event should be triggered;
8. when an event is triggered the student receives information about specific tasks to be conducted at their location; and
9. if the system determines that the student is not performing adequately then alternatives sources of information are provided based on the context of the individual student.

An Example Lesson:

One lesson currently in the curriculum focuses on the use of a GPS to collect data. The purpose of the lab is to show how GPS works, to demonstrate the concepts of precision and accuracy, and to provide data for later exercises. In this exercise each student is required to visit objects located on The University of Iowa campus and follow the directions pushed to them when they enter within a certain distance of those objects. In this situation, the PDA directs them to collect 6 GPS coordinates for that object. The sample points are transmitted to the server for subsequent analysis in our GIS laboratory. Fig. 25.5 shows a digital orthophoto image of campus with the spatial trigger locations as circles. In the center of each circle is an object whose location must be captured. The actual locations of these objects are known, and during the in-laboratory segment of the lesson students produce maps that illustrate the accuracy and precision of the points they collected. They are also asked to examine the magnitude of the observed differences with respect to positional dilution of precision (PDOP), the number of fixed satellites and other aspects of GPS data collection.

Fig. 25.6 shows the GPS tracking data collected during one GPS lab. These GPS locations are automatically sent to the server every three seconds from student PDAs and are used to determine when students enter a

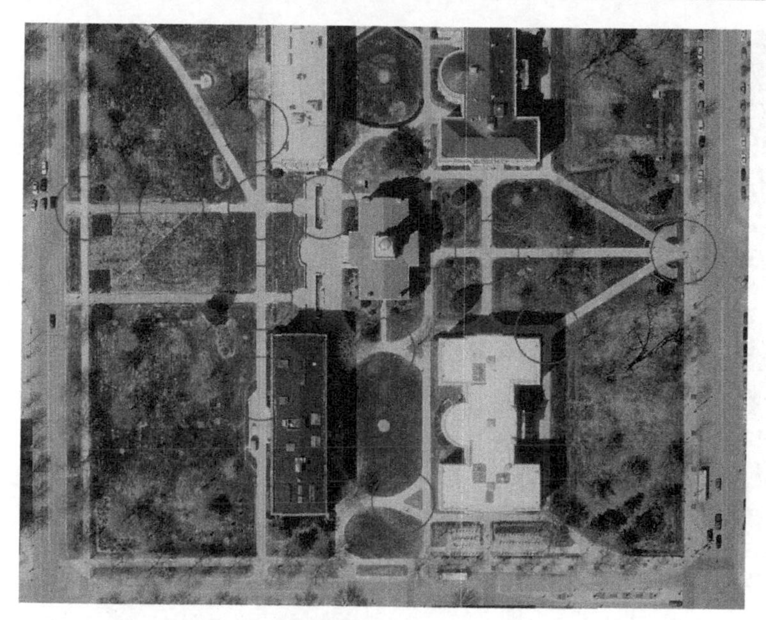

Fig. 25.5. Digital orthophoto with circular spatial triggers.

spatial trigger. The light colored circles indicate tracking locations that are not located within a spatial trigger, whereas the dark colored circles indicate those locations that fall within a trigger zone. The size of the circle indicates the PDOP value associated with each sample; larger circles indicate high PDOP values and thus decreased positional precision. We collect approximately 17,000 to 18,000 GPS coordinates during a one week lab. A powerful feature of the system is that the locations of spatial triggers can be easily moved. The system, therefore, dynamically moves the spatial triggers if the field activity requires it. For example, the instructor could create a geocaching game in which the goals, or triggers, vary their location based on the performance or age-group of the student. Contextual triggers can also be dynamically changed as part of a variable in a SQL query statement. Event triggers are merely records in the database and can be altered, deleted, inserted, or associated with a particular user or group of users.

25.5 Discussion and Conclusions

The mobile geographic education system provides a new educational tool for GIScience and a variety of other disciplines. It provides students with a

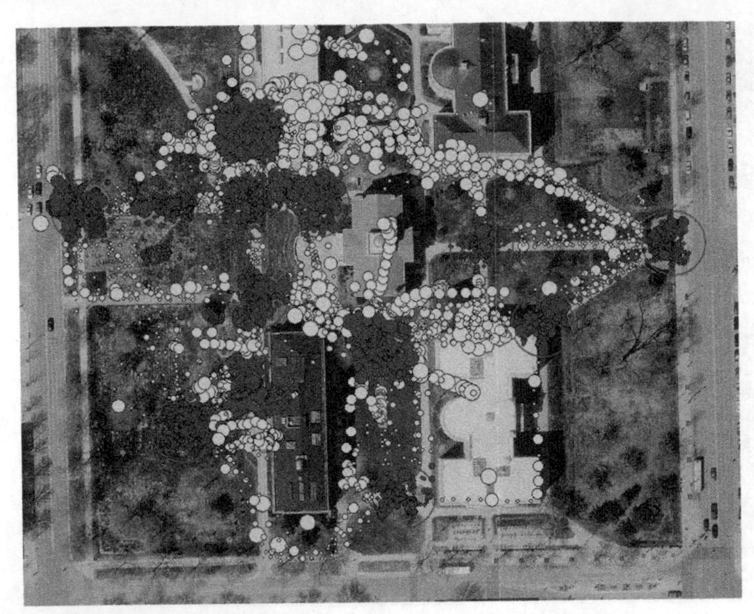

Fig. 25.6. Digital orthophoto with GPS points collected during several sections of a laboratory exercise.

new educational experience that promotes field-based studies while retaining access to external knowledge repositories and computationally intensive geoprocessing capabilities. The combination of the experiential, the abstracted, and the representational experience will help students understand complex geographical phenomena that operate at multiple spatial and temporal scales.

References

Armstrong MP, Bennett DA (2005) A manifesto on mobile computing in geographic education. The Professional Geographer, vol 57(4), pp 506-515

Egenhofer M, Herring J (1990) A mathematical framework for the definition of topological relationships. In: Proceedings of the 4th International Symposium on Spatial Data Handling, Zurich, Switzerland, pp 803-813

Index